The Mobile Audience

Architecture Technology Culture 5

Editorial Address:
Prof. Dr. Klaus Benesch
Department of English/American Studies
University of Munich
Schellingstrasse 3/VG
80799 Munich

The Mobile Audience
Media Art and Mobile Technologies

Edited by Martin Rieser
With an Introduction by Howard Rheingold

Rodopi

Amsterdam - New York, NY 2011

Cover illustration: ©Umbrella.net by Katherine Moriwaki 2004.

Cover design: Inge Baeten

The paper on which this book is printed meets the requirements of "ISO 9706: 1994, Information and documentation - Paper for documents - Requirements for permanence."

ISBN: 978-90-420-3127-2
E-Book ISBN: 978-90-420-3128-9

Printed in The Netherlands

Acknowledgements

The editor would like to thank the following institutors and individuals for their help and collaboration in researching this book:

Mobile Bristol: Dr Jo Reid/ Dr Constance Fleuriot / Dr Richard Hull
Bristol University: Jon Dovey
Dunne and Raby: Fiona Raby
Watershed Media Centre/Clark's Digital Bursary: Stanza and SquidSoup
Blast Theory: Matt Adams/Nick Tandavanitj
The Mixed Reality Lab Nottingham University: Professor Steve Benford
MIT Media Lab Europe, Human Connectedness group: Cati Vaucelle,
Arianna Bassoli, Valentina Nisi
Trinity College Dublin: Dr Linda Doyle, Jonah Brucker-Cohen, Katherine Moriwaki
Urban Tapestries/Proboscis: Giles Lane
Land Design: Peter Higgins
UIAH Helsinki: Andrew Patterson
The ARKI Group: Arena/Heidi Tikka
Interactive Institute Gotenberg: Margot Jacob
Future Physical: Debbi Lander
University of the Arts/Central St Martins:
Trisha Austin, Sharon Baurley and Lisa Stead

Preface

When I started researching this book at the end of 2004, Mobile artworks utilising new technologies such as GPS were a recent innovation. Since then Web 2.0 phenomena including mobile gaming applications and mobile entertainment have blossomed with enormous rapidity and new language has evolved to describe them: "Pervasive Media," "Locative Media," "Mobile art," "Wearable Computing," "Peer-to-peer" and so on are now commonplace terms. Together with these words, have emerged those to describe human behaviour in this mobile ecology, words as exotic as "Flashmobs" or "Human Swarms," "Swarm Intelligence" etc. This is a field developing too rapidly to adequately capture in a medium as traditional as a book.

What this necessarily represents then is an attempt to chart the underlying debates and document the early experiments in the field. I have been blogging new examples, since the inception of the project, anticipating the rapidity of change in mobile artworks, and readers can find ample updated references on http://www.mobileaudience.blogspot.com. Further detailed online references are well documented on excellent weblog sites such as http://www.we-make-money-not-art.com and also on Turbulence's now discontinued research site of Networked_Performance, to be found at http://transition.turbulence.org/blog/. As a measure of this rapidity of change, until midway through my researches the very best source of documentation was Dr. Reinhold Grether's network research site http://www.netzwissenschaft.de/, to which at the time of writing was attached the following sad postscript: "Abandoning the Internet in 2003 and the cell phone network in 2006, my netzwissenschaft link page is completely out of date. It's time to bid farewell." So while the book cannot be a definitive survey, it represents the first comprehensive attempt to map the field and lay down a number of markers for the serious understanding of how, as art, this pervasive medium relates to its audience.

Contents:

2.2 Understanding Public Spatialisation

2.3 The Creative User

Section 3: Case Studies
3.1 Locative Art

3.2 The Creative User: The User as Co-creator

3.3 Wearable Computing

Section 4: Artist Interviews
4.1 Locative

4.2 Wearables

List of Illustrations

Introduction
Howard Rheingold

Ten thousand years ago, nomadic hunter-gatherers became less mobile and what we now call "civilised," when they settled in river plains, practiced agriculture and built cities. Humans had been creating external expressions of our inner lives—one way of defining "art"—for tens of thousands of years before that. And now, with the prospect of a majority of the human species walking around with powerful, untethered computation and communication devices that know where they are: it looks like culture is already in the early stages of yet another leap. I suspect that our definitions and mental models of "audience" and "art" are going to warp and morph. As the examples in this book testify, the leading edges are already pushing art into multiple separate dimensions.

I've long been interested in the un-provable hypothesis that Leonard Shlain proposed in Art and Physics: Artists conceive and represent the world in new ways, changing the culture's consciousness in a way that makes it possible for scientists to understand, predict, and control the world in new ways. Shlain claimed that without perspective painting, Newton's physics would not have been possible, and that Impressionism not only preceded, but also set the stage for quantum physics. Now that millions of computers are joining into voluntary computation collectives (SETI@home, fightAIDS@home, etc.), entire populations are experimenting with performance art, via flash mobs, and participating in augmented reality games, locative learning experiments, collective mapping and geotagging—the untethering of instant global communication, significant computing power is joined to new location-aware capabilities. Who can doubt that new arts, new performances, kind kinds of audience and performers, and, if Shlain is right, entire new worldviews are already emerging?

The avant-garde have been annotating, inscribing, mapping, but the rest of the world has already changed major cultural and social conventions, as we appropriate the capabilities of mobile communication to the needs of our lives. For example, the "softening of time," as sociologist Rich Ling has termed it: everywhere, particularly among those younger than 25 years old, people no longer arrange to meet at 5:15pm at Fifth and Main. They agree to meet downtown after work and negotiate everything else through text messages. In recent years dashboard GPS systems have been changing urban way finding in vehicles. Affordable multimedia real-time mapping capabilities are only beginning to find their way into the hands of the populations; who will, as they always do, appropriate and mutate their media to suit human needs, desires, and institutions.

Cities today are the shape that elevators, railroads, steel, and fixed line telephones made possible. The way people use cities today is only beginning to transform as a result of the devices we carry. The meaning of public art, of urban crowds as performers and audiences, of physical conveyance, have not yet begun to show the profound changes that certainly are taking place. It is a time when art is not only reflecting but influencing the most important step in the evolution of any communication technology—the literacies that people invent, transmit, and evolve when a new technology makes new means of discovering, representing, and transmitting knowledge possible.

Overview
Martin Rieser

"Locative media emerged over the last half decade as a response to the de-corporealized, screen-based experience of net art, claiming the world beyond either gallery or computer screen as its territory. Initially coined as a title for a workshop hosted by RIXC, an electronic art and media center in Latvia during 2002, the term is derived from the "locative" noun case in the Latvian language, which indicates location and vaguely corresponds to the English prepositions "in," "on," "at" and, "by"."[1]

This is a book based on primary field research, its principal aims being to interrogate and explain the visual languages and critical discourses growing up around the new art and narrative forms which are adapting to mobile technologies, in particular through examining audience response. The book forms a study of these emergent uses of mobile, wearable and wire-free technologies, which have moved the audience for screen-based work out of the gallery and cinema into public spaces and geographies, with a particular focus on forms of experimental art works using narrative in its new spatialised or "locative" incarnations. This continues arguments on reception rehearsed in *Rieser, M, Zapp, A, New Screen Media: Cinema/Art/ Narrative (BFI, London, 2002)*, but examines and contextualizes more recent and developing work founded on mobile technologies.

Contexts

Screen cultures today are dominated by narrative and by its modes of framing. Dispersed forms of interaction raise a whole new series of intriguing questions on the nature of narrative and communication, particularly in relation to an audience's modes of participation and reception. The convergence of mobile technologies and ubiquitous computing are creating a world where information-rich environments may be mapped directly onto urban topologies. This creates an opportunity to pose a series of questions around changing concepts of *space and place* for a wide range

---

[1] Tuters, Marc, and Kazys Varnelis. *Beyond Locative Media*. <http://networkedpublics.org/ locative_media/ beyond_locative_media> (accessed 12/03/08)

of traditional disciplines ranging from Anthropology, Art and Architecture to Cultural and Media Studies, Fashion and Graphic design.

Over the past few years, current developments in ubiquitous computing technologies have been explored in parallel by organisations in Britain and Europe such as the *Mixed Reality Lab at Nottingham University* and the *Mobile Bristol Project* (jointly between Bristol University and Hewlett Packard European Research Labs, Bristol). The author negotiated collaborative engagement with many of the case study subjects. This book examines industry futures at the leading edge of technology and aesthetics, where environments of production and reception currently lack definition. This new and experimental work, which has so far been undertaken in the arena of interactive public art or in spatialised interaction through mobile technologies, is in pressing need of exploration, definition and documentation for the benefit of a wider audience.

Dispersal, Narrative and Technological Convergence

The essays in this volume ask in what ways can the new modes of audience engagement and participation in dispersed or mobile interactive art works, be evaluated? Emergent technologies of interaction and the changing nature of public interactive engagement present a radical challenge to Western narrative and its vehicles and traditions. Boundaries between established forms (i.e. games and cinema) are thrown into question and the very concept of creative authorship becomes problematic. In this context, the book explores the creative potential of interactive art and narrative forms in a public and dispersed environments. Whilst other emerging technologies are already redefining existing forms of screen-based exhibition and reception (interactive television and digital cinema), they still tie down the audience in relation to the screen. Locative technology blurs the borders between physical and virtual space, leading to the redefinition of the concept of the virtual from that of *simulation* to that of *augmentation*. While the book focuses on new forms of public and mobile art, where interaction has become spatialised, it evaluates new the modes of audience engagement and participation in such works. through both subjective and objective strategies.

Participatory Media and Form

The texts also question if there are other ways that these ubiquitous interaction technologies can satisfy the demand for participation and create viable art forms, which move beyond the merely sensational? The nature of audience interaction responds to a socio-cultural dynamic that, although far from being quantified, demonstrates a desire for a greater degree of 'participation', evidenced in popular broadcast television e.g. *Big Brother* and its interactive outlets and in the meteoric expansion of social networking on sites such as MySpace and Facebook. Such examples, however, fall far short of the requirements of serious art.

What then is the potential for the emergence new visual and auditory languages and strategies of narration? The book analyses and redefines the languages required to enable the realisation of effective interactive narratised art forms in urban and site-specific environments. Through understanding these new and radical forms of experiment, it attempts to map changes in sociability and communication patterns and new forms of collaboration.

How can this extension of interactive technology from the fixed installation to real urban geographies radically alter the modes of audience participation and reception? If the physical space overlaps the space of diegesis, can this emergent space for art and performance create new perceptions of space and place in an audience? We appear to need a redefinition of the concept of physical space (including hybrid environments): where through such technologies, a new perception of urban space is emerging which is not simply visual, but is both conceptually driven and nascent with multiple embodied potentials.

Much reflection on Locative media art has been premature, for as Drew Hemment observes:

"It is too early to offer a topology of locative media arts, however, or to tie the field down with strict definitions or borders. While artists such as Masaki Fujihata (JP), Teri Rueb (CN) and Stefan Schemat (DE) have been producing work in this area for many years, more widely there have been only a handful of fully realised locative art works, with many projects remaining in the beta-stage, if not still on the drawing board. We have not yet reached the point at which the technology disappears—all too often the tendency is to focus on the technology and tools rather than the art or content."[2]

The waters have been further muddied by the convenient way in which artist's projects have often aligned with the consumer research interests of the mobile phone companies, where yesterday's locative project becomes tomorrow's "killer app." While the tagging of urban space is a process enabled by the commercial concerns of big software players such as Microsoft and Google, it will probably only be when that process meets the next generation of GPS 'smartphone' mobiles that the really interesting art works will begin to emerge:

"Mike Liebhold of the Institute for the Future (IFF) regards "geohackers, locative media artists, and psychogeographers" as key players in developing the "geospatial web," in where the web becomes tagged with geospatial information, a development

---

[2] See <http://www.drewhemment.com/2004/locative_arts.html> (accessed 12/03/08)

that he sees as having "enormous unharvested business opportunities" and believes that this context-aware computing will emerge as the "third great wave of modern digital technology."[3]

An early example based on individual issues of sustainable lifestyles, was Katherine Moriwaki's *Inside Outside* pollution-sensing handbag.[4] Social uses of technology are always beyond prediction. Christian Nold, for example, has definitely found a new way to exploit the personal context of the technology with his emotion mapping and bio-sensing in relation to location.[5] I too am sure such affective hybridity combined with the collective construction and augmentation of site-specific knowledge through wiki-like interfaces gives an evolving future for locative art.

Which brings me to a further question relating to the art itself. Much of what is named 'Locative Art' is not really art, but rather games or spatial documentary or simply demonstrators of advanced toolsets that happen to use this technology. The potential is clearly there, but art has a different function to these uses—and when it is truly present you can smell and see it from afar, altering us through its reinvention of the world. This brings us back to my earlier question about the pleasures and modes of user experience and how we can distinguish these from other media art forms or genres of work.

Mobile devices already appear performative in their nature, with public space interpenetrating our private concerns, so that any conversation has its willing or unwilling eavesdroppers. Add to this the potential for social interaction, crudely demonstrated by *Flashmobs* and in more sophisticated ways by mobile gaming, and you have a case of new technology creating adaptive social behaviours, which contain strong performative elements.

"The mobile games industry has long been the poor relation of the PC and console markets, but a combination of new technology, services and investment is fuelling optimism that mainstream adoption is not too far off. Ask anyone to name a mobile phone game and the most common response will be Snake or Tetris. And while the classic Russian puzzler is the world's most played and downloaded mobile game it is not an accurate reflection of the industry."[6]

Far more demanding games are already being played using mobile technologies

---

[3] Quoted in *Beyond Locative Media* by Marc Tuters and Kazys Varnelis. See also "The Geospatial Web: A Call to Action- What We Still Need to Build for an Insanely Cool Open Geospatial Web" by Mike Liebhold, Senior Researcher, The Institute for the Future. <http://lists.burri.to/pipermail/geowanking/2005-May/001536.html> (accessed 12/03/08)

[4] See <http://www.kakirine.com/> (accessed 12/03/08)

[5] See <http://www.biomapping.net/> (accessed 12/03/08)

[6] Waters, Darren, "How mobile got its game on" Technology editor, BBC News website, San Francisco. <http://news.bbc.co.uk/1/hi/technology/7254123.stm> (accessed 12/03/08)

such as *Catchbob*[7] and Blast Theory's *Uncle Roy All Around You* (explored in several places in this volume), which combine internet and mobile technologies, where the City and the Internet were regarded as related stages on which we play, regardless of the specific context. Steve Benford of Nottingham University now talks of "seamful" media where players have learnt to exploit GPS "shadows" (where tall buildings block satellite triangulation) to their own advantage during game play, describing how such unforeseen effects of the technology encouraged new kinds of movement through the city.[8]

The failure of such works is often in terms of misapplied contextual practice: I once tested Valentina Nisi's *Media Portrait of the Liberties* in Dublin before the demise of MIT's MediaLabEurope.[9] We had gone about a block when the local youths began stoning us. The technology was certainly impressive, but this new form of public art was alien even to the children of the collective contributors to the artwork. When participating in Blast Theory's *Uncle Roy All Around You*, I reflected on how the game's format had reduced the richness of the city to a few textual clues and a dangerous process of frantic searching, with users crossing roads with even less awareness than the average iPod listener. This emotional distancing through the mapped interface seems in direct contradiction to the affective engagement sought by artists—and remains a circle to be squared.

Spatial annotation has emerged in the last three years as a major Internet phenomenon, particularly with the growth of Google Maps and social photosharing sites such as Flickr. In spatial annotation projects like *Yellow Arrow*[10] and *Neighbornode*,[11] cities are increasingly being treated as surfaces on which individuals can inscribe annotation, and which will ultimately become repositories of collective memory. While such story-telling projects give new social and cultural readings of space, allowing private narratives to become public and subject to reinterpretation, they rely heavily on commercial mapping models and their associated biases.

Satnav systems tend to reduce our world to roads between A and B. The specific tagging potential of the locative can certainly overlay this reductive idea of space with all the richness of personal experience, but that depends on the framework provided and the context set by the artist, and in many projects this is so loosely

---

[7] See <http://www.epfl.ch/research/catchbob/> (accessed 12/03/08)

[8] Performing Space - Arts and Humanities Research Council Seminar at Nottingham Trent University (February 2008)

[9] See <http://www.valentinanisi.com/liberties.html> (accessed 12/03/08)

[10] See <http://yellowarrow.net/index2.php> (accessed 12/03/08)

[11] See <http://www.neighbornode.net/> (accessed 12/03/08)

drawn that we simply achieve a kind of public palimpsest.

Mark Tuters has perceptively identified how such annotation and tracing fits into the legacy of Situationism, which Locative Media has claimed as a philosophical base from its inception.

> "Roughly, these two types of locative media' Annotative and Tracing' correspond to two archetypal poles winding their way through late 20th century art, critical art and phenomenology, perhaps otherwise figured as the twin Situationist practices of détournment and the derive."[12]

Situationism in Locative media resists easy definition, but may best be represented says Tuters, by one of Deleuze and Guattari's maps which distinguish between annotation and tracing:

> "Maps can be torn, reversed, adapted to montages of every kind, taken in hand by an individual, a group or a social formation. It can be drawn on a wall, conceived of as a work of art, constructed as a political action or as a meditation... Contrary to a tracing, which always returns to the 'same', a map has multiple entrances." [Deleuze and Guattari, 25-26]

The increasing importance of maps in defining space within these projects should not blind us to the fact that mapping is not a neutral process, but always has been a highly selective and subjective one, in which can be embedded various (invisible) ideological assumptions. Many GPS mapping projects tend to forget this and even revel in the act of remapping without context.

Media artist, Coco Fusco, also launched a headlong attack on new media practices associated with networks and mapping, declaring:

> "It is as if more than four decades of postmodern critique of the Cartesian subject had suddenly evaporated...In the name of a politics of global connectedness, artists and activists too often substitute an abstract 'connectedness' for any real engagement with people in other places or even in their own locale."[13]

---

[12] Tuters, Marc, and Kazys Varnelis. *Beyond Locative Media.* <http://networkedpublics.org/locative_media/beyond_locative_media> (accessed 12/03/08)

[13] Fusco, Coco. "Questioning the Frame: Thoughts about maps and spatial logic in the global present." <http://www.inthesetimes.com/article/1750/> (accessed 12/03/08)

A Taxonomy of Mobile Media Art

Locative and Mobile Media Art remains an emergent field for both practice and associated theory. This book attempts to map this new diversity from its roots through to its many genres and modes of interaction. While it is possible to create taxonomies of the field (and several people have already done so), it seems fated to remain merely an interesting exercise, a net where rare specimens continue to flutter and evade definition.

Accuracy is resisted by complexity, as can be seen in the table above, drawn from a variety of sources and reading across the page. It can be permutated to fit most cases, depending on primary and secondary emphasis, but of course categories elide and blur in many works. An example might be "Drift" by Teri Rueb, which is explored later in this volume (See Section 2.2). Clearly an "individual" *mode of interaction,* and in *genre* a mix of "singular directed journey" mixed with "serendipitous individual discovery." But is it based on a "situated," "arbitrary" or "ubiquitous" *conceptual model*? The *sensory interaction* is predominantly "audio', but "embodied interaction" (sense of orientation) is vital to its meaning. It also based on a mixture of *spatial concepts*, being a "distributed narrative" using "landscape as interface"; but can we define its effects of interaction using the definitions established by Dovey and Fleuriot? (See Section 2.1) We could probably map all of those definitions across a temporal graph of user experience during the work, so that all conditions would at some point be true of some aspect of the project.

Conceptual Models:	Situated: Close fit to actual environment	Mapped/ Geographic: Seeming fit to represented environment	Linear: Single pathway between points	Omnipresent/ Ubiquitous: Distributed and constant	Arbitary Triggered by specific user conditions
Sensory interaction:	Textual	Predominantly Audio	Mixed Reality	Predominantly Visual	Embodied
Modes of interaction:	Individual	Group Present - Collaborative	Group Present - Competitive	Group distributed-Collaborative/ collective intelligence	Group distributed-Competitive
Genres:	Singular Directed Journey	Serendipitous Individual Discovery	Individual Game	Group Game	Individuals Interacting
Spatial Concepts:	Hertzian/ Invisible Space	Space Annotation/ Geographical	Layered Space/ Landscape as Interface	Distributed Spatialised Narrative	Social Spaces
Effects of Interaction: (Dovey/ Fleuriot)	Pleasurable Discomfort	'Magic moments'	Synaesthetic Confusion	Response Enhancement	Deep Immersion/ concentration

A Taxonomy of Locative Media Art

Research Methods

These took account of the contextualisation and interpretation of work samples from the specificity of the audience's views and interacting behaviours. Research visits to other case studies, specifically in Bristol, London, Nottingham and East England, Dublin (MIT), Oslo and Helsinki (UIAH) and North America (Funded through RAE monies) were used to produce detailed documentation. The interrogation of case studies through interviews with makers and their audience was made directly or through questionnaires. The analysis of audience reaction through subjective testimony and its positioning against theoretical positions drawn from an interdisciplinary approach has be the guiding spirit of this work.

Structure

The book is divided into four sections:

Section 1 Towards Hybridity, A Critical History, examines the genesis and development of audience mobility and digital artefacts and art works. In **Section 2 Critical Issues** in contemporary understandings of mobile forms of work are unpacked from a variety of viewpoints: through definitions of language and practice, through spatial understanding and through the role of the 'Creative User'. These categories are revisited in **Section 3 Case Studies** which also interrogates and places the works into three main categories: *Spatial or Locative Works*, where location awareness or geographic proximity dominates; *The Creative User and Play and Improvisation*, where collaborative, ludic, telematic and social works are examined: and *Wearables*, where the body itself becomes platform and vehicle of expression. **Section 4: Artist Interviews** records interviews with some of the pioneers of mobile and wearable art, covering issues as diverse as new media politics, surveillance, new concepts of public space and the role of technology in art.

Content in Detail

Examining the sections in more detail—**Section 1: A Critical history of Audience Mobility** traces the origins of current mobile art practices, covering the story of early media experimentation and artists' use of screen interfaces using various wire-free technologies. These range from sensor-based interactive video installations and Expanded Cinema in the 70s, 80s and 90s, through the public art and interactive architecture of artists such as Christian Moeller, Toshio Iwai, Diller and Scofidio and Rafael Lozano-Hemmer, to the present maturing of mobile and wearable devices and the use of such new media in emergent practice.

In **Pockets of Plenty,** Erkki Huhtamo creates a history of the precursors of mobile devices by examining older technologies, their social uses and their meanings, ranging from Nadar's balloon photography to the Gameboy. Huhtamo argues that, so far, media histories have been based on the idea of media as "fixed coordinate systems," equally in public or private spaces, where traditionally the user was locked into a fixed location before beginning the communication. This is illustrated by "the telegraph office; the phone booth; the cinema theatre; the television set in the living room corner; and the desktop computer on the office desk or on the bedroom table." Even Marshall McLuhan, "whose prophetic insights about the new extensions of man anticipated many future developments," had little to say about mobile media. In this paper Huhtamo attempts to redress this imbalance in the media discourse, by identifying the astonishing diversity of mobile technologies preceding the ones of the present day.

In **The Temporal and Spatial Design of Video and Film-based Installation Art in the 60s and 70s** Dr Susanne Jaschko examines the temporal and spatial cinematic installations and video installations of the late 1960s and early 1970s. The starting point is the seminal installation work of the Vasulkas, which are put into context against installation works of their contemporaries. Essential characteristics of this first decade of spatial work in video and film are demonstrated, especially with regard to the role of the recipients in the overall conception of the artistic works. The unique integration of the recipients in the artwork is also examined in relation to the participatory and performance aspects of contemporary media works.

My Forgotten Histories of Interactive Space is a critical framing of the history of interactive public art, including pioneers such as Ichnatowicz, EAT, Myron Krueger and their inheritors. In its applications as public art, interactive work has frequently attempted to gain critical purchase by exploiting the tension created in an audience between the virtual electronic space and the physical location. Consistent themes and uses for electronic art in public contexts were established early in the 20th century, where threads of similar practice may be traced through from Dada, Futurism and Constructivism and the Bauhaus to the present day. Tatlin was playing with motorised architecture in his Monument to the Third International in 1920. The distant relationship between artist and architect has also created problems in the proper integration of public art in cities, a problem compounded in mobile digital art into public spaces. It is no accident that some of the early successful examples of actual or potential public art works using new technologies have frequently been produced by architecturally trained artists, most notably Shaw, Moeller and Diller and Scofidio.

Adriana de Souza e Silva's **Art by Telephone: From Static to Mobile Interfaces** investigates art works that use telephones as interfaces. Considering telephones as *telepresent* technology, it focuses on the point of transition from the fixed to the mobile telephone, exploring how artistic practices change when the component of mobility is added. She outlines how location awareness capability transforms cell phones into far more than voice-only devices. The consequences are examined in artistic experiences that bring the medium into public spaces, transforming them into ludic, and collective interfaces, pointing toward how mobile technologies can be used in the future. From a broader perspective, this study addresses how art mediated by technology deals with the connection between physical and digital spaces.

Section 2 Critical Contexts begins with **Mobile/Audience: Thinking the Contradictions** by Sean Cubitt and Mary Griffiths. The paper tackles the questions of novelty and the democratic potential of mobility, through selected contemporary examples of aesthetic practice, moving towards an inventory of mobile user dispositions, and the social futures, which they may engender. Mobile technologies have a history, which meshes theoretically with, yet often works against, three other phenomena: the diverse *histories of the screen* in its many forms, the notion of the *public/private gallery space*, and the expectations generated by *public art* located in civic spaces.

This chapter also seeks to map some of the points of similarity in those histories and theoretical prioritisations, and make visible the points of divergence, with the following subheadings in mind: "connectivity; content; temporality; space; privacy;

publics; power; consumption; distribution; visibility; desire; place; identity; and play." They argue that anyone theorising fast moving and converging technologies and their uses by artists, needs to unpack the terms *mobile* and *audience*, and think plurally and flexibly about capital, consumption and power. They ask key questions: "Is this individuated and customised technology more prone to producing user anxieties and re-inscriptions of consumption than it is the active moments of participation in, and recognition of, community? Should the construction of subjectivities take centre stage as the object of mobile research analysis? Are there ways of re conceptualising the desire for Peer-to-Peer connectivity, and community, within democratic frameworks?"

In **Towards a Language of Mobile Media** Jon Dovey and Constance Fleuriot document five seminar discussions at Mobile Bristol held between November and December 2004, aimed at refining the language we use to describe and understand our experience of pervasive, mobile and located media applications. The aim of the authors was to contribute to the development of a common discursive framework for description, understanding and production of mobile and locative media. This research project was prompted by the perception that (the necessary) interdisciplinarity of collaborative authoring teams lacked an agreed set of terms upon which to proceed. This definitional difficulty was additionally compounded by the blurring of previously distinct categories of media and reception which mobile media bring about e.g. spaces of reception lose their borders as media reception becomes more and more pervasive, and media genres similarly start to slip, as reception practices change.

Dr Beryl Graham, founder of CRUMB (Curatorial Resource for Upstart Media Bliss) uses **Snapshots from Curating Mobility** to outline the problems of mobile art works and their curation, drawing on the online discussion archives of CRUMB. She bases her discourse on the notion of a located third space between the audience and the artwork, which is identified as one of the exciting potentials promised by mobile media. Illustrating ways in which what these potentials, which need substantial experience and skill, can be fulfilled; she discusses the nature of engagement as a complex area, attained mostly by the content of the artwork, but also by the characteristic of interaction and cites Rafael Lozano Hemmer's *relational architecture*, Nicolas Bourriaud's *relational aesthetics* and a longer history of participative works including live art, community art, art activism, and Situationism.

The next part **Understanding Space** looks at the issues and languages defining space in relation the mobile artwork. It begins with **Beyond Mapping: New strategies for Meaning in Locative Artworks**, in which I propose an approach to mobile and locative structures, which goes beyond those of a simple mapped correlation between location and story. Narrative is examined in relation to mental maps, architecture and sacred and ritual spaces, revealing that the embedding of narrative in public and constructed space is an ancient practice. That Artists learn to understand these vocabularies of space seems essential in the development of more meaningful and engaging mobile artworks.

Anke Jacob provides a critical overview of interactive architecture and reactive physical surfaces in public spaces in **Digital Media and Architecture—An Observation** and tracks the transformation of the lived environment. She examines the impact of new media and digital technology by looking at transformable surfaces, the types of interactivity connected to architecture, and new display technologies.

In **Urban Screens as the Visualization Zone of the City's Invisible Communication Sphere**, Mirjam Struppek sees the emergence of Urban Screens as the *visualization zone* of the city's invisible communication sphere and as a unique arena for exchange of rituals and communication. Struppek contends that such media facades challenge traditional ideas of development in our urban society, which is now in a constant process of renewal. Citing a variety of contemporary examples, she questions how Urban Screens can contribute both to the idea of Public space, both in moderating urban planning in a city of *free players*, and in developing as a communication medium.

The next section **The Creative User** is intended to focus on aspects of audience as co-creator. It begins with **Future Physical: The Creative User and Theme of Response-ABILITY** by Debbi Lander. This critical essay explores the role of the public user/audience as a Creative User: someone who combines creativity with consumption. It adopts a user focus and draws on the research, observations, analysis and recommendations published in an Arts Council commissioned report and DVD publication, *The Creative User—the new market for interactive digital art*. Lander concludes on the reactions of audience to the ambitious collaborative art works commissioned by Future Physical / Shinkensen in mobile and wearable formats

In **'A Fracture in Reality': Networked Narratives as Imaginary Fields of Action and Dislocation**, Andrea Zapp explores the role of the network in the construction of narrative spaces for open participatory creation, transforming into "imaginary places of being." In her understanding networks are a place of relationship and co-dependency, not simply of communication, "as an open resource of a participatory order." She contends that the Net is "a comparatively unique cosmos of invented identities, partakers, and accomplices in joint forces, hidden in the endless labyrinth of homepages, chatrooms, and communities." These cinematic and performative elements of the Net offer an unceasing platform for user experimentation and innovation. By its nature the Net, (and by implication Mobile networks) provides a framing for "unforeseen content" which should be regarded warily as *works in progress*. We are now looking at the "effects of immediate *real-time* exchange," as opposed to compiled or edited information. She emphasises the transient nature of such content, resulting from textual and visual dialogue between physically remote online participants. As such she provides a valuable template by which to understand remote transactions in gaming and other mobile forms of dramatic or narrative participation.

The third major section is that of **Artist Case Studies**. These are primarily to give an in depth understanding of a broad variety of works both in terms of the taxonomy outlined in this overview and as a record of the development of seminal exemplars in the field.

In **Part 1 Locative Art**, Jo Reid and Richard Hull ask **"What Makes Mediascapes Compelling?"** This joint article on the findings of the RIOT! Project conducted in Queens Square Bristol in by Mobile Bristol team (Hewlett Packard/ Bristol University). The Mobile Bristol authoring environment had been used to develop a range of location-triggered mediascapes such as a Ferryboat tour, an educational game called *Savannah*, situated digital stories, a nature walk, *CitiTag* and many others. What these mediascapes shared was the same simple interaction paradigm of using movement to trigger different media depending on location. In this chapter they discuss the role of the mobile audience in these locative mediascapes by drawing on a quantitative and qualitative analysis of public reaction to a rich mediascape called Riot! 1831 deployed in a public field trial lasting three weeks, using an interactive locative drama based on the Bristol riots of 1831.

Valentina Nisi and Glorianna Davenport describe two further seminal projects constructed at MIT's MediaLabEurope in Dublin, namely **Hop Story** and **Media Tales of the Liberties**. This is a joint description and evaluation of two works, exploring sound and video locative experiences. MediaLabEurope suggested itself as an ideal venue for experiments in place-based narrative. MediaLabEurope (2000-2004) occupied a renovated hopstore building that has been historically part of the Guinness brewery complex on James's street, Dublin 4. Situated in the heart of the Liberties area, the hopstore served as an historical and anecdotal reference point for the neighbourhoods. However, the everyday life of MediaLabEurope was significantly different from the traditionally brewery business of the Guinness complex. Open 24 hours a day, the building hosted researchers exploring the potential of advanced technologies and visitors who came to the Dublin lab to taste the future. Every few months the lab would host a daylong event. On those days the researchers would demonstrate their current projects to the visitors. These situations provided the opportunity to implement and evaluate some challenging ideas about location-embedded stories.

In **Loca: Location Oriented Critical Arts** by Drew Hemment, John Evans, Mika Raento and Theo Humphries, a major public presence for Locative art is described. Hemment examines the genesis and development of *Loca*, a project foregrounds secondary characteristics of mobile communications, such as the ability to locate consumer mobile devices in real-time and near real-time, and the kinds of peer-to-peer pervasive surveillance that is possible as a result. Loca explored the shifting nature of surveillance as it ceased to be the preserve of governmental or commercial bureaucracies. He asks what happens when it is easy for everyone to track everyone, when surveillance can be affected by consumer level technology within peer-to-peer networks without being routed through a central point?

In describing the Sky Ear project in **Invisible Topographies**, Usman Haque contends that the overlapping territories of art and architecture have developed, in large part, because of technological developments, which have upset conventional understandings of spatiality. The Mobile technology, through which we conduct our daily lives and businesses, has made us far more aware of the electromagnetic environment that envelops us. He feels that the traditional dichotomies between audiences and performers, designers and users, architects and occupants are less evident than they used to be. He writes "Wireless technologies in particular have challenged our relationship to designed space because they encourage us to think not of static silent structures that surround us, but rather of fluid dynamic fields beyond the edge of our natural perception." These are fields within which "we are all consumers and all contributors." Mobile communication has relied equally on scientific and cultural evolution and has prompted a distinct shift both in the way we relate to space and the way we relate to each other. Haque describes the evolution of Sky Ear—a major mobile artwork, which visualises these electromagnetic flows.

In **Wifi-Hog: The Battle for Ownership in Public Wireless Space**, Jonah Brucker-Cohen's controversial agit-prop work on the ownership and provenance of publicly accessible wifi networks is described in depth. *Wifi-Hog* is a tool that enables control over a specified network by someone who is not the networks administrator and looks specifically at what happens when these seemingly open networks are made exclusive and competitive. Since these networks exist as private, public, and corporate monitored services, there is also confusion about rights ownership over networks in public spaces, thus *Wifi-Hog* is specifically reacting to the lack of an acceptable usage policy of wireless networks. As mobile technology has entered public space and brought private conversations and interactions along with it, an interesting rift was forming between what is deemed acceptable usage. In a sense, Wifi-Hog exists as a tactical media tool for controlling and subverting this claim of ownership and regulation over free spectrum, by allowing a means of control to come from a third party.

In the second part of this section, **The Creative User** begins with Paul Sermon writing in **Puppeteers Performers or Avatars: A Perceptual Difference in Telematic Space** on the displacement of related spaces and their reconnection through telematics. This enabling of collaborative narrative construction is argued for in this examination of Sermon's own visual practice. His work in the field of telematic arts explores the emergence of user-determined narrative between remote participants who are brought together within a shared telepresent environment. Through the use of live chroma-keying and videoconferencing technology these divided audience participants enter a video installation and initially suppose they entering a passive space—sitting, standing or sometimes lying within it. Their presence within the space is recorded live on video camera and mapped in real-time, via a chroma-key video mixer, with an identical camera view of another participant in an identical installation space—combining two shots of live action by replacing a blue or green back drop in one image with the image of the other. The two spaces which can be any geographical distance apart are linked via an internet videoconference connection, making it possible to combine these telematic installations and there performing audiences between almost any location in the world.

This is essentially how all his installation projects function, but what is most surprising for the intended viewer is that they form an integral part within these telematic experiments, which simply wouldn't function without their presence and forced participation within it. The audience participant rapidly becomes a performer, or at best an actor within these spaces, by observing their body within a telepresent space represented on self-view video monitors in front of them. The user/actor ascends a rapid learning curve and begins to control and choreograph their human avatar representation of themselves in a new telematic space, in combination with another physically remote role-playing user. The Narrative that unfolds here would appear to be self-determined by the user, but what is essential in such experiments is the architecture of this installation. As an artist he is both designer of the environment and director of the narrative, which the artist determines through the social and political context in which he chooses to play out these telepresent encounters.

Mobile Feelings was a mobile art project where users could send and receive body data over a wireless communication network. Christa Sommerer and Laurent Mignonneau used specially designed Mobile Feelings devices to allow remote users to feel each other's heartbeat signals and breath over distance. The system explores novel forms of intuitive and non-verbal communications that go beyond the conventional transmission of voice, sounds and images used in standard mobile communication. Mobile Feelings enables intuitive bodily communication between remote users by exploring the emotional quality of touch and breath as some of the less explored communication senses.

In the **The Living Room** Victoria Fang created the first RFID-driven narrative space. *The Living Room* was a narrative puzzle installation. Players took part by moving panels with LCD monitors in an effort to solve a 'whodunit' murder mystery. When all three panels were positioned correctly, video memories are triggered with which one is able to unravel the mystery. Through the position change of the panels, one's experience with the space changes in every scene. Ultimately, the installation functions as a giant physical puzzle that houses a narrative puzzle. *The Living Room* is a spatial movie experience—the user is in front of the movie, between the movie, around the movie, and physically shifting the movie. While the movie itself is primarily linear, players' participation to progress the fragmented narrative results in changing their surroundings, both physically and in terms of narrative. Each segment of the film forces players to interact with the space in a way that compliments the content of the scene they are watching.

Arianna Bassoli uses **tunA and the Power of Proximity** to demonstrate how new proximity-based ICTs are introducing an overlap between the virtual and the physical space where people currently interact, generating the oxymoron of mediated face-to-face communication. Short-range wireless technologies are partly responsible of this process of remediation, where socio-technical networks can be created among co-located strangers and acquaintances, sharing digital resources and communicating through mobile devices. Bluetooth and WiFi do not only increase the ubiquitousness of ICTs and give access to the World Wide Web, but they also provide a link between people who are at a relatively short distance from each other, opening up a whole range of unexplored potentialities in terms of human-human interactions. Bassoli's early experiment with shared and exchanged mobile music, developed at MIT Dublin is examined in depth.

The third part of the case-studies section addresses **Play and Improvisation**, as essential aspects of the new technology. Margot Jacobs in **Engagement with the Everyday** describes a number of projects undertaken in Gotenberg in collaboration with Play Studios, using unusual game-based strategies for public engagement with mobile and locative sound works.

Cati Vaucelle in her essay **"Between improvisation and Publication"** examines how from an early age, we play, learn and exchange ideas about our identity using stories; and test our hypothesis about the world using toys, telling stories and acting in the world. She believes that with mobile technology, we are moving into a creative and collaborative world in which images and sounds can mix with local language. Her shared movie-making devices can engage people in exploring multi-dimensional approaches to expressing and exchanging point of view on their environment. She envisages a future where people can create movies about how their lives interact with everyday means of interaction such as commonly used technological objects; for example, computer game consoles, cell phones, and video cameras, and these objects can themselves serve as vehicles for manipulating personal media to co-construct video games, movies, and songs. She asks: "How might creative people engage debates with their spectators about the technological environment we live in?"

Anthony Rowe examines four gallery-based projects by **Squidsoup** aimed at developing creative audience interaction through play, improvisation and collaboration. This paper discusses the processes of creative development in four projects by Squidsoup: *Altzero* (1999-2003), *Come Closer* (2004-5), *Freq2* (2006) and *Driftnet* (2006-7). Each piece presented users with a structured audiovisual composition that could be manipulated, probed and explored in different ways. Together they illustrate a search for novel, intuitive and subtle forms of interaction where the processes of interaction are creative, rewarding and (as far as possible) understood by those taking part. This is an investigation not only into direct user interaction, but also into the participant's interpretation of their experience; what it makes them feel and think, and what meanings they ascribe to their experience and the work.

The fourth part of this section: **Wearable Computing** is selectively examined as a distinct area of practice. It begins with **The Emotional Wardrobe** by Lisa Stead, Petar Goulev, Caroline Evans and Ebrahim Mamdani, which examines the new alliance emerging through the integration of electronic technology and smart materials on the body. This study addresses the integration of technology with clothing from a fashion perspective and examines its expressive and interactive potential. It proposes the concept of 'The Emotional Wardrobe': clothing that represents and stimulates emotional response through the interface of technology. It asks if fashion can offer a more personal and provocative definition of self, which actively involves the wearer in a mutable aesthetic identity.

In **Social Fashioning and Active Conduits** Katherine Moriwaki describes her explorations in wearable interaction and collaborations with Jonah Brucker-Cohen. In the early days of the Internet the dematerialization of place allowed for the development of a digitally embodied aesthetic, where virtual equivalents of physical objects were believed to be capable of replacing their material counterparts. The physical world, with its spatial and temporal limitations fell out of vogue, unable to compete with the chimerical seduction of the electronic agora). However, as ubiquitous computing technologies have emerged in the third wave of the digital revolution, the importance of the physical and the local has acquired renewed relevance as emerging technologies are increasingly incorporated into the everyday environment. New communications infrastructures stress decentralized and temporary networks, which support mobile and distributed computing needs. More importantly, these emerging network communications technologies have the capability to reinforce and supplement existing social and behavioural structures allowing for deeper resonances between the virtual and physical. This paper will present examples of socially fashioned networks, or networks which use social behaviour and human mobility as the means for establishing a network infrastructure between collocated individuals. The various projects discussed represent social fashioning at different degrees of behavioral and technical resolution. They function as lenses through which spatial-temporal constraints are focused and applied in order to create interrogative disruptions in everyday experience. Through this process individual and group relationships in public and urban space are challenged and re-imagined using everyday objects and activities.

In Wunderkammer: **Wearables as an Artistic Strategy, Laura Belloff** examines how hybrid space is conditioned by the use of mobile and wireless technologies. Examining various theorists and researchers who have scrutinized this space, she positions her practice. Amongst others she quotes Timo Kopomaa on the concept of a *third space*, Anthony Townsend on *phonespace*, and Adriana de Souza e Silva who has defined the more general concept of *hybrid space* which is formed from a merge of physical and virtual spaces, and is discussed elsewhere in this volume.

In **Section Four** of the book: **Artist Interviews,** I cover in depth the conceptual and technological development of the work of major global players in this new territory, from the early locative works of **Dunne and Raby** and **Teri Rueb** to the ambitious work of **Blast Theory** supported by **MediaLab at Nottingham University**. Drew Hemment examines the **Politics of Mobility**, and experiments in **Wearable Technology** and their social effects are teased out in interviews with Berzowka and Lovejoy.

The first interview **Trace, The Choreography of Everyday Movement and Drift** covers Rueb's major art works dating from *Trace* 1996 to *Drift* 2004. It examines her interest in creating experiences of space and time that are multiple, overlapping and even sometimes contradictory as is her concept of narrative interpreted as by a sculptor through use of space. With **Flirt and Mset** Fiona Raby discussing the genesis and developmental problems of the pioneering mobile works of Dunne and Raby. It focuses on their early Finnish trials looking at future technologies using WAP such as Flirt and Stampede, through later experiments for MSET such as *Pixel Kissing* and *The Garden in your Pocket* concluding with *Lazy Crow*-an early use of Landsat data.

Blast Theory is an interview with **Matt Adams** examining all the major works by Blast Theory, their evolution, execution and the various types of public engagement anticipated and observed. It raises a number of issues, which are to do with the ludic aspects of relationships between real and virtual worlds, and how a group, they came to the decision that this was the area of discourse in which they were most interested. The central challenge of works like *Uncle Roy All Around You* and *I like Frank* is how to combine two spaces-the online world and the space of the street player out there in the city and how to deal with the very divergent sets of expectation and experience that those two spaces engender.

Mixed Reality Lab is a detailed interview with **Professor Steve Benford** on the collaboration with Blast Theory by the Mixed Reality Labs at Nottingham University and the lessons in development of meaning and interaction in locative works. It examines the full range of work developed by Blast Theory. The nature of the social interaction and necessary refinement of technologies as the partnership progressed.

The Politics of Mobility is an interview **Drew Hemment**, which rehearses the major arguments around emergent technologies of mobility and the freedom of the individual, surveillance and its ethical and moral dilemmas. He argues that in many ways we are seeing a fundamental shift: that we ourselves have become the drivers of the surveillance society because it is a secondary effect of so many things we otherwise value. The basis of the Panopticon was that we would not know when we were being observed or not, and so would come to act as if we were being observed at all times. With network technologies comes a whole new set of variables that govern this same scenario. We now leave data trails behind us that will last potentially forever.

In **Wearables: Heart on Your Sleeve**, Annie Lovejoy covers her early experiments in locative sound with Mobile Bristol and her subsequent use of wearable technology in collaboration with the Bristol University Wearables group. She believes that the aesthetic visualisation of data has potential for wider shared experience. Locative media can move beyond cartographic documentary style narrative. Her work is generated through interactions with sites or situations; technological tools are party to this, but not central. She is currently attempting to articulate a non-media specific overview of this approach for a web space. "Interfaces of location and memory" is a helpful phrase: interface—as a point of connection between things, location—the site, situation, context or position of something and memory—as experiential, collective, associative, knowledge base, data and archiving.

Memory Rich Garments and Social Interaction with **Joey Berzowska** focuses on her own researches in wearable computing from XS Labs and later Concordia University. In this 2005 interview she describes the social drivers for her uses of interaction. On the one hand portable technologies are changing behaviour, but on the other current portable technologies are developed in a vacuum for the consumer electronics industry. She emphasises that the body is a more appropriate space for these intimate conversations and communications. She wants to force people to consider how we are redefining intimacy and to bring this back to the body and locations on the body through her experimental works, which are cited in detail.

Martin Rieser 2008

Bibliography

Deleuze, Gilles, and Félix Guattari. *On the Line*. New York: Semiotext(e),1983.

Section 1: Towards Hybridity

A critical history of audience mobility

Pockets of Plenty: An Archaeology of Mobile Media

Erkki Huhtamo

A history of the precursors of mobile devices examining their technologies, social uses and meanings (This text is based on a keynote address given at the 12th International Symposium of Electronic Art 2004 in Helsinki, August 2004. Although written for this publication, it contains material published in the ISEA 2004. 12th International Symposium on Electronic Art catalogue, edited by Tapio Makela and Mare Tralla, Helsinki: m-cult, 2004, 23-26.)

Take any standard media history, say, Brian Winston's *Media Technology and Society, A History: From the Telegraph to the Internet* (1998), and look for words like "mobile," "portable," "wearable," or "nomadic." Most likely they are missing. At most, you may find a few references to them in the final chapter dedicated to recent and future developments.[1] Media histories have so far been based on the idea of media as fixed coordinate systems—signals penetrate walls, and broadcasting blankets huge virtual territories, but the material nodes of the network, the equipment used as transmitters and receivers, are found in fixed locations. Whether situated in public or private spaces, to use media we have to sit down in front of the device, switch it on, adjust the controls, and only then begin our communication. When the session is over, we switch it off and leave the media machine in its place, possibly closing its cover or doors (as in the case of many early TV sets). Think about all the devices kept on a table or a stand of some sort. Think about the telegraph office, the ham radio operator's den, the phone booth, the cinema theatre, the television set in the living-room corner, and even about the "desktop" [sic] computer in the office cubicle or on the bedroom sidetable. Even Marshall McLuhan, whose prophetic insights about the "new extensions of man" anticipated many future developments, had little to say about "mobile media."

---

[1] See for example Asa Briggs and Peter Burke, *A Social History of Media from Gutenberg to the Internet*. Cambridge: Polity Press, 2002. Less than two pages of the 374-page book have been dedicated to mobile phones (306-307 + references on pages 284, 285. Mobile media does not play a role in Armand Mattelart's *The Invention of Communication*, translated by Susan Emanuel, Minneapolis & London: University of Minnesota Press, 1996 [orig.1994].

There seems to be a remarkable contrast between media histories and the mediated realities we live in now. Particularly for those belonging to the younger generations, the word "media" is practically identical with "things mobile." Many people hardly use "location-fixed" media machines any longer. The desktop computer—a sensational novelty in the 1980s—has often been replaced by the laptop computer, which is constantly gaining power, while losing weight and shrinking in size. Yet often even the laptop is considered too bulky and uncomfortable. Many tiny devices that combine the functions of the telephone, the personal computer, the electronic notebook, the music player, the camera, the information terminal and the game machine have appeared on the market; and this is only the beginning. Unlike the media machines of the past, such devices fit into the pocket or the handbag. They go where their users go, unless they are dropped on the sidewalk, forgotten on the bar desk or left in the restroom of a high-speed train. They are media for the "dromological" society (Paul Virilio) that is constantly on the move. Mobile media are linked with any imaginable forms of urban mobility from skateboards and private cars to taxis, subway trains and aeroplanes, not to forget the persisting "proto-motion" of walking. Even if we did not count the futuristic ultra high-tech gadgets seen on the pages of Wired (but rarely elsewhere), the constant presence of more mundane mobile media—mobile phones, digital cameras, car radios, PDA's, pagers, Gameboys, iPods and iPhones—haunts us as we move through urban "non-places" [Augé 1995].[2]

Why have media historians neglected the mobile media? Of course, the answer could be so simple that it hardly needs to be stated: until recently they did not exist. However, the issue is far more complex. As media-archaeologists and cultural historians of technology have purported to demonstrate, the history of media does not comprise only material things—those gadgets that were invented, built, sold, used and finally disposed of—only to be replaced by others. It also includes the discourses that anticipate them, accompany them and contribute to the formation of their social roles and cultural meanings. The 'mobile media' may not have played a widespread role in actual communication practices until recently (although this assumption could be contested), but the 'idea' certainly did not appear from nowhere. Mobile media led discursive lives before the material ones; they may even have existed as the shadows cast by other applications—traveling with the users in guises, fulfilling functions, but rarely perceived for what they were. As it often happens, we needed the explosion of mobile media in the marketplace and the public consciousness before we began to see anything in these shadows. Still, we

[2] Perhaps surprisingly, even Augé does little to integrate mobile media into his thinking, even though this would seem a logical step to take.

should not let our latter day schemata (pre-) determine what we perceive; claiming to have detected the outlines of Alexandrian iPods or Brabantian Blackberrys in the obscurity of the past would be highly suspect, to say the least.

Excavating, identifying and connecting traces left behind by both material and discursive developments are tasks for media archaeology. Often the relationship between the material and the discursive is highly complex; no easy synchronism can be expected. Any attempt at developing "archaeology of mobile media" must be ready to face complicated questions. Do contemporary mobile media really have predecessors or pre-forms? Wouldn't it be more sensible to assume, on the contrary, that a rupture between immobile and mobile media has occurred? Shouldn't we concentrate our efforts on locating its coordinates? Could there be both ruptures and continuities? Even if we came to the conclusion that there indeed have been predecessors that are worth noting, what would we gain from such an insight? Wouldn't its value be mostly antiquarian, useful for historians or collectors, but hardly for the protagonists and agents of today's mobile media world? It is one of the premises of the media-archaeological approach that culture does not merely progress along linear paths, but also constantly recycles its elements, sometimes over long stretches of time. As 'rupturist' and anti-passeist as it may seem; today's mobile media culture may nevertheless contain elements that have already been active within earlier cultural formations. Understanding such elements and the ways they function within varying cultural fabrics can give us clues about the present condition, including its obsessions and excesses. In such a 'spirit,' the present article will undertake a test-excavation of this largely unexplored terrain.

The Desire for Mobile Media

Based on the strategies of usage, three types of mobile media can be identified: portable, wearable and vehicle-mounted. While this division doesn't have any direct historical relevance, it can serve as a tool for analysis. "Portable" refers to things that are carried around, such as a transistor radio or a laptop computer. To use such a device, the owner often stops and places it on a support, such as a table. This is not always the case: a ghetto blaster can also be used in motion. "Wearable" means things that are attached to the user's body in a more rigorous sense than "portable" things, and used in motion. The Walkman and the iPod are typical wearables, although a mobile phone kept in the pocket and used with a headset (a common sight in the cyborg state of California) qualifies as a wearable as well. Perhaps the most common piece of wearable technology, although it is not a "medium" in its basic form, is the wristwatch. A pocket watch qualifies as well, as it becomes part of the owner's clothing, attached to a belt or kept in a dedicated pocket. "Vehicle-mounted" refers to devices that are either built-in parts of the vehicle—like the car radio or the in-flight entertainment system—or regularly used in the car or some other vehicle. Of course, these categories and the relationships between them are not always sharply defined. An iPod can be placed in a docking station that now often replaces the domestic AV-system. A mobile phone could also be classified as a portable, depending on usage. When it comes to vehicle-mounted devices, the borderline is blurred as well, because mobile phones and iPods are increasingly replacing (or supporting) permanent vehicle-mounted fixtures. They are used when the vehicle is in motion, but removed when the owner steps out—thus the "vehicle-mounted" transforms itself fluidly into "portable" or "wearable."[3] In spite of the indistinctiveness of these categories, they are useful aids for explaining the historical 'logic' of mobile media.

It could be claimed that the emergence of mobile media pre-supposes a "desire for mobile media." It often first appears in the intuitions of a few visionaries, manifesting itself in fiction, illustrations, or proposals by inventors, often decades or even centuries before the technical prerequisites come to exist. Sometimes the desire is first fulfilled by another device that will later be associated with other uses. A good example is the telephone, which was not immediately identified as a medium for two-way voice conversations. Telephone concerts were organized, and subscriber-based services that anticipated broadcasting—like the Hungarian Telefon Hirmondo—established. That telephony has never been exclusively reserved for two-way voice communications is proved by the fact that especially in Japan picking up the signal of broadcast television channels has become one of the uses of the mobile phone [Briggs 40-65] and [Marvin 222-231]. In the 19th century, the desire for "broadcasting" was also manifested by other phenomena, including the mass-appeal of illustrated magazines, and stereoscopic photographs and the popularity of giant panoramas depicting recent news-events in the form of wrap-around 'virtual realities'. In *Burning with Desire* Geoffrey Battchen has made an interesting effort to map the coordinates for the emerging desire for photography in the late 18th and early 19th century, pointing out the complexity of the endeavour. Doing something similar for mobile media is not an easy task either—at least if one wanted to look further back than the early 20th century.

Michael Brian Schiffer has shown that the introduction of radio broadcasting in the 1920s inspired almost immediately a variety of designs for portable radio sets, many of them more or less impractical—pinkie ring radios, billfold radios, purse radios etc. [Schiffer 80]. As it frequently happens in media history, these efforts to exploit the "radio craze" and to spark a desire for mobile media had already been

[3] One of the reasons for this usage is fear of crime: these days it is not recommended to leave anything valuable inside the car.

anticipated by the insights of visionary figures like Nikola Tesla and Lee De Forest. Tesla had famously described "[a]n inexpensive receiver, not bigger than a watch, [that] will enable [the user] to listen anywhere, on land or sea, to a speech delivered or music played in some other place, however distant" [Tesla 66].[4] De Forest had even displayed a "Wireless Auto" as a publicity stunt for mobile media in 1904 [Schiffer 23]. As Schiffer's invaluable mapping of the hitherto neglected history of the portable radio shows, it did not appear overnight. The first portable transistor radios, introduced in the mid 1950s, may have seemed a real breakthrough, but the phenomenon had been gathering force for decades. Beside the transistor, social, cultural and economic factors—the emergence of youth cultures, the new consumerist lifestyle—more liberal attitudes toward personal mobility—all contributed to the illusion of its sudden appearance.

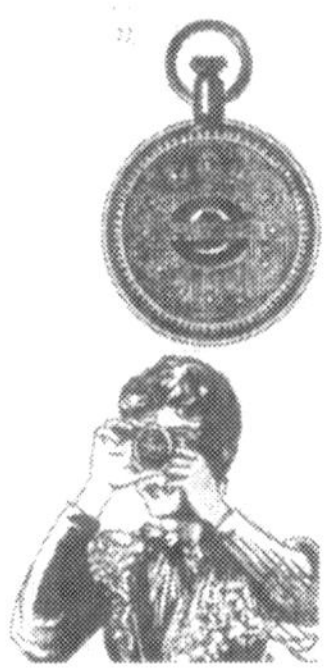

(Figure 1: Panorama Watch)

Another manifestation of a desire for mobile media was the outburst of amateur photography in the late 19th century. Indeed, the easy-to-use box cameras loaded with celluloid roll film could well be described as the first true mobile medium. Introduced in the 1880s, their use soon reached mass proportions, with thousand of amateurs roaming public spaces in search for a subject. Up until then photography had been an occupation for professionals. When amateur photography made its breakthrough (thanks to technological improvements, and the business acumen of George Eastman, the founder of Kodak), it rapidly activated the non-professionals' dormant desire to photograph. The amateurs were encouraged to "snap" every aspect of public life, including unanticipated events, like traffic accidents. Eastman Kodak and the other manufacturers presented amateur photography as an easy, democratic and modern hobby, suited also for women and children [Collins]. What is interesting is that the ideals promoted by the photographic industry were embraced anything but wholeheartedly by the late Victorian society. There were frequent complaints about "The Camera Epidemic." Amateur photographers, including women, were accused of transgressing existing

[4] In his autobiographic writings Tesla quotes from a 'technical statement' describing his "'World-System' of wireless transmission from around 1900. I have not been able to locate the original source for the quotation.

social rules, particularly those related with privacy and decency [Jay 218-240].[5] Their activities were felt to develop into a kind of distributed panopticon—anybody could become a target at any moment. Demands were made for legislation to regulate the use of cameras in public spaces.[6]

An integral aspect of the late 19th and the early 20th century camera craze, was the candid camera. Numerous types of candid cameras, disguised as bags, walking sticks, hats and even pocket watches, were brought to the market. Some of them, like C.P. Stirn's "Concealed Vest Camera" that shot through a buttonhole, enjoyed a commercial success.[7] Whether hiding the camera was primarily a protective reaction against the negative attitudes toward amateur photographers, the ultimate weapon of the snapshot-voyeurist or a token of the surveillance sneaking into 'liberal' societies, is an open question. What is clear is that contrary to the descriptions in corporate discourses and standard histories of Photography; amateur photography wasn't received with unqualified enthusiasm. It was a site for discursive controversy. It is tempting to relate the debate about the snapshot craze—"another terror to the modern existence"—to the ongoing argument about the use of mobile phones, and particularly camera phones, in public spaces. While loud ring-tones have been considered a nuisance from the beginning of mobile telephony, and legislative measures have already been taken in some countries to limit their use in spaces like trains and concert halls, the existence of camera phones may prove to be a more difficult issue. The stories about Japanese 'otaku' taking snapshots from under the

---

[5] The words "The Camera Epidemic" were used in an article in the New York Times in 1884 (quot., 227). Caricaturists often interpreted the intrusions by amateur photographers as sexually motivated, but the camera seemed to have a de-humanizing effect on both sexes. In a telling cartoon (reproduced in Jay, 239) a group of ladies are seen pointing their cameras at a man hanging from a branch of a tree, struggling for his life. Instead of terror or empathy, the faces of the ladies show excitement about the photogenic event; none of them makes the slightest effort to run for rescue. Obviously this refers to changes in the woman's role. Photography club excursions gave women a way of moving through public spaces without their chaperones, and perhaps even exercising their newly acquired power on men, with photography serving as an excuse (and a weapon). However, other caricatures show men in exactly similar actions (Jay, 223, 231).

[6] "The rights of private property, personal liberty, and personal security—birthrights, all of them, of American citizens—are distinctly inconsistent with the unlicensed use of the instantaneous process." Reprinted from an unnamed American periodical by The Amateur Photographer, 25 September 1885, 379 (quoted in Jay, 228).

[7] Stirn's add in the April issue of Scribners Magazine 1889 boasted: "The only camera invisible to the eye; carried concealed under the coat or vest; takes 6 sharp pictures without a change of plate; always ready and in focus. Instantaneous Pictures, stationary or moving, caught unknown to the object. No Tourist, Artist or Student, Amateur or Professional, should be without this Camera."

schoolgirls' skirts in crowded subway trains are only one, albeit an extreme, example [Ito, Okabe, Matsuko, 308-9, 208-11]. Camera phones can function as candid cameras, and their images uploaded on a website in an instant.[8] Although the late 19th century camera craze and the current camera phone craze may seem to have absolutely nothing to do with each other, it could be argued that both have activated the same 'topos', a stereotypical reaction model, to negotiate the role of new technology in a social and cultural setting. According to media archaeology, such 'topoi' accompany and regulate the development of media culture [Huhtamo, Electronic Culture, 296-303, 425-427].

Imaginaries about Mobile Media

Would it be possible to trace the desire for mobile media even further back in time? There is nothing self-evident in the connection between "mobile" and "media." Although communication is a behavioral need shared by all human beings (as well as animals, although not mainly through code-based learning), the idea of moving around with devices that perform this function or perhaps replace the normal auditory environment with a "soundtrack" is cultural rather than universal. It emerges when certain social, cultural and economic conditions are met. As Flichy has noted, proposals about communication at a distance by means of "handy" magnetic devices can already be found from the 17th century [Flichy 17-18]. Observing a magnetic sliver of stone vibrate was assumed to transmit the sentiments of one's lover separated by distance. The belief in the mystical power of magnetism was not far from the idea of telepathic communication. Similar ideas were still proposed in the next century, but again only associated with intimate communication. It was during the French Revolution that the political (and a little later, financial) meaning of telecommunications became clearly understood. However, the means to realize this politicized desire were fixed lines of observation towers crossing the landscape ("the optical telegraph"), followed some decades later by the poles and wires of the electric telegraph. Mobile media was not part of this development; nor had desperate lovers any access to the system. Even horse-mounted messengers, an ancient communication "medium" on the move, had only transported messages between fixed points; the riders had no access to communication devices *en route*.

Obviously the cultural, social and mental circumstances in pre-industrial and pre-urban societies would have prevented any widespread imaginary about mobile media, even if they had been technically viable. The advantages of mobile phones

---

[8] As Ito, Okabe and Matsuda point out "[a]lthough keitai cameras are all designed with a shutter sound that cannot be turned off, users can mute the sound with a finger or a piece of clothing." (Ibid., 309.)

would not have occurred to the inhabitants of a village, who stayed within a limited radius from their homes all their lives and communicated face-to-face with familiar people. As Alain Corbin has observed, the chimes of the village bells created "a territorial identity for individuals living in range of its sound" [Corbin 117]. The community defined by these sounds was what mattered, not the alien communities "beyond." According to similar reasoning, the Walkman would not have made much sense to the landed gentry of the ancien régime, who would have found no reason to seclude themselves aurally from their immediate auditory environment. Living in guarded country estates, surrounded by bird song, engaged in "polite" chatter with peers, listening and performing music only as a coded element of their social routines, free from the dangers and the escalating noise of the city—the aristocrats would not have known what to do with the Walkman, at least before the shouts of the revolutionary "mob" began getting louder. With 19th century industrialization and urbanization devices like the mobile phone and the Walkman would have begun to make sense, although they still seem to have been largely missing from the cultural imaginary. Interpreting Baudelaire's (and Benjamin's) flaneur as a kind of "human candid camera," engaged in 'stealing' impressions (and occasionally being 'caught' by a fleeting gaze) while hiding in the crowd, would be tempting, but it would also mean forcing a mould from a later era onto earlier circumstances.

In our quest for traces of mobile media, it might make sense to change direction slightly. Instead of looking for anticipations of portable devices meant explicitly for communication, recording and / or playback, we could focus on the symbolic meanings of everyday objects [Bull 138-143].[9] Objects are never merely functional—they gain private meanings as part of their owners' and users' life histories. Conventions of displaying or hiding one's personal things can be highly significant. Of course, such idiosyncratic meanings are only possible within the wider signifying practices of the culture of an era. While performing their 'primary' material functions, objects also serve as discursive objects that can be both highly complex and unstable. The daguerreotype was unprecedented in its ability to preserve and represent an almost uncanny likeness of a human. Soon after its introduction in 1839, a massive industry emerged around it. Contemporaries described the daguerreotype as an "image fixed on the mirror," which opens a whole spectre of meanings [Csikszentmihalyi and Rochberg-Halton 143].[10] Gazing at the shiny silver

[9] For a critique of the notion of flaneur as it has been applied to later cultural contexts within urban and media studies, see Michael Bull.

[10] There are many kinds of portable objects that could be discussed here from travel altars to pocketbooks, the invention of which radically changed the mobility of the written word. Analyzing these cases must be left to another occasion.

nitrate surface of the daguerreotype, the looker saw his/ her own face (or part of it) superimposed on the Figure depicted in the picture. This created a communication of a kind, which it would be tempting to describe as a "metaphysical picture phoning," particularly because daguerreotypes were commonly enclosed in hinged boxes that could be carried in the pocket.

Once again, such an interpretation risks superimposing future ideas onto a past that did not have the "picture phone" in its imaginary (it appeared only in the late 19th century, with the emerging discourse on "tele-vision"). However, one might counter this argument by claiming that the imaginary around the daguerreotype is actually part of the longer discursive tradition of seeing mirrors (including hand-held ones) as virtual gateways to other realities [Battchen 40].[11] Another category of objects that has been neglected by media scholarship, but which has significance from the media archaeological point-of-view, are women's fans and hand-screens. These seemingly superfluous objects have a long and varied history [Huhtamo, Iconics, 31-82]. Their importance goes beyond their roles as collectables or decorative arts. Fans and hand-screens had practical functions, guarding the bearers from heat or intruding gazes. However, they also became symbolic manifestations of the owner's social status, and means of erotic play. Barely perceptible movements of the fan could be used for secret communication. There were codes "languages" of the fan. Throughout centuries, enormous varieties of pictorial fans have been produced, ranging from delicate miniature artworks to cheap mass-produced giveaways [Bruno 134]. Some fans bear political slogans, while others have concert programs or maps printed on them—a folding souvenir fan for the 1867 Universal Exhibition in Paris has a bird's-eye-view of the exposition grounds on one side and a clearly marked map of the principal gates and routes on the other; the temptation to name it as a 19th century GPS device is great [Payen-Appenzeller]. There are fans that have been designed as novelty attractions, incorporating 'technology' like translucent 'dioramic' images or even movable panoramic picture rolls.[12] It could be argued that all this makes pictorial fans and hand-screens a portable "medium" of a kind, although only in an oblique sense, through issues of usage and symbolic meaning. Still, these issues are not unfamiliar

---

[11] According to Battchen, the professor of chemistry Richard W. Habersham remarked "how grand it would be if we could invent a method of fixing the image on the mirror," recollecting his collaboration with his student Samuel B. Morse in the early 19th century. Morse became a pioneer in both telegraphy and daguerreotypy. As Battchen remarks, Habersham obviously wanted to add his name to the list of the innovators who had felt the desire to photograph before Daguerre and Talbot. For us, the expression he uses is interesting.

[12] I have seen two slightly different designs of this fan. One was sold on eBay, and the other by the French collector and toy seller, Francois Binetruy. I keep pictures of both fans in my archive.

in the context of current mobile media either. Covers that can be exchanged, stickers and imaginative carrying cases—together with the aural attraction of idiosyncratic ringtones—convert mobile phones into symbolic attractors resembling the fans of earlier centuries.[13]

Wearables: Fantasies and Realities

For the media-archaeologist, one of the interesting issues raised by wearable media is the interplay between overt and covert usages. Regardless of the 'primary' function of the device, it is important to note how we carry it and explain the meaning of this habit. I have seen men who carry their mobile phones visibly in a holder on their belt much like a revolver—is this just a coincidence? Or does it, perhaps unawares, carry connotations of power and masculinity? Historically one extreme in 'wearing the technology' overtly are stunts where a person (usually a woman) practically dresses up in technology, turning it into a body-mounted spectacle. This topos was frequently evoked in the 19th century, obviously to celebrate new technologies—in fancy balls (as well as in illustrated fashion prints), women wore camera-hats and dresses covered by photographic prints or by (functioning!) light bulbs.[14] Such extravagant creations could well be linked to the on-going discussion about the representations of technology as the (feminine) other, although this issue cannot be elaborated on here [Ferguson 28-29].[15] Still, it is tempting to claim that there may well be a link between these 19th century wearable stunts and the many "smart garments" displayed in recent years at venues like Siggraph's Cyber Fashion Show. Using "active" fabrics, flexible displays, or media technologies like fiber optics or LCD panels sewn into the dress seems a celebratory gesture as well, with—for the time being—relatively limited practical uses in everyday life [Huyssen].

---

[13] Examples of both types are in the author's collection. They are French and date from the 1820s or 30s.

[14] In Japan the tradition of using imaginative straps with plush toys, miniatures and amulets of all kinds attached to the cell phone can be linked to the old tradition of netsuke, as professor Machiko Kusahara from Waseda University, Tokyo, has taught me. Netsuke were miniature sculptures that had a practical function: because Japanese kimonos had no pockets, thinks had to be carried by attaching them to the sash belt by means of strings and a netsuke. Netsuke also had symbolic meanings to the bearer and could communicate certain meanings to others as well.

[15] The last mentioned idea was resurrected by the Japanese Gutai artist Atsuko Tanaka in her work Electric Dress (1956). Tanaka performed herself in the dress that was covered by hundreds of blinking light bulbs, emulating the traditions of the Japanese marriage ceremony. This work has been seen as prophetic in its way of combining the body and 'dangerous' technology. It is not clear whether she knew about the 19th-century anticipations. A working replica of the work was displayed at the Documenta 07 in Kassel, 2007.

In the covert end of the spectrum we find phenomena like "ero-tech" and "psy-tech." These are devices carried around, but hidden from other people's gazes. "Ero-tech" contains large varieties of erotic objects, often with mechanical or optical features, meant to raise sexual desire when discretely displayed for someone. Such devices include peep-viewers with miniature pornographic photographs [McQuaid]. "Spy-tech" refers to the huge tradition of secret devices designed to record conversations, take snap-shots, transmit and receive secret messages and even kill people. Candid cameras—not to say anything about candid guns—are related to this tradition as well. The history of spy-tech fluctuates interestingly between real spy gadgets and those imagined in detective stories, films (James Bond) and television series (Men from U.N.C.L.E.), said to have influenced each other [Bourgeron].[16] The two-way wristwatch radio used by Chester Gould's comic-strip hero Dick Tracy in the 1940s inspired companies to produce working models, although they were often little more than toys. Eventually even the US Army Signal Corps produced one with a short wire antenna that ran up the sleeve "comic-strip style."[17] Spy-tech is simultaneously overt and covert. In spite of its widely known discursive presence, it is at the same time exclusive, carving out secret channels for "private" communication.

If conventions of use and symbolic meanings are accepted as worthy of critical attention, a device like the wristwatch deserves attention as a predecessor to wearable media. According to a well-known story, the French jeweller and clock-maker Louis Cartier invented this now ubiquitous device in 1904 for the Brazilian aviation pioneer Alberto Santos-Dumont, who found it difficult to use his pocket watch to check the time while steering his aircraft [Schiffer, 154-155, 174-175]. The wristwatch was the perfect device to the high-speed technological environment of an aeroplane or a motorcar, where intense concentration is required, and one false move of the arm may be fatal. Although fascinating, the story about Santos-Dumont's wristwatch is not complete: the device had been invented much earlier, but its popularity grew slowly, because it was considered feminine—it was associated with the habit of wearing bracelets [Winters 52, 113-119].[18] The technology-studded profile of Santos-Dumont may have convinced some men about

---

[16] In Japan, erotic netsule were also popular among men.

[17] See the interesting web site *Spy-Fi Archives*. <http://www.cia.gov/spy_fi/index.html>

[18] Cartier's Santos watch is still in production. About the early history of wristwatches, see Carlene E. Stephens: On Time. How America HAS learned to Live by the Clock, Boston, New York and London: Bulfinch Press / Smithsonian Institution, 2002, 133-135. Stephens does not mention the Cartier & Santos-Dumont episode, although she has published a picture of an aerial photographer wearing a wristwatch in 1918 (p.134).

the masculine potential of the device. However, more convincing proof of its usefulness and newly found masculinity was gathered in factories and offices where the employees spent much of their time controlling a machine, and eventually in the trenches of the two world wars, where having a wristwatch could become a matter of life and death. Placed on the axis between the overt and the covert, the status of the wristwatch seems to be constantly shifting, depending on the movements and the length of the sleeve. Its predecessor, the pocket watch, seems to belong closer to the covert end. A pocket watch is attached to the clothes (hanging from chains and often kept in a special pocket), and taken out only sporadically; the user stares at its "screen" and puts it back.[19] This evokes some of the ways in which mobile phones are used.

When David Sarnoff, one of the pioneers of radio broadcasting, introduced in 1922 his vision about the portable radio, he used the watch as the model. According to Sarnoff, the radio should have as its ideal "the watch carried by a lady or a gentleman, which is not only serviceable but ornamental as well."[20] Although it is not clear whether he meant a wristwatch or a pocket watch, Sarnoff's idea of the portable radio as a both useful and neatly designed personal utility, resonated within media culture. The St. Louis-based jeweler J.A. Key introduced a radio set modeled after the pocket watch, and even a radio "pinkie ring" was proposed by another inventor [Schiffer 67-68].[21] As this shows, the design for Dick Tracy's a voice activated wrist-mounted videophone did not appear out of nowhere. Although such novelties have often remained little more than publicity stunts, the interplay between real R&D and discursive fantasies is extremely interesting, and an essential aspect of an "archaeology of mobile media." It also shows that wrist-mounted television-sets, telephones, digital cameras, music players and other devices that are regularly promoted as the newest of the new in Wired and elsewhere, are based on a well-established tradition [Schiffer, 80].[22] It may seem curious that merely changing the appearance or adding some new functionality can wipe away the awareness about

---

[19] According to Stephens, "as late as 1943, wristwatches were still sometimes called 'bracelet watches' or 'wristlets,' recalling feminine jewelry." (op.cit., 135).

[20] Around 1900 "dioramic" picture viewers in the shape of the pocket watch were produced in France. The devices contained a rotating disc with minuscule images. Georges Demeny, Etienne-Jules Marey's former assistant even invented a pocket Fonoscope, a movie camera resembling a pocketwatch. It took a short moving image sequence on a rotating sensitized disc, and could also be used for viewing developed (positive) discs. The only know example in the collection of the Musée du Cinéma of the French Cinématheque. I would like to thank its keeper Laurent Mannoni for displaying the device for me.

[21] According to Schiffer, Sarnoff suggested that the apparatus could also function as a flashlight!

[22] The "pinkie ring" receiver was invented by Alfred G. Rinehart.

this, again and again. Yet, this may be the very point: more often than not, novelty is guided by the invisible "hand of tradition."

Another aspect of wearable media, situated somewhere along the overt-covert continuum, still needs to be named: hands-free interfaces, most often associated with a head-mounted headphone-microphone combination connected with a mobile phone in the pocket or placed in a stand next to the driver.[23] In places like California this combination has come to accompany the movements of countless citizens—many of them concerned about radiation, others just following fashion—in supermarkets, cafes, and in and out of the car. Considering the pervasiveness of this outfit, turned into a technological harness worn from morning until the evening, it is tempting to characterize its users as 'cyborgs'. Although a full treatment of this issue would take too much space here, certain manifestations of the 'cyborg topos' can be mentioned briefly. In the nineteenth century, cartoonists often presented the early photographers as "a new species," partly human, partly technological: the camera, with its one large "Cyclops' eye" (the lens) had replaced the photographer's head, hidden under the hood. The (mis)adventures of the "Elephans Photographicus" as it looked for its optical preys were fantasized about and laughed at.[24] Later in the century the cyborg appeared in offices, tapping a typewriter and wearing the headset of a dictating machine, and at telephone exchanges, where young female cyborgs were forced to spend hours "bondaged" to the switchboard, wearing a headphones / microphone combination [Henisch and Henisch, 29]. How the 'cyborg' became mobile is not yet quite clear, although s/he was certainly vehicle-mounted (sitting on the pilot's seat, for example) long before beginning to walk and run donning a Walkman.

Conclusion

The excavations undertaken in this article are tentative. Little has been said about vehicle-mounted mobile media, a topic for another article [Gold and Koch, 68-70,114; Lupton].[25] "Mobile media" has been treated more of less as a lump, as if crucial differences did not exist between their various forms. Although many different functions are converging in the latest super gadgets, it would still be important to

---

[23] The "Play: Fetish" section of *Wired*, Vol.12, No 8, (August 2004) present, among other things, a wristwatch that includes a 1.5-inch TV set from the Japanese company NHJ.

[24] To which extent the popularity of this interface could have been influenced by the use of wireless head and neck-mounted microphones by popstars like Madonna is an interesting question. The situation also somehow resembles the ways of useing some early candid cameras. They were wearables hidden underneath the clothes and secretly operated with a wire from the pocket.

[25] Working women were forced to become "cyber" also when they were submitted to physiological tests measuring the efficacy of work with body-mounted interface.

reflect on the relationships between devices meant for voice or text communication, image or sound reproduction, or just "playback." A candid camera is very different from a radiophone or a portable cassette player; such differences are necessarily reflected in the conventions of usage as well. Nothing has been said about the content that the various mobile media transmit. However, I do agree with Michael Bull when he writes "the structural placing and role of the personal stereo in everyday experience can be analytically distinguished from that which is listened to" [The Auditory Culture Reader, 357-374].[26] Mobile media devices can be treated as "apparata," that are partly technological, partly psychological, partly cultural. The devices themselves incorporate certain "built-in" modes of usage that are then negotiated—perhaps embraced, perhaps contested—by the users themselves. It is possible to explain the functioning as such an "apparatus" without performing an analysis of the content transmitted by the system. We glance at the "screen" of our watch to see the time or the date; we can get the same information from the mobile phone screen as well, but we also use it for many other purposes. As apparata, these devices can be compared because of their "interfaces" and their conventions of usage, including the looks they "encourage" [Bull 14].

Contemporary mobile media seems to have been born as if through parthenogenesis. Both the gadgets and the social formations that surround them seem to have appeared very recently, as if out of the blue. This idea is embraced by the mobile media users themselves (particularly the younger ones), who eagerly grasp the opportunity of being part of something new and unprecedented, while living their everyday lives. Although there are certainly many mobile users whose relationships to their devices are practical and mundane, there are also those, for whom the act of using them contains surplus value that exceeds (or at least parallels) the value of the communication itself. Seeing the mobile media as the newest of the new is in the interests of the manufacturers and retailers, who regularly parade the latest mobile phone or iPod models in front of their customers, always with new features, new designs or more processor power (as a counterpoint, there are also stripped down, easy-to-use models to persuade even the prejudiced elderly to join the world of mobile communications). By performing a series of test-excavations as a prelude to "archaeology of mobile media," this article has shown that the novelty of mobile media is more about ideology than about factual reality. Although it is incontestable that mobile communications have reached truly massive dimensions only recently, this has been anticipated on various levels during long

---

[26] About connections between cars, media and visuality, see Anne Friedberg, "Urban mobility and cinematic visuality: the screens of Los Angeles - endless cinema or private telematics." *Journal of Visual Culture*, Vol. 1, No 2 (2002), 183-204.

periods of time. Excavating the earlier manifestations will give us new insights to the current—and perhaps future—forms and uses of mobile media.

Bibliography

Augé, Marc. *Non-Places. Introduction to an Anthropology of Supermodernity.* Trans. John Howe, London: Verso, 1995.

Battchen, Geoffrey. *Burning with Desire: The Conception of Photography.* Cambridge, Mass.: The MIT Press, 1997.

Bourgeron, Jean-Pierre. *Les Masques d'Eros: Les objets érotiques de collection à systèmes.* Paris: Les Editions de l'Amateur, 1985.

Briggs, Asa, and Peter Burke. *A Social History of Media from Gutenberg to the Internet.* Cambridge: Polity Press, 2002.

Briggs, Asa. "The Pleasure Telephone: A Chapter in the Prehistory of the Media." *The Social Impact of the Telephone.* Ed. Ithiel de Sola Pool. Cambridge, Mass.: The MIT Press, 1977.

Bruno, Giuliana. *Atlas of Emotion: Journeys in Art, Architecture, and Film.* London: Verso, 2002.

Bull, Michael. *Sounding Out the City: Personal Stereos and the Management of Everyday Life.* Oxford and New York: Berg, 2000.

– – –. "Soundscapes of the Car: A Critical Study of Automobile Habitation." *The Auditory Culture Reader.* Ed. Michael Bull and Les Back. Oxford and New York: Berg, 2003.

Friedberg, Anne. "Urban mobility and cinematic visuality: the screens of Los Angeles—endless cinema or private telematics." *Journal of Visual Culture*, Vol. 1, No 2, 2002.

Collins, Douglas. *The Story of Kodak.* New York: Abrams, 1990.

Corbin, Alain. "The Auditory Markers of the Village." *The Auditory Culture Reader.* Ed. Michael Bull and Les Back. Oxford and New York: Berg, 2003

Csikszentmihalyi, Mihaly and Eugene Rochberg-Halton. *The Meaning of Things: Domestic Symbols and the Self.* Cambridge (UK): Cambridge University Press, 1981

Flichy, Patrice. *Une histoire de la communication moderne. Espace public et vie privée.* Paris: Éditions de la Découverte, 1991.

Ferguson, Russell, ed. *Between Performance and the Object, 1949-1979.* New York: Thames & Hudson, 1998.

Henisch, Heinz K., and Bridget A. Henisch: *Positive Pleasures. Early Photography and Humor.* University Park: Pennsylvania University Press, 1998.

Gold, Helmut and Annette Koch, eds. *Fräulein vom Amt.* Munich: Prestel-Verlag, 1993.

Huhtamo, Erkki. "From Kaleidoscomaniac to Cybernerd. Towards an Archeology of the Media." *Electronic Culture.* Ed. Timothy Druckrey. New York: Aperture 1996.

– – –. "Elements of Screenology: Toward an Archaeology of the Screen." *ICONICS: International Studies of the Modern Image*, Vol.7 (2004 Tokyo: The Japan Society of Image Arts and Sciences.)

Ito, Mizuko, Daisuke Okabe, and Misa Matsuda, eds. *Personal, Portable, Pedestrian. Mobile Phones in Japanese Life.* Cambridge, Mass.: The MIT Press, 2005.

Huyssen, Andreas. *After the Great Divide: Modernism, Mass Culture, Postmodernism.* Bloomington: Indiana University Press, 1986.

Jay, Bill. *Cyanide & Spirits. An Inside-Out View of Early Photography.* Munich: Nazraeli Press, 1991.

Lupton, Ellen. *Mechanical Brides. Women and Machines from Home to Office.* Princeton: Princeton Architectural Press, 1996.

Marvin, Carolyn. *When Old Technologies were New. Thinking About Electric Communication in the Late Nineteenth Century.* New York & Oxford: Oxford University Press, 1988.

McQuaid, Mathilda. *Extreme Textiles: Designing for High Performance.* Princeton: Princeton Architectural Press, 2005.

Mattelart, Armand. *The Invention of Communication.* Trans. Susan Emanuel. Minneapolis & London: University of Minnesota Press, 1996.

Payen-Appenzeller, Pascal. *Fancy Fans.* Paris: L'Aventurine, 2000.

Schiffer, Michael Brian. *The Portable Radio in American Life.* Tucson & London: The University of Arizona Press, 1991.

Tesla, Nikola. *Moji pronalasci / My Inventions.* Ed. Branimira Valic. Zagreb: Skolska Knjiga, 1987.

Winters, Nancy. *Man Flies. The Story of Alberto Santos-Dumont Master of the Balloon.* Hopewell, New Jersey: The Ecco Press, 1997.

Winston, Brian. *Media Technology and Society, A History: From the Telegraph to the Internet.* London and New York: Routledge, 1998.

The Temporal and Spatial Design of Video and Film-based Installation Art in the 60s and 70s: Their Inherent Perception Processes and Effects on the Perceivers' Actions

Susanne Jaschko

This text is an excerpt from a paper which was researched on the basis of a grant by the Fondation Daniel Langlois. Hence focus has been laid on the documents archived at the Foundation.

I. New Territories and Transitory Places

The shaping of temporal and spatial *topoi* of installation media works of the late 1960s and 1970s was accompanied by the occupation of new spatial and contextual territories. The search for new spaces and different publics for art characterized this period. This resulted from a fundamentally new determination of the function and formats of artistic production. Many artists working with video and computers in a performative and installative way did not inevitably define their creations as artworks. More often their method of production and the way their works were conveyed was based on the principle of experiment and was connected to a utopian concept of their respective medium. Video in particular was viewed as a democratising medium, which theoretically enabled everyone to produce something creative. It was also seen as a medium of the expression and the mediation for a new type of experiential culture, the central point of which was the broadening of perception and consciousness—a culture of experience, which was influenced by the drug culture of the time.

In accordance with this image of an avant-garde experiential culture; computer and other electronic media were artistically utilized for the creation of new, immersive experiential spaces and compositions. Already before the video and computer era, dynamic environments had been created, which had been orientated by a bodily and sensory experience of space. An important example in relation of this was the legendary *Philips Pavilion* (1958) for the World Exhibition in Brussels that was created by Iannis Xenakis and Le Corbusier. In following years, Xenakis created a series of *Polytopes*, buildings which were similar to pavilions and which were conceived of as dynamic systems and time-based art. The *Poème Electronique* in the *Philips Pavilion* already used the newest technologies developed by Philips in order to cause an immersive sensory experience in the interior by using moving image, projections of colour images and spatial tones [Sterken, 262-273]. A large number of temporary architectural works were created in this tradition, in the framework of the following world exhibitions, such as the Pepsi-Cola Pavilion, which was carried out by the group Experiments in Art and Technology (E.A.T.) for the World Exhibition of 1970 in Osaka.

The close co-operation of artists and engineers from Bell Laboratories in the context of E.A.T. resulted in a spectacular building, which stimulated every sense. Its futuristic conception was mediated from a distance by artificial clouds of fog, which continuously covered the hemisphere. In the interior, visitors experienced a spectacle of light and images, which were transformed many times by projections, mirrors and lasers, as well as stationary and mobile sources of sound.

The concept of an immersive, multimedia utilization of large interiors like the World Exhibition pavilions is found again in some video festivals of the early 1970s. With the flowering of video art and video culture came a desire to reach a wide audience. On the one hand, the practical motives for the organization of festivals lay in the lack of a distribution network; this was only created bit-by-bit through the establishment of initiatives like Electronic Art Intermix and art video libraries at the beginning of the 1970s. On the other hand, festivals with their independence from an institutional, spatial framework offered an ideal opportunity to capture the innovative, experimental character of video culture in their scenography. They were by no means inevitably and wholly conceived of as pure art festivals, but, at least partly, as a platform for information with a widespread socio-political impact.

> "The intended audience is anyone with an interest in the future of society—the public, media people, and industry can all learn from this interchange. The festival proposed here is an attempt to explain the new video technology by showing what it can do."[1]

The planned *Video Festival, an exposition for understanding the medium* (which was not actually carried out, but was supposed to have taken place from the 25th-29th of May 1972 in two inflatable indoor tennis halls in New York City) was in its conception one of the largest and most inclusive video festivals of its time. The festival concept came into being in the collaborative development work of the *Alternative Media Project*, where Steina and Woody Vasulka were also consultants. The concept, which runs to 30 pages, is a rare document of its kind, as it describes all the elements of the festival and the organization of the content in detail. The *Video Festival* was conceived of as a transitory space for experience, where it was planned that the visitor would be integrated into a totality of picture and sound

[1] Media Project. Video Festival. Proposal. New York City, April 20, 1971. 8. VAS B22-C9.

media as he moved through the different environments and temporarily lingered in them: large-format video projections were planned to be shown in different parts of both indoor tennis centres, projected on screens or on objects like a 20-foot high cube and accompanied by regular electronic sound.[2] This video and acoustic environment was supposed to create an all-encompassing impression. The so-called "Video Environments," in their various forms (and partly situated in an interior with furniture and objects) were conceived of for the presentation of videotapes. A matrix of monitors was also supposed to be used in this spatial context, either as a giant wall of monitors or grouped in smaller units.[3]

A closed-circuit video sculpture, recorded the visitors in real time from different perspectives and relays. Its recording on monitors was supposed to stimulate the visitor to interact with their own image. In a second version of this installation, a time delay in the relaying of the recording was planned. In addition to the so-called "Controlled Interaction situations," it was planned that the video recordings from further closed-circuit cameras would furnish the whole projection and monitor environment with "Free Organic Interactions," by the visitors, camera operators and technicians who were in the space. Other areas in the interconnecting halls were to serve as places to demonstrate video technology and future television technology: places for acting out or for relaxing.

The plan to hold the festival in two indoor tennis centres was undoubtedly a product of a conscious decision in favour of a place which was publicly effective and which was not associated with art. Like the Vasulkas, the majority of the artists, curators and organizers were interested in finding the most "uncontaminated,"[4] new and temporary places. The creators of the *Annual Avant Garde Festival* worked in this way, the festival being innovative and experimental in terms of form and content, and which started in 1963, and began presenting video works in 1967. The festival introduced contemporary artistic works, which overlapped in terms of genre and media in new spatial contexts. The decision to use mobile venues like ships (1967 and 1972) or freight cars (1973) and the festival event in the form of a parade (1968) was in accordance with the process-related, mobile and transitory character of the festival.

---

[2] These data as well as the descriptions and quotations occurring in this section are from Media Project. Video Festival. Proposal. New York City, April 20, 1971. VAS B22-C9.

[3] It refers to the Video-Matrix of the American Can Company, which is discussed at length later in this section.

[4] "We showed in churches, we showed in state houses, so we had our own policies to forget the official word that was very corrupt, you see. (...) So we had it uncontaminated." Woody Vasulka in the interview, which the author conducted on July 28, 2005. In this interview, Steina and Woody Vasulka participated.

E.A.T. organized a series of media performances as early as 1966, the *9 Evenings* in the Armory, a building which served as a military institution, as the clubhouse of the National Guard, and where the legendary Armoury Show took place in 1913. In *9 Evenings*, the audience in the performance space was completely surrounded by image and sound media and was part of a series of multi-media performances.

Television as a space of experience was fascinating for artists, primarily because of its capability of reaching a mass-audience. In a similar manner to the festivals, where artistic production was presented in non-artistic spatial contexts, artists created artistic works for television, a non-artistic context, and hoped by broadcasting them to create a lasting effect on the viewers. The break in the perception habits and the undermining of expectations were the main intention of artists like Douglas Davis who worked with the medium of television. In a number of videotapes created especially for broadcasting on the television networks, he presented television as an object and released the viewer from his passive role by challenging him to participate in an active way.

There was a strong belief that video was an important part of a newly formed Cable-TV landscape and it was reinforced by the initial openness of the broadcasting stations. The video pioneers saw the role of the cable stations as being that of educators of consciousness and the supporters of local culture [Ryan 12]. Moreover, cable television was moving towards the dream of a two-way-communication between broadcasting stations and receivers, the latter likewise becoming a station:

> "Its two-way capability can revitalize the democratic process in heavily populated areas. For example, multiple input video dialogues on subjects of local concern could prove to be a revival of the town meeting in areas where such gatherings are no longer possible."[5]

One of the many professed utopias was for the decentralization that would come at long last from the older channel networks which had been inaccessible to the public, by building up a computer-based programme archive, where it was possible for everyone to submit their video productions. Even if this dream remained unattainable, the utopia of cable television which offered public access to groups and individuals did in fact become reality for a relatively short time at the beginning of the 1970s.

---

[5] Alternative Media Project. Video Festival. Proposal. New York City, April 20, 1971. 5. VAS B22-C9.

The invention of the Porta-Pak, the first manageable video system which was freely available on the open market, was the basis for a concept of society which produces moving images and which uses video as a natural medium of information and communication. Even if today, in times of digital video on mobile phones and cameras, the beginnings of this movement seem comparatively modest; the effect of amateur recordings of everyday and unusual situations, in contrast to the professional and mostly conservative products of official forms of media, must have had a massive effect on the public.

The *Video Festival*, which in the end was not carried out, reflected in its planning stages all the claims and hopes which were linked to the Porta-Pak. Large-screen projections would have shown videos of unusual sights such as heart transplants. It was planned that many hundred video productions, ranging from interviews to feature films on video, would be shown on monitors. Thus, new territories in terms of content, which were until that point either the preserve of professional media producers or were not covered by the contemporary media, went hand in hand with the utilization of new spaces, which to date had not been used for cultural purposes.

Many new spaces, places for exhibiting and producing were created, which were above all dedicated to creative work in video. One of these centres, and definitely one of the most effective ones, was *The Kitchen* in New York, which was founded by Steina and Woody Vasulka in 1967 in order to give the video community, who frequented their apartment and studio, a new refuge. The development of *The Kitchen* is representative for a series of such initiatives, mostly by artists, which sought to furnish artistic experiments with new media. Alongside these artist initiatives, the communal libraries played an important role in the propagation of video, as a place of distribution of video works on cassette and as a place to rent video equipment [Hill 27-36]. Without doubt, the special feature of *The Kitchen* was the openness towards a wide spectrum of activities, which took place without curatorial and administrative authority in the spaces of the former kitchen of the Hotel Central in New York City. In this, *The Kitchen* differed substantially from traditional art institutions like galleries and museums, which had devoted themselves to a much narrower definition of art. In including bland urban spaces on the map of art, a mostly critical, and, at the very least, a distanced position was expressed with regard to institutional venues and their programmes:

> "The art gallery was something neither of us knew anything about. Coming from music and film, it seemed quite a distant possibility, to feel at home there and additionally, we understood video to be an activity, not art a priori. At that time, video was not "above" ground so to speak. The only serious exhibit that I had seen until then, was the "television as a creative medium" at Howard Wise's

gallery, so for me, instead of trying to establish myself and my activity within the art galleries, it became obvious, that I could simply bypass the art scene altogether. The conditions were correct for that. Everyone was trying to bypass whatever establishment channels there existed in all directions, from politics to art."[6]

The establishment of video initiatives, video centres, groups like E.A.T. or platforms like *The Kitchen* served a mainly pragmatic purpose: to emancipate the artists and make them self-sufficient by offering unhindered access to equipment and exhibition spaces. In view of the experimental and performative quality of most of the artistic works, there was a great need for presentations with short-term programming and technical infrastructure. One result of this policy of establishing their own platforms for artistic creation and presentation was that a new, parallel scene came into being, which had only a limited exchange with other art and culture scenes. It would nevertheless be wrong to assume that the traditional art institutions had not begun working with this new media art. Already as early as the mid to late 1960s, some museums and galleries became involved and exhibited this new art themselves or supported the many video festivals, which were mostly organized by artist and co-operative initiatives.

Public space as a place of production and exhibition also came into the artists' focus due to the growing manageability and mobility of the equipment. While video productions were possible in the public sphere due to the mobility and convenience of the video camera, and were often carried out, the public sphere as a platform for the presentation of artistic media works came with practical problems of its own. Witness Steina on this point in an interview with Russell Connor:

"I would like very much to do those kinds of things—installations—in a public. It is very difficult … it is hard to put it there, in a public space. It is hard to maintain. It is an investment, but what is interesting to me is to interact with it, to walk around and be part of the whole thing. But in the meantime I have no other means to do it but here at home."[7]

---

[6] This quotation originates from an original document of the Vasulkas, which was written in several versions. It seems to be a draft for a statement of the Vasulkas on the beginnings and goals of The Kitchen, which was published by The Kitchen in 1978. The published statement no longer contains this passage. VAS B 34-C6-2 dll433/1 to 2

[7] Connor, Russell. Vasulka. Manuscript. Interview report. VAS B32-C10-01

In retrospect, Steina also expressed criticism about the public sphere as a place of exhibition and thus put her position of that time into perspective:

> "I never wanted to convert anybody. The idea of the innocent passer-by by chance looking at my work and loving it, I never bought into that. I was more into that concept having it somewhere, where people would come for the purpose. In that sense, gallery is nice that people come for the purpose of knowing and learning. But it has a lot of other problems."[8]

Characteristic of the conception of the public sphere as an ideal platform for art and agitation is the argument over the presentation of the *Modular Video Matrix* in that public sphere. Originally produced for the American Can Company in collaboration with Woody Vasulka,[9] the *Modular Video Matrix* was composed of a closed-circuit system of 56 television monitors, seven live cameras and seven video recorders, encased in a modular system of steel and plexiglass. In its original form, which was created by Ira Schneider, Frank Gillette, Paul Ryan and John Riley, it was used for presentation purposes for the products of the American Can Company at trade fairs, even if that was not in line with the intentions of the four creators.[10] Characteristically, an argument flared up later over the use of the matrix between Videosphere, a charitable foundation which had supported experimental video work and *Automation House*, a project of the American Foundation on Automation and Employment, which had renovated an old building in New York City into an highly-technological building, into which E.A.T, among others, had moved and was actively giving it an artistic form. While *Automation House* saw the matrix as a permanent installation used for artistic experimentation in the house, in Videosphere's eyes, the matrix, due to its mobile character, should primarily be utilized in different places: "in churches, satellite street showings, community centers, etc."[11] According to

---

[8] Steina in the interview with the author on July 28, 2005.

[9] This responsible contribution by Woody Vasulka to this project is taken from a confidential appraisal that Harvey Lloyd wrote on December 28,1970. The appraisal served in support of Woody Vasulkas application for a scholarship from the John Simon Guggenheim Memorial Foundation. VAS B21-C2.

[10] This quotation originates from an original document of the Vasulkas, which was written in several versions. It seems to be a draft for a statement of the Vasulkas on the beginnings and goals of The Kitchen, which was published by The Kitchen in 1978. The published statement no longer contains this passage. VAS B 34-C6-2 dll433/1 de 2

[11] The artists expressed themselves as follows: "This Matrix was designed for The American Can Corporation and contrary to the desire of the designers the software was assembled by Harvey Lloyd Productions. Software for the Matrix presented at Industrial Trade Shows consisted mainly of bald-headed men touting American Can products intermixed with men (live camera) gawking as cheesecake hostesses." Radical Software. Volume II. Number Video and Environment. Winter 1973. 18-19. <http://www.radicalsoftware.org/volume2nr5/pdf/VOLUME2NR5_art05.pdf>

Videosphere's plans, the matrix should have been used for the distribution of video productions:

> "We are aware that there is a multitude of software shelved with no chance of distribution. There is no tool of distribution. By design the *American Can Video Matrix* is organically the perfect system to fill this void. The responsibility of the artist today is not only to experiment but to communicate."[12]

There were many reasons behind the numerous attempts to bring artistic video works into the public sphere. As well as the new understanding of the artists' role as communicators and distributors of their work, and the utopian idea of an art which had the power to change society through enlightenment and direct confrontation, there was another goal: to escape the dark, fictional space of the cinema and to set it against these open and real spaces. The use of television monitors enabled public and semi-public space to be captured in installations, since television monitors, in contrast to projections, are independent of the light conditions; moreover they are an off-the-shelf-product and were substantially cheaper, easier to manage and more mobile than the projectors of the time. Surely this fact was another reason behind the video artists' predominant use of the television monitor in the 1960s and 1970s.[13]

II. Spatial and perceptive immersion

> *"In the early days of video, everything was an installation or an "environment," as we used to call it."*[14]

In the initial phase of working with video, the artists did not make any specific differentiation between various formats and forms of expression such as installation, performance or environmental. Indeed, rigid categorization was to be overcome and replaced by new concepts of presentation and interaction. In this way, new terms like "environment" were welcomed and used as a description for various spatial formats and sizes. However, this investigation differentiates between the terms sculpture, installation and environment, in order to highlight the different spatial

---

[12] This quotation is taken from a letter, which Videosphere addressed to the American Can Company on April 1, 1971, in order to clarify their intended purpose for the matrix and distance themselves from their former co-operation partner Automation House. VAS B22-C9.

[13] Ibid.

[14] A further, relevant reason was naturally the television culture, which unfolded a substantial social effect in this time, on which artist such as Nam June Paik in its work reported. See for this the paragraph "The audiovisual presence of the equipment."

complexities of the individual works. In the following, the term *environment* describes a complex, three-dimensional form of artwork that generally can be walked into and that can fill a whole space.

> "In contrast to an *installation*, an *environment* means the creation of a self-contained construction, which fits into a larger spatial context, but is experienced primarily as a spatial unity with more or less open boundaries. The term *installation* describes a spatial design, which has a relation to the surrounding space without creating the space itself. This spatial relation can be specific, in that the installation is created for a specific space and relates to the characteristics of that space, or it can be unspecific, in the case of a work which could theoretically be presented in different spaces and does not inevitably rely on the characteristics of a particular space.
> The term *sculpture*, furthermore, is related to a three-dimensional creation, which, relative to the *installation* and e*nvironment*, only has a limited spatial extension, which in this sense does not extensively utilize space or determine that space. A *sculpture* can also be created for a specific spatial context, but usually this is not the case. Therefore, a *sculpture* in most cases does not have any specific relation to the space it occupies. Obviously, these terms only serve as practical simplifications for these spatial-temporal constructions, which are in part extremely complex, especially when the artwork is spatially and temporally extended by e.g. projections, light or sound."[15]

László Moholy-Nagy's *Light-Space Modulator* (1923-30) was an early spatial media composition due to the use of light and shadow projections, or rather, it was a "light prop" (*Lichtrequisit*)[16] which worked on the one hand as a sculpture and on the other hand also as an installation, when the back of the box was removed and the movable light and shadow play could extend into the room. This work of Moholy-Nagy's was a temporal composition too, as the lights were activated at intervals and the other elements, such as the stencils, were in continual motion over a period of time, due to the mechanical construction, and thus over the course of time ever new coloured light and shadow plays were created. This work also had a strong performative character due to this special temporal construction; in fact the "light prop" was presented to an audience. In this respect, Jean Tinguely's performative

---

[15] Vasulka, Steina. "Artist's Note. My Love Affair with Art: Video and Installation Work." Leonardo, Vol. 28, pp. 15-18, 1995. VAS B2-C9-15; Dossier d18944;

[16] László Moholy-Nagy, quoted from: Medien Kunst Netz/ Media Art Net. <http://www.medienkunstnetz.de/works/licht-raum-modulator/>

machine-sculptures of the 1960s, seen in their most radical form in the self-destructive machine *Homage to New York* (1960), were clearly related to Maholy-Nagy's light apparatus. Although these kind of machine-based works can be called the definite forerunners of kinetic video and computer-based installations, a vast paradigm shift took place in the early film and video installations and environments: the recipients were given new roles, the performative element was freed from it's temporal rigidity by the use of loops and by the individualisation of the experiential processes, and the spatial creation aimed at immerse and include the recipient completely.

One of the most interesting phenomena connected with this, and in relation to the current emergence of dynamic image walls in public or semi-public spaces is the *Video Matrix Display* (1974), which the Vasulkas designed for the Albright-Knox Art Gallery in Buffalo, New York, but which was not, in the end, executed. The source of the design lies in Woody's work on the *Video Matrix* of the American Can Company four years previously. A matrix of at least 40 television monitors was planned, which would make up a kind of passageway; whereby the walls, ceiling and floor would consist of monitors. The sound from the speakers, which were built into at least four places of the construction, would have surrounded them acoustically in the same way as the moving pictures on the monitors.

> "These monitors would be connected to a synchronous program selector which would carry a programme designed by the artist and which would distribute simultaneously to the monitors up to three different video programmes from loop-cassette playback desks. Because this exhibiting system is modular and therefore can be rearranged for any number of different programmes once it is set up, many video artists would be able to present their existing and future works, and original works could be commissioned by the gallery. The creation of this video matrix display would provide, in effect, "wall space" for video art, gallery space for changing exhibitions of electronic environments."[17]

The Vasulkas planned a programme for the first recording for the environment, which was to convey two phenomena they had explored: the "image flow experience," which was the result of the continual flow of images over the matrix in one direction, and the "colour and textural saturation." The horizontal movement of the image, that is to say, a frame, which had been made possible by video—a phenomenon that had been discovered by the Vasulkas—was explored in a whole

[17] Vasulka, Steina and Woody. Proposal for a Video Matrix Display (1974). VAS BI-C5-I; Dossier d19303.

series of installations. The sound in these works was generated by the image or, conversely, the sound generated the image:

> "Shapes and forms skid, roll, and metamorphose across multiple screens like sound travelling through geometric space to our ear. In these matrices, the Vasulkas reduce the image and sound to their bare essentials in order to examine the essence of the electronic image and sound—the signal."[18]

The experiencing of the *Video Matrix Display*, constructed chiefly as a physical space, is closely linked to the body's physical movement in and through the space. The horizontal movement of the images, for example, from the entrance of the space in the direction of the exit could be synchronous to the perceiver's bodily movements in the same direction and so enhance his perception of his own movements. The designing of the matrix as an image space in which the individual pictures move from one display to the next and the abstract aesthetics of the video pictures chosen, as well as the close processual association of sound and image, demonstrates the relationship with yet another completely different, contemporary and immersive environment: the multimedia club environment in which video has become the "mirror ball of the electronic era"[19] and sound and image reciprocally transform. The contemporary club environment usually consists of a row of screens that are arranged to form a space onto which moving digital images are projected in a synesthetic relation, of varying intensity to electronic music, which is typically produced live or at least mixed live. Images multiply themselves on the screens surrounding the visitors and merge together with the sound into a total spatial, temporal and physical experience. Jan Rohlf writes about this:

> "Decisive is the momentary, the unforeseeable, the experience of the instant in which the terror and the inconsequence of chronological time collapses." [Rohlf 54]

Interestingly enough, the concepts of the *Video Matrix Display* and the matrix of the American Can Company are already aimed at the creation of a system that was as open as possible and accessible to artists, who could work with established parameters such as the technological and architectonical design of the system—just as is done today within the context of the electronic clubs. The Vasulkas were

---

[18] Spielmann, Yvonne. Video and Computer: The Aesthetics of Steina and Woody Vasulka. Acrobat (PDF) file. English and French text. Text written in 2003. Date of reference: June 7, 2004. 4. <http://www.fondation-langlois.org/media/activites/vasulka/Spielmann_EN.pdf>

[19] Ibid.

unable to carry out the *Video Matrix Display* in the Albright-Knox Gallery; however, the concept of a highly immersive image space was apparently common among the artists of the time:

> "Some of the dreams of Frank Gillette, thinking about the first news of flexible flat TV screens, were being able to construct a tunnel that you could crawl through and have your image all around you."[20]

The Vasulkas made use of their multi-screens at the beginning of the 70s to fashion installations that were less immersive, yet still space-forming, in which the impression of motion through space was evoked by horizontal drift:

> "Our concern right from the beginning was how you put pictures in space. We worked with the monitor only as a reference, but our major medium was the matrix...we were always interested, how things work in space."[21]

As part of the eighth Annual Avant-Garde Festival of New York (1971), the Vasulkas presented the installation *Video Kinetic Environment*, which consisted of a number of television monitors arranged in a circle with the picture horizontally drifting from one monitor to the other during the 12 hours the festival lasted. The festival took place in the Armory building, which E.A.T. had already used for their performance evenings in 1966. The installation adapted to the open space there and was visually limited only by the circle formed by the monitors. The work was only one of many in the hall-like space, so that visual and acoustic isolation would only have been made possible through the construction of an enclosed inner space. As it was, the installation was rather permeable and because of its being embedded in the whole audio-visual structure, was unable to develop the immersive effect that the matrix installations normally had.

> "We would set up a lot of monitors, 10-15, (...) and then let the images drift from one to the other, these kinds of things. We used all audio synthesized music.

---

[20] Frank Gilette and Ira Schneider were probably the first who publicly displayed multi-screen installations. Their closed-circuit installation Wipe Cycle, which was shown in the exhibition TV as a creative medium, consisted of a wall of new monitors on which four cycles of moving images drifted from one monitor to another.

[21] Jud Yalkut in: Yalkut, Jud. Part Three. Open Circuits: The New Video Abstractionists. THE KITCHEN: An Image and Sound Laboratory: A Rap with Woody and Steina Vasulka, Shridhar Bapat and Dimitri Devyatkin. Radio interview. <http://www.vasulka.org/Kitchen/K_Audio.html>

Because of the time, people just sat there for 3 or 4 hours. They didn't care; they would sit or lie on the floor. There would be a fair amount of marijuana smoke in the air. Some people would be cross-legged and rolling back and forth. It was casual." [22]

The idea of an immersive space in which pictures and sounds more or less timelessly flow such that the perceiver (in a relaxed manner) could absorb them over a longer period of time, represents a reoccurring concept of the ideal *space of experience*, which is also considered a further development of theatre and cinema. Stan Vanderbeek realised his utopia of the ideal theatre in 1966 with *Movie-Drome,* in which the audience was meant to perceive the flow of images and sounds whilst lying on their backs in a domed building. A worldwide series of such constructions and the development of a universal, international pictorial language were planned. Many artists of this period made use of the possibility of creating new spaces through the medium of pictures. The breaking down of spatial and temporal linearity was the focus of these works, which also suggest a distant relationship to the panoramas of the 19th century.[23]

III. An Almost Kinetic Mobility

The inclusion of the third dimension in artistic design demands a form of reception that goes beyond the act of seeing (or/and hearing): spatial exploration. In the early period of civilization, with the creation of large sculptures and cult sites and their structuring for viewing from multiple perspectives, the mobility of the recipient was taken into consideration or even provoked through their design. The perceiver either moved around the sculpture or cult site or cultural site. In contrast to this exploratory and free mobility of the perceiver, most of the early installations and environments based on timed-based media set a relative rigid, spatially restricted boundary for the movement of the perceiver through the space, and so aimed at an intensive bodily awareness. The *Live Taped Video Corridor* (1969/70) by Bruce Nauman is an early, and the probably the most discussed, example of this type of work [Jaschko 32-35]. Only one of a total of five corridors of varying widths is passable due to its dimensions. The observer experiences this space by entering it, squeezing his way through it, and thereby approaches two monitors at the end of the corridor. The observer sees his own image in real-time on one of the monitors

---

[22] Woody Vasulka in an interview with the author on July, 28, 2005.

[23] STEINA-May, 1995. Interview by Chris Hill. Transcription by Julia Dzwonkoski. Edited by Chris Hill. Acrobat (PDF) file. <http://vasulka.net.zkm.de/archive/Contributors/ChrisHill/InterviewSteina.pdf> 3.

recorded by a video camera above the entrance, allowing merely a rear-view of himself when he is facing the monitor.

In this work, in comparison with other art forms, some fundamental changes in recipient mobility in time-based media are implied in the relation between art work and perceiver; that is, the interaction through perception and the alterations in the state of the art work elicited by that (inter) action. The movement of the perceiver in relationship to the physical components of the artwork is a spatial and temporal dimensioned, individually executed, and subjective process of reception. The play with spatial perception, the breaking down of habits of perception and behaviour, and the offering of new ways to experience reality were the goals of the designers of such perception processes.

In many of the media, large-scale works that were created in the first decade of video art, the movement of the perceiver and the resulting change in the state of the art work was an essential element. The movements of the perceiver: that could range from walking through the space to simple bodily movements in place, are performative actions that make the perceiver part of the artwork and usually lead to performative immersion. It can be said of this first decade of video installations and environments, although the spatial qualities and the forms of movement offered the perceiver were of a variable nature, that the movement of the perceiver was an elementary component of many works. In comparison with current mobile art projects, however, their spatial and temporal ambition was of a relative minor degree.

Two Sides to Every Story by Michael Snow for instance is concerned with the construction of a space and the action that takes place in it. The installation's picture material is a kind of performance, which, on the basis of the dramaturgy, possesses the hybrid character of a theatrical play and an experiment. Two cameras positioned opposite each other recorded a performer standing in front of a screen, which she works on in various ways over a period of time [Iles 254]. The installation consists of an aluminium-coated screen placed in the centre of the room upon which one recording of the performance is projected on one side and the second on the other. The perceiver must walk around the screen in order to be able to experience the performance from both camera perspectives. The common audio plane of both projectors reunites the spatial area divided by the screen. The directions to the performer given by Michael Snow, who acts as a director and who is also visible in one camera perspective, describe the actions of the performer, who, depending on the position of the observer, *cannot* be seen, and so indirectly have the effect of an incitement to move to a position from whence the particular action *can* be seen.[24]

On the basis of the relative minor variance, the limited measure and the repetition of physical movements offered the perceiver, one could speak in the context of these early media installations of a "kinetic mobility" of the perceiver. In many cases there actually exists an obvious connection between the forms of motion of the machines used and the movements of the perceivers within the installation or in the environment. What is more, artists have performed the kinetic motion of mechanic tools in video works. In Steina's video experiments of the 70s, the kinetic movement of the perceivers as well as Steina's own, almost kinetic motions in interplay with the tools, is a central subject. In the complex (experimental) arrangements of *Orbital Obsession* (1975-77) and *Switch! Monitor! Drift!* (1976), a permanent, circular motion is performed one time by Steina (*Orbital Obsession*), and another time (*Switch! Monitor! Drift!*) by use of a rotating camera. In *Allvision* two cameras circle around a mirror ball, whereupon the perceiver instinctively moves likewise in a circle around the kinetic sculpture. The same motions around one's own axis, or in a circular motion, are a feature of the *Finch College Project* and the linear video work related to it, *Mirror* (1969). Bruce Nauman made the motions around one's own axis as a destabilising, disturbing factor in spatial perception, the conceptual focus of his film installation *Spinning Spheres* (1970).

In a similar manner the linear movements of the perceiver towards a projection or monitor resemble those set out in many installation video works, a kinetic backwards and forwards motion comparable to a zoom. Thus the perceiver in the closed-circuit installation *Mem* (1975) by Peter Campus finds himself facing a projection, which is showing him. The picture captured by the camera is projected at a sharp angle on the wall. If the perceiver moves in the direction of the projection and so into the camera's field, his picture drifts across the wall at an angle, though unexpected, yet logical to his movement. If the perceiver approaches the camera, his image, contrary to his expectations, becomes smaller.

It should be noted that the artists were in no case concerned with emulating kinetic motion. Rather the installations and environments and the "kinetic mobility" of the perceiver set forth in them are playful explorations of their given technological possibilities. The mobility of the artist, machine and perceiver appears limited against the backdrop of the urban and spatial experience of the Situationists in the 50s and 60s and today's degree of mobility in recording and communications

---

[24] The complex work consists of several action phases. A complete description of the work and an interactive simulation of the installation can be found on the DVD *Digital Snow*.

technology and the works created with it. A reason why exterior space was rarely examined is that this wasn't a focus of interest as its own subject, but rather the physical and cognitive experience of space, which was precipitated through the media.

Bibliography

Hill, Chris. "Attention! Production! Audience!: Performing Video in its First Decade." *Rewind: Video Art and Alternative Media in the United States*. Ed. Chris Hill. Chicago: Video Data Bank.

Iles, Chrissie. "Video und Film Space." *Space, Site, Intervention. Situating Installation Art*. Ed. Erika Suderburg. Minneapolis: University of Minnesota Press, 2000.

Jaschko, Susanne. "Space-Time Correlations Focused in Film Objects and Interactive Video." *ISEA Papers*, Nagoya/Japan 2002.

Rohlf, Jan. "Die Diskokugel der elektronischen Ära. Klang/Bewegte Bilder im Clubkontext." transmediale 99, internationales medienkunstfestival berlin. Berliner Kulturveranstaltungs-GmbH, Berlin, 1999.

Ryan, Paul. "Cable Television. The Raw and the Overcooked." *Radical Software* 1.1 (1970).

Sterken, Sven. "Towards a Space-Time Art: Iannis Xenakis's Polytopes." *Perspectives of New Music* 39.2 (2001).

Forgotten Histories of Interactive Space

Martin Rieser

> "The nomad, nomad space, is localised and not delimited. What is both limited and limiting is striated space, the relative global: it is limited in its parts, which are assigned constant directions, are oriented in relation to one another, divisible by boundaries, and can interlink; what is limiting (limes or wall, and no longer boundary) is this aggregate in relation to the smooth spaces it "contains," whose growth it slows or prevents, and which it restricts or places outside. Even when the nomad sustains its effects, he does not belong to this relative global, where one passes from one point to another, from one region to another. Rather, he is in a local absolute, an absolute that is manifested locally, and engendered in a series of local operations of varying orientations: desert, steppe, ice, sea [and salt-lake]." [Deleuze & Guattari 382]

A Brief History of Technology, Public Art and Performance

In its application to public spaces, interactive work has frequently attempted to gain critical purchase by exploiting the tension created in an audience between its electronic space and its physical or mechanical one. Consistent themes and uses for electronic art in public contexts were established early in the 20th century and threads of similar practice may be traced through from Dada, Futurism and Constructivism and the Bauhaus to the present day. Tatlin was playing with motorised architecture in his "Monument to the Third international" in 1920. The distant relationship between artist and architect has also created problems in the proper integration of public art in cities, let alone that of mobile digital art into public spaces. It is no accident that some of the more successful examples of actual or potential public art works using new technology have frequently been produced by architecturally trained artists, most notably Shaw, Moeller and Diller and Scofidio.

In researching the origins of interactive work inspiring audience mobility, I was profoundly embarrassed to discover a common tendency, including my own, to repeat experiments achieved 40 years ago. So much of what we think of as innovative was explored in some way during that period. EAT—Experiments in Art and Technology—was founded in 1966: By 1969 it boasted a worldwide membership of 3000 artists and 3000 engineers. Famously, Rauschenberg and the group staged a series of large-scale public events and installations called *9 Evenings*. The most notable was "Open Score" at the NY Armoury in 1966. A Tennis match with rackets containing tiny FM transmitters ensured that each time a ball was hit, a light would extinguish to an amplified sound: eventually the match ended in total darkness. 500 volunteers then entered the court and went through a number of behaviours in the darkness, while their cctv infra-red images were the only thing the audience could see on huge projection screens-a form of "darkness visible."

Rauschenberg played with technology in the round in "Soundings," collaboration with Bell labs. Rauschenberg installed sound sensors to control a lighting rig. Audience talk set at four frequencies controlled the light falling on the huge multiple-mirrored plexiglass panels, which had been silk-screened with images of wooden chairs. The movement of lights made the chairs appear to move in random patterns in association with audience sound. Rauschenberg's *Solstice,* another interactive piece that was created through E.A.T., was a series of sliding glass doors with silkscreen images that opened and closed as you walked through. *Mud Muse* also used technology for its nine by twelve foot basin, which was filled with "drillers mud" (bentonite mixed with water) that bubbled and gurgled continuously. The technological limitations of the pieces and the vagueness of the art intention, as opposed to the exploration of audience interactions, combined to drive Rauschenberg back towards more traditional methods. With his move in 1970 from New York to Captiva, an island off the Gulf Coast of Florida, Rauschenberg favoured an abstract idiom and the use of natural materials.[1]

EAT's seduction by technology was to culminate in the Expo 1970 Pavilion in Osaka. Sponsored by Pepsi, it was an attempt to create a "living responsive environment," a non-hierarchical theatre space. It was a reprogrammable space with a giant "mirror room" full of interactive sound areas, a giant fog sculpture and motorised exterior sculptural elements or "floats" by Robert Breer. This extraordinary project brought together sixty-three American and Japanese artists, engineers and scientists. Innovative work was done with individual wireless handsets, for use within the mirror room, and with programmed laser displays. Visitors were responsible for their own experiences. Combining the world of Fluxus and the Happening it was little more artistically than "son et lumiere," but pushed the boundaries of technological art some distance, particularly in relation to audience mobility.[2]

Robert Kluver. Co-founder of EAT, worked with Robert Breer, Robert Whitman, Frosty Myers and David Tudor on the first design. The roof of the geodesic dome was covered by a water vapor cloud sculpture, designed by Fujiko Nakaya, capable of generating a 6 foot thick 150 foot diameter area of fog.[3] It wasn't until 2002 that

---

[1] See Alexandra Sprowls, "Time, Space and Rauschenberg." <http://www.csuchico.edu/art/contrapposto/contrapposto00/pages/themeperspective/sprowls.html> (September 2005)

[2] See also <http://www.artmuseum.net/w2vr/archives/Kluver/09_Soundings.html> (September 2005)

[3] This correlates almost exactly with to the most famous project of architects Diller and Scofidio, three decades later the *Blur Building* 2002 consisted of a fog of mist (made of filtered lake water shot

a similar architectural project was attempted successfully in Diller and Scofidio's Blur building.

On the terrace were seven of Robert Breer's Floats; six-foot high slow moving sculptures which emitted sound. When they hit an obstacle or were pushed they would reverse direction. Upstairs the main space of the Pavilion was a 90-foot diameter 210-degree spherical mirror made of aluminized Mylar. Visitors could see themselves reflected in this through an optical effect producing a real image resembling that of a hologram. Because of the size of the mirror, a spectator looking at an image could walk around and see it from all sides. The sound systems in the dome were truly innovative and delivered spatial sound via a huge array of speakers and through mobile handsets, resembling early mobile phones. Kluver describes this in detail:

> "David Tudor designed the sound system as an "instrument" with 32 inputs and 37 speakers arranged in a in a rhombic grid on the surface of the dome behind the mirror. Sound could be moved at varying speeds linearly across the dome and in circles around the dome. Sound could be shifted abruptly from any one speaker to any other speaker, creating point sources of sound. All speakers were capable of giving out the same sound. The lights and sound could either be pre-programmed or controlled in real time by the artists from a console at one side of the dome.
>
> The floor was divided into 10 areas made up of different materials, such as astro-turf, rough wood, slate, tile, asphalt. Through handsets, ... visitors could hear specific sounds on each different floor material. On the tile floor: horses hooves and shattering glass; on the astroturf: ducks, frogs, cicadas and lions roaring. These sounds were transmitted from wire loops embedded in the floor.
>
> Twenty 100-turn wire coils or loops 1 foot in diameter were embedded under each of the floor sections and were fed by tape recorders. The low frequency magnetic field they generated was picked up and amplified through handsets the visitors carried. The innovation in this system was the use of a large number of coils for each area to obtain an even distribution of the sound and not have sound spill over to another area." [Kluver]

After the financial over-stretch of the Pepsi Pavilion, EAT became increasingly inactive and the Seventies were a relatively quiescent time for the digital arts.

---

through 31,000 fog nozzles) around a steel structure, with the result appearing to be a massive white cloud floating on a lake. The building was created for Swiss Expo 2002, held off the shore of Lake Neuchatel at Yverdon-les-Bains in Switzerland.

Robotics and Audience

The engineer turned artist, Edward Ichnatowicz, produced one of the most successful works of computer art of the late Sixties and early Seventies, in terms of a fully realised interactive installation. The giant public piece, which performed a seminal role in the realisation of what was artistically possible with computing and robotics, was the Senster.[4] It was an active metaphor playing on an audience's techno-fear and its simultaneous ability to remotely control the products of nightmare fantasy. Installed in 1971 at the Phillips' industrial exhibition 'Evoluon' at Eindhoven, it represented an extremely ambitious technical and artistic feat. About fifteen feet long and 8 feet high, the Senster consisted of six independent electro-hydraulic servo systems based on the articulation of a lobster's claw, allowing six degrees of freedom. The Senster had a "head" with four sensitive microphones, which enabled the direction of the sound to be computed and also a close range radar device, which detected movement. The Senster would move toward sounds, but would shy away from loud noises, or if you tried to touch it. The whole was controlled in real-time by a digital computer which sent feedback from the movement and sounds of visitors to the Evoluon, so that the servos could reposition the head anywhere within 1,000 cubic feet of space within a couple of seconds.

Using a "predictor," the programme put the machine through a complex series of accelerations and decelerations for the maximum efficiency of motion. The net result was convincingly lifelike in its movements. Unlike the automata of earlier ages, *The Senster* didn't try to conceal its inner workings, nevertheless the public's response was to treat it as if it were a wild animal. Apparently according to contemporary accounts the complicated acoustics of the hall and the unpredictable behaviour of the public made *The Senster's* movements seem a lot more sophisticated than they actually were. It soon became obvious that it was that behaviour and not anything in its appearance, which was responsible for the impact *The Senster* had on its audience. *The Senster*, which worked on so many levels of meaning, was never surpassed as a robotic piece. It was unveiled in 1970 and remained on permanent show until 1974 when it was dismantled, seemingly, because the exhibit was attracting people away from the rest of the exhibition! [Benthall 80-83][5]

Another early and monumentally large public installation exhibiting the same techno-enthusiasms as Pavilion was Michael Hayden's *Arc en Ciel* (1978), consisting of many-hooped fluorescent tubes arching above the platforms of the Yorkdale subway station in Toronto. The piece measured 570 feet in length and was

---

[4] The Senster. <http://www.dse.nl/~evoluon/senster-e.htm> (September 2005)

[5] For a complete archive see <http://www.senster.com/ihnatowicz/senster/> (September 2005)

responsive to infra-red radiation. The arrival or departure of a train caused the piece to ripple with sweeps of multicoloured light following the passage of the warm subway train. 20000 effects were programmed on a cycle of 40 hours. The sensors could even respond to the movements of individuals on the platforms.

These permanent interactive works remained uncommon through the 1980s and 1990s. Designed for a Silicon Valley cinema in San Jose, California *Jump Cuts* (1996) by Diller and Scofidio and Paul Lewis (with technical support by Ben Rubin) placed a row of twelve large liquid crystal display panels on the theatre's glass facade. Mounted in the lobby, a matching set of projectors cycled through live images of people on theatre escalators, trailers of movies supplied monthly by United Artists on laser disks, and a transparent mode where the theatre lobby could be viewed through the panels, alternating the real view and the view transmitted live from inside the building, When the direct view was cut off by the liquid crystal, it was replaced by its video substitute. Because the field of view of each camera overlapped the next, there was a stuttering effect to the crowd images. Before a person fully disappeared from one screen, they appeared on the next, just as the title implied. To quote the artists:

> "The difference between liveness in broadcast and liveness in surveillance is the difference between the act of witnessing and the act of monitoring. Whether an illicit or official gaze, the ubiquity of surveillance cameras has left few shadow zones of privacy"[6]

Abandoned since 2000 by United Artists, both the complex and laser technology are now obsolete.

A Third Space

At the same time as these early and grandiloquent projects were being thrust into the public eye, Myron Krueger was patiently mapping the territory by quietly developing a language of interaction. Since 1974 his efforts have been focused on the development of 'Videoplace'. It was created to give the computer control over the relationship between the participant's image and the objects in the graphic screen. It could coordinate the movement of a graphic object with the actions of the participant. A series of simulations could be programmed based on any action and Videoplace offered over 50 compositions and interactions (including Critter, Individual Medley, Fractal, Finger Painting, Digital Drawing, Body Surfacing, Replay,

---

[6] See <http://www.prototypo.com/Essays/Essays1/001_3.htm> (September 2005)

among others). For example when the participant's silhouette pushed a graphic object-the computer could choose to move either the object or the silhouette.

It was an early attempt to create a wire-free projection environment, capable of responding to each participant differently. The most famous of these was the 'critter'—a small circular figure, which avoids contact with a projected image of the participant. Krueger's outstanding achievement lay in the foresight he showed in examining possible types of unencumbered interaction, many of which inform both telematic and installation practice today. Interacting at an exhibition in Duisburg with a reconfigured version of this piece, I was struck by how the playspace seemed capable of reducing many adults to their five year old selves.[7]

In the following extract from an interview in Ctheory magazine Krueger explains his early rejection of VR as an encumbering technology:

"CTHEORY:

Today, many new realities being built by developing and speculative technologies are being confused with having exclusively immersive properties. However, all along, you conceived of Virtual Reality (which you then called "Artificial Reality") as being primarily composed of external and not immersive properties. You also discussed the possibility that extroverted realities could be harmonized with the "real" rather than harvesting it. One could walk in physical space and experience a prototype to what Michael Heim now calls "Exo-Virtuality" (EVR). Are we then seeing a gradual return to your original idea?

MYRON KRUEGER:

In 1970, I considered HMDs (Head Mounted Displays) and rejected them because I thought whatever benefit they provided in visual immersion was offset by the encumbering paraphernalia, which I felt would distance participants from the world they were supposed to feel immersed in. When I pondered what the ultimate experience would feel like, I decided that it should be indistinguishable from real experience. It would not be separated from reality by a process of suiting up, wearing gear, and being tethered to a computer by unseen wires. Instead of an alien planet accessed through an airlock, it would be like a doorway to a fantasy world that you could enter simply by attending to it.

Rather than limiting your participation to a single hand-held 3D pointing device, your image would appear in the world and every action of your body could be

---

[7] InterAct: Schlüsselwerke Interaktiver Kunst, Wilhelm Lehmbruck Museum Duisburg 1997

responded to instantaneously. Whereas the HMD folks thought that 3D scenery was the essence of reality, I felt that the degree of physical involvement was the measure of immersion. Instead of being concerned about the stagecraft, I focused on the play."[8]

Haptic Interfaces

This brings me to a crucial subject of this paper: the development of physical interaction in public space. Intimacy or crowd collaboration are both legitimate modes for the experience of public art. While Jeffrey Shaw is not an example of a public artist he is a wonderful source of examples of appropriate physical and mobile interfacing to works, which could easily be placed in public contexts. Perhaps because of his background in architecture Shaw has always included strong physical elements for interaction in all his works.

His famous piece The Legible City[9] 1989, combined a highly physical interface employing motionless motion, combined with virtual reality. The 'City' is a computer controlled and projected virtual urban landscape made up of solid three-dimensional letters that form words and sentences instead of buildings along the sides of the streets. The architecture of text replaced exactly the positions of buildings in a plan of the real cities (New York and Amsterdam). This spatial transformation of narrative was literal in every sense. Bicycling through this city of words is a journey of reading, choosing a direction is a choice of text and meaning. The image of the city is projected on a large video screen in front of the bicycle, which is fixed like an exercise-bike.

His less well known "Revolution" 1990[10] was an interactive videodisk installation which allowed the user to turn the 'mill of history', tracing 200 years of turbulent history from 1789 to 1989, The considerable physical effort required to turn the installation was enough on its own to give "gravitas" to the content, demonstrating a perfect synchronisation of metaphor and haptic interface.

---

[8] *Jeremy Turner* Myron Krueger Live Ctheory: <http://ctheory.net/articles.aspx?id=328> (September 2005)

[9] The Manhattan version (1989) of this work comprises eight separate fictional story lines in the form of monologues by ex-Mayor Koch, Frank Lloyd Wright, Donald Trump, a tour guide, a confidence trickster, an ambassador and a taxi-driver. Each story line has a specific letter colour so that the bicyclist can choose one or another to follow the path of a particular narration. In the Amsterdam (1990) and Karlsruhe (1991) versions all the letters are scaled so that they have the same proportion and location as the actual buildings which they replace, resulting in a transformed but exact representation of the actual architectural appearance of these cities.

[10] Documented at <http://www.medienkunstnetz.de/works/revolution/> (September 2005)

Collaborative works

While Shaw's works required a single user, another experimental interface (which prefigured multi-participatory public works), was created at the Banff Centre in Canada by Perry Hoberman. *Bar Code Hotel* [11] was an interactive environment for multiple participants. An entire room was covered with printed bar code symbols. An installation was created in which every surface could become a responsive object, making up an immersive interface that could be used simultaneously by a number of people to control and respond to a projected real-time computer-generated three-dimensional world. Each "guest," who checked into the *Bar Code Hotel* was given a bar code wand. Because each wand could be distinguished by the system as a separate input device, each guest had their own consistent identity and personality in the computer-generated world. And since the interface was the room itself, guests could interact not only with the computer-generated world, but with each other as well. The objects in *Bar Code Hotel* were based on a variety of familiar and inanimate things from everyday experience: eyeglasses, hats, suitcases, paperclips, boots, and so on. Thus the co-dependence of our two universes was established through the simplest piece of supermarket technology.

Early uses of Public Space

This sense of play, curiosity, and inventiveness is reminiscent too of the fairground attraction and in many ways his approach mirrored that of Toshio Iwai whose entire 'oeuvre' including his public art is based on play. In *Another Time, Another Space* created at Shinjuku Station in Tokyo in 1993. Toshio Iwai made an electronic hall of mirrors using a tree structure of video screens. The installation featured 15 video cameras, 30 computers, 30 video monitors, and a videodisk recorder. The comings and goings of people through the station were filmed by the cameras, and manipulated in real-time by the computer to deform shape, time reference, and showing a different time-space environment in each movement. Video processing software reflected back crowds like fields of wheat where algorithms interpreted successive layers of crowd as wave-like motions. Sober-suited businessmen leapt and cavorted in front of these magic mirrors.

> "I used the "Another Time, Another Space" system to create an experimental event as part of an NHK television program. People passing in front of Shinjuku Station were photographed by a video camera and the images were altered and projected onto the giant Alta Vision screen across the street. It caused a much

---

[11] Barcode Motel Banff Centre for the Performing Arts 1992; See <http://www.itofisher.com/PEOPLE/PERRY/BarCodeHotel/> (September 2005)

larger commotion than we expected. The moment the image appeared on the screen, hundreds of people started gathering in front of the station and waving their hands and moving their bodies as they watched their images on the screen. In that moment the big screen that everyone had been taking for granted suddenly became a giant interactive event."[12]

In contrast to this rather formal and monumental project, there were a number of attempts to create interactive architectural spaces by British artists. For example Simon Biggs with his installation "Heaven," commissioned by the European Media Art Festival 1993 for a projection onto the ceiling of the Dominikanerkirche, Osnabrück, Germany, 18 metres above the viewers heads. The Cathedral used remote visual sensing techniques to track the viewer. Each viewer was allocated an angel (or demon, depending on location), which followed the position of the viewer on the floor analogously on the ceiling. The viewers actions controlled not only the behaviour of the angels/demons, but also a large range of other images, which were dynamically composed on the ceiling used audience movement to alter virtual architectural features such as angels and gargoyles projected onto the roof space. In the 1993 *River Crossings* public art project, Susan Collin's *Tunnel*, similarly mapped responsive soundscapes, and video projections into a pedestrian tunnel under the Thames, triggered by audience movement.

Interactive Art and Architecture

If we look next at the possible fusion of physical architecture and public art works, we see another discrepancy. Materials technology in the 1990s began to deliver the means for artist-architect collaborations, which might finally realise some of the 1960s dreams of adaptive or "liquid" architecture. Dreams of groups like Cedric Price's Archigram and later visionaries like John Fraser.[13] The development of electro-heliological fluids which transform from liquid to solid state at the passing of a current, piezo-electrical ceramic which can change colour to order, SMA-shape memory alloys which act like muscles and liquid crystal glass, paint and inks that respond to tiny electrical or temperature changes allow a building or artwork to behave in a biological manner. New research in nanotechnology combined with artificial life programming implies self-repairing and "living" systems grown around human needs.

---

[12] Personal e-mail interview Toshio Iwai and Martin Rieser 1997

[13] For details of John Frazer's Evolutionary Architecture see <http://www.aaschool.ac.uk/publications/ea/intro.html> (September 2005)

However, realistically, even at the basic level of combining existing architectural materials with digital artwork, very little was being achieved until the end of the 1990s, although the techniques were already in place, more permanent integration of such work in public contexts remained elusive. Right into the 1990s Public art tended still to rely on the same electro-mechanical mechanisms developed fifty-odd years ago by EAT.

An early exception to this curious reluctance to engage with new materials is Christian Moeller.[14] His pioneering work points the way, with buildings such as the ZeilGalerie in Frankfurt (1992) which changes colour at night according to wind direction and speed, while a sine wave of light ripples its length governed by ambient noise from the street. People gathered at night to clap and create sounds that alter the wave. How seriously one should take such interaction as art is another question.

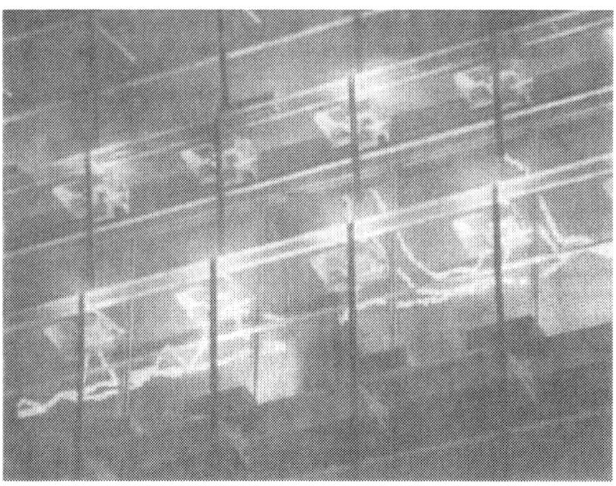

Figure 2: Zeil Gallerie Frankfurt, Christian Moeller

Moeller's more thoughtful gallery piece Electronic Mirror confounded our narcissism with a distance sensor and electro-responsive LC glass, clouding-over our image on close approach so we are literally swallowed by the glass like digital prisoners-shadowing the original myth of Narcissus.

In *Space Balance* (Ars Electronica 1992) a virtual interior architecture mirrors the hydraulic tipping of the viewing platform. The participants could roll virtual balls, which clicked as they collided controlled by the movement of their bodyweights on the platform. A similar device was used in *The Virtual cage* in Frankfurt in 1993. The

[14] Moeller, Christian. Interaktive Architektur. Galerie fur Architektur und Raum, Berlin 1994.

viewer danced on the platform in relation to a virtual swarm that interacted with the viewer's movements. This use of a tilting floor was also developed by Grahame Weinbren as a way of allowing audience participation in his interactive films.[15]

It sometimes seems there are as many types of interactive digital art as there are artists, but that the same forms are reinvented and reappear with successive technologies. As we have seen successful practice for audience mobility must always place content and meaning above mere technology. However, if it fails to engage with the full potential of those new technologies, it fails to find the new form and meaning for which all art ultimately aims. The old voices may be saying the same things, but, as always, only the new voices can be heard, by the tired ears of the "Public."

To quote Regina Cornwell:

"These explorations are crucial to how the world can be re-drawn and viewed in an art whose power is in its open-endedness and polyphony. And for the participant the installation too is hard work. To be meaningfully experienced demands time and serious attention." [Graham 171]

Bibliography

Benthall, Jonathan. *Science and Technology in Art Today.* London: Thames and Hudson, 1972.

Deleuze, Gilles, and Félix Guattari. *A Thousand Plateaus. Capitalism and Schizophrenia.* London: Continuum, 2002.

Graham, Beryl. "Playing with Yourself: Pleasure and Interactive Art." *Fractal Dreams.* Ed. Jon Dovey. London: Lawrence and Wishart, 1996. 154-79.

Kluver, Billy. *Pavilion: Experiments in Art and Technology.* Boston: Dutton, NY, 1972.

Moeller, Christian. *Interaktive Architektur.* Berlin: Galerie fur Architektur und Raum, 1994

[15] Interview in "ctheory" 1995

Art by Telephone: From Static to Mobile Interfaces [1]

Adriana de Souza e Silva

This paper investigates how the artistic approach toward telephones changes when they become mobile. The transition occurs mainly because cell phones are no longer used solely as two-way voice communication devices. Mobility and location awareness transform them into social and ludic technologies, giving us the ability to merge physical and digital spaces, and to find one's relative position in the globe. Examples of artworks with fixed and mobile telephones help to clarify this transition.

Reviewing artworks with telephone handsets[2] helps us to remember how the device has previously been used as an artistic interface, and to imagine new approaches when this interface becomes mobile.

1. Some early experiments on telephone-based art

Very early on, some artists started to use telecommunication media to develop projects. Experimenting with remote-controlled creation may have been the first use of telephones to produce art. Considered one of the first to create a telepresence piece, László Moholy Nagy experimented using the telephone to transmit directions for fabricating enamel tile paintings. He wrote:

> "In 1922 I ordered by telephone from a sign factory five paintings in porcelain enamel. I had the factory's color chart before me and I sketched my paintings on graph paper. At the other end of the telephone the factory supervisor had the same kind of paper, divided into squares. He took down the dictated shapes in the correct position. (It was like playing chess by correspondence.) One of the

---

[1] This article was originally published at Leonardo Electronic Almanac (LEA), volume 12, issue 10 (October 2004).

[2] Other telecommunication media, such as satellites, slow scan TV, and even those which use the telephone network, like modems, videophones, and telefacsimiles, are outside the scope of this review. Some examples of telepresence art with satellites, and the transmission of data via modem, faxes, and slow scan TV are "Hole in Space" (1980) from Kit Galloway and Sherrie Rabinowitz, and "The World in 24 Hours"(1982) from Robert Adrian. Other examples can be found in Stephen Wilson's *Information Arts* (2002), Frank Popper's *Art of the Electronic Age* (1993), and Heidi Grundman's Art Telecommunications (1984).

pictures was delivered in three different sizes, so that I could study the subtle differences in the color relations caused by the enlargement and reduction."[3]

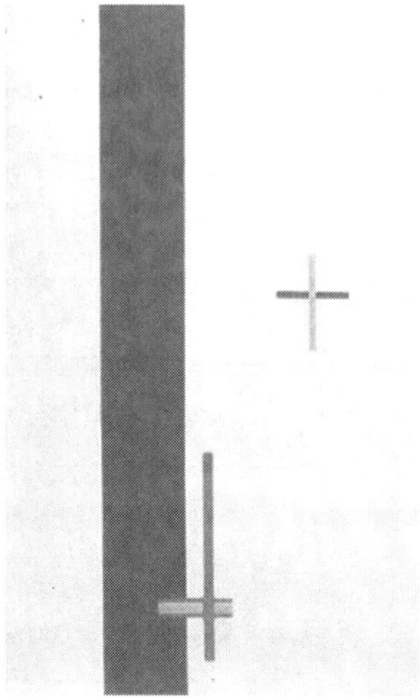

Figure 3: Laszlo Moholy-Nagy, Telephone Painting or Em 3, enamel on steel, 24x15 cm, 1922. (Reproduction, BauhausArchiv, Berlin. Copyright, Hattula Moholy-Nagy.) Permanent Collection of the Museum of Modern Art, New York. This is one of the five enamel paintings that Moholy-Nagy ordered by telephone.

Eduardo Kac suggests that nobody knows whether Moholy-Nagy's story is true or not, because his wife stated that in fact she ordered the paintings in person. Moholy-Nagy's work, whether actual or apocryphal, demonstrates that the artist could be removed from the location of artmaking.

In 1969 the Chicago Museum of Contemporary Art organized an exhibition called *Art by Telephone* that somehow repeated Moholy-Nagy's experiment. Thirty-six artists were asked to place a phone call to the museum and to instruct museum staff

---

[3] Moholy-Nagy Apud Eduardo Kac, "Aspects of the Aesthetics of Telecommunications," in John Grimes & Gray Lorig (eds.), *Siggraph Visual Proceedings,* New York: ACM, pp. 47-57 (1992) <http://www.ekac.org/Telecom.Paper.Siggraph.html> (23 July 2007).

about what their contribution to the show would be. The museum then produced the pieces and displayed them. The telephone as a new artistic medium was not explored as a creative medium in *Art by Telephone*; it was only used as a remote interface to accomplish something that could be done, for example, if the artist went to the museum and talked to the curator.

Kac says that Robert Huot accomplished one of the few creative uses of the technology by an artist in this exhibition. The artist:

> "potentially involved all visitors of the museum and attempted to generate unexpected first meetings by employing chance and anonymity. Twenty-six cities in America were chosen, each starting with a letter of the alphabet, and twenty-six men named Arthur were selected, one in each city. Each Arthur's last name was the first listing under the initial letter of the city (Arthur Bacon, in Baltimore, for instance). The Museum displayed a list of all cities and names, and invited visitors to call and ask for "Art." The work was the unexpected conversation between "Art" and the visitor, and its development totally up to them." [Kac 3]

Huot's piece presents the artist as the creator of a context, in which the visitor participates in the creative process. Here the telephone is used to turn art making into a social experience. Generally, up until the end of the 1990's artworks that used telephone handsets were almost all restricted to calling another party, using the phone's ring as an artistic element, and recording voice messages.

More recent pieces that employed the telephone include works developed by the Disembodied Art Gallery, a British Group that explores conceptual and telecommunication-based art. For instance, *Babble* was a telepresence-art installation created in 1993 that received over 70 voice contributions from the United States, Australia, Japan, and Europe. Callers telephoned a U.K. number and could record poetry, stories, and thoughts on an answering machine. Then these messages were collected and replayed automatically to visitors of the gallery whenever a member of the public entered the installation room. *Temporary Line* (1993/94), another piece by the Disembodied Art Gallery, was an audio-reactive sculpture constructed from telephone handsets. Whenever a member of the public walked close to the sculpture, the sound of whispering voices would dart around the sculpture, from telephone handset to telephone handset, at random around the feet of the visitor.

The common idea behind most of these projects is not so much to explore synchronous communication, as is considered the general use of the telephone, but to investigate the use of a recorded presence or voice as a past presence. Telephones have been considered the ultimate virtual medium, because they eradicated the distance between disembodied voices [Wilson 489]. Therefore they transformed the pure element of voice into presence: an absent presence. The above-mentioned artists used the communication technology to emphasize not only this removal in space, but also a removal in time, by bringing past recorded voices into the immediate present.

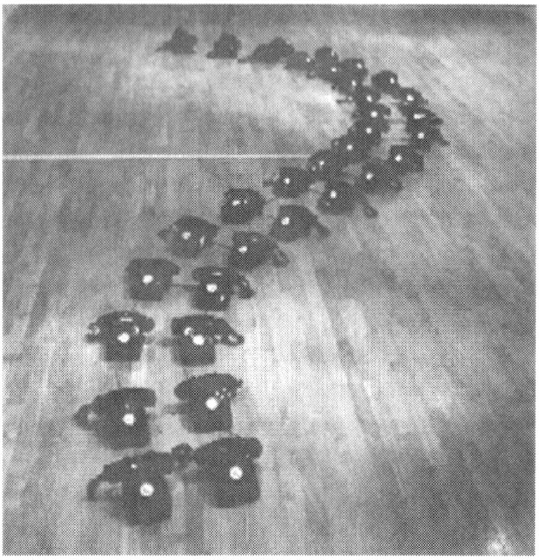

Figure 4: View of Temporary Line telephone installation (1993/94). (Photograph by Mark Monk-Terry)

Heath Bunting, a contributor to the Disembodied Art Gallery, created a piece that incorporated the use of the Internet to reach a more spatially distributed audience. The 1994 piece entitled *Kings Cross Phone-in* scattered numbers of the telephone kiosks around Kings Cross British Rail station using the Internet and asked whoever found them to choose a number and call it at a specific time and chat with whoever picked up the phone. Likewise, in December 1996, StallPlaat created *The Answering-Machine Solution C.D.*, a large collection of 30-second tracks that could be used as answering machine messages to celebrate their 100th CD release. Keith de Mendonça, from the Disembodied Art Gallery, provided the front cover for the CD and an answering machine message.

Also using telephones in public space, Stephen Wilson created a telephone-based project called *Is Anyone There?* (1992) during one week in San Francisco. In the project, a computer-based system with digitized voice capability made hourly calls to five ringing pay phones on the streets with the aim of involving whoever answered the call in a conversation about life in the city. The system used intelligent response programming to engage passers by into a short dialog. The conversations were then digitally stored in a database and accessible through an installation, which included a database of these recorded calls. In the gallery, the installation changed randomly to a real time mode that placed live calls to the pay phones, linking viewers with a real person on the street. With this piece, Wilson tried to explore random communication between unknown people, placing the user as a content creator. Furthermore, he looked into possible developments for artificial intelligent systems, by analysing dialogs between computers and humans.

It is possible to perceive two characteristics in the above-mentioned works. First, although some have been performed in urban spaces, such as *Is Anyone There?* they were still connected to a fixed place, like a pay-phone. Second, the pieces mostly transmitted voice and stored voice messages. This scenario changes with mobile phones.

2. Mobile phones: Bringing the interface into public spaces

In the last ten years, cell phones have become highly popular among telecommunication technologies, exceeding the number of fixed landlines and personal computers. Cell phone ownership has increased much more rapidly than PC ownership partly because of its relative affordability. This gap is markedly larger outside the United States, especially in developing countries (de Souza e Silva, 2007; Castells, 2007). In places where fixed telephone lines are expensive and not so widespread, the difference is substantial. For example, Africa has been named by the International Telecommunications Union (ITU, 2007) "the least wired region in the world," due to the lack of wired infrastructure and the rising number of cell phones in the continent. As of the end of 2006 (ITU, 2007), the continent had 3.15% fixed landline penetration rate, and 4.72% Internet user rate, versus 20.94% mobile subscribers. Some countries in Latin America have a similar information and communication technologies (ICTs) distribution. For example, as of 2007, Brazil had 20.7% penetration rate of fixed landlines, and 17% of Internet users. On the other hand, the country achieved 56.45% rate of cell phone users, and the highest number of mobile phones in the whole Latin America (about 106 million) [6] . Even in developed countries, like Japan, we can clearly see the popularity of mobile phones over other ICTs, such as fixed lines and PCs. Japan had, as of 2005, 66.59% Internet users, and 45.32% fixed landlines, as opposed to 75% mobile phone subscribers. Finally, many countries in the world, such as Israel, Austria, Czech Republic, Italy and the United Kingdom already exceeded 100% penetration rate of mobile phones (ITU, 2007).

The large number of cell phones in use worldwide makes them significant social and communication tools. Moreover, the use of mobile phones as artistic interfaces both reaffirms their popularity, as well as suggests new uses and unexpected uses for the technology. The artistic use of mobile communication interfaces is like an arrow pointing toward two directions. First it draws our attention back to past telephone-based artworks. Second, it foresees new uses for the mobile interface, mostly not predicted by technology developers. Although in the United States and in most countries of Latin America, the cell phone is mostly used as a two-way voice communication device, much like a "mobile telephone," use of the mobile Internet, Short Message Service (SMS), camera phones, and location-based services, observed mostly in Asian and Scandinavian countries, transform the mobile into more than a telephone. In these countries, we can witness cell phones playing the role of micro-computers, identity items, and more importantly, collective (many-to-many) communication devices. It is when mobile phones are socially and culturally appropriated that creative uses of the technology emerge, leading to a shift not only in our social relationships, but also in our perception of space.

Telecommunications-based art is primarily concerned with connecting distant and contiguous spaces. According to Frank Popper [Popper 127], communication art has six main characteristics:

it stages physical presence at distance,
it telescopes the immediate and the delayed,
it focuses on the playfulness of interactivity,
it combines memory and real time,
it promotes planetary communication, and
it encourages a detailed study of human social groupings.

In a broader sense, it can be said that telecommunication art not only foresees new developments for existing technologies, but also changes our perception of space. It focuses on the relationships between participants, rather than on the creation of material objects, in a situation where the author is the context provider, not the content creator. This configuration of telecommunication art also includes telephone-based artworks. There is a critical difference, however, between artworks that use fixed telephones as interfaces and those that use cell phones as the main focus of artistic exploration.

While the fixed telephone connects specific places, cell phones connect people who move through urban spaces. Mobility adds another layer to our perception of connectivity, since it allows people to be connected to each other, instead of only places. Moreover, when cell phones include location awareness (through a Global Positioning System (GPS) device or triangulation of radio waves), people can also be connected to places and to each other depending on their relative position in space.

This shift adds a fundamental difference on the possibilities of artworks with mobile technologies when compared to fixed telephones.

When cell phones arose, as in the early days of the telephone [Gitelman & Pingree 143],[4] they were regarded as media to transmit messages, and generally only urgent messages. Even now, cell phones are still viewed in many parts of the world as mobile telephones, that is, a telephone that can be carried around, used mostly for voice conversations. However, the incorporation of new functions such as text messaging (SMS), multimedia messaging (MMS), and location-based services contribute to the creation of new meanings for the mobile interface. The cell phone's potential for making new art is explicitly highlighted in cities with dense populations, because there is more potential for people to interact with each other. The emergence of nomadic technology devices allows whole cities to be used as a "responsive surface," or as a game board. It is as though the urban space has become a map of itself, a place for interaction and long-distance contact, without the need for a restricted or fixed space.

The following works use cell phones as promoters of collective and social actions in public spaces. They envision the phone no longer as only a voice transmission device, but also as a musical instrument, a game controller, and location-aware device. [5]

3.1. *Dialtones*: a telesymphony.

In the Ars Electronica 2001 in Linz, Austria, Golan Levin and the Ars Electronica Festival used the audience's cell phones to create a music concert at the Brucknerhaus Auditorium. Prior to the concert, members of the audience could register their cell phone numbers in kiosks, after which they would be assigned a seat in the auditorium and have a set of ringtones downloaded to their phones. Knowing each person's position in the auditorium and their respective ringtones, the computer could call them individually and produce a musical symphony, which was eventually a product of collective authorship.

---

[4] Writing about the first days of the telephone in Lancaster, England, around 1910: "Telephone company advertising in the village weekly newspapers amplified these themes by emphasizing the value of the telephone in times of emergency: accidents, fires, illness, stolen horses, mad dogs, robbers, and threatening weather."

[5] Other examples of artworks with cell phones can be found at Stephen Wilson's compilation under "Telecommunications: Telephone Art/Cell Phone" <http://mercury.sfsu.edu/~infoarts/links/wilson.artlinks2.html> (23 July 2007), and in Golan Levin's list "An Informal Catalogue of Mobile Phone Performances, Installations and Artworks" <http://www.flong.com/telesymphony/related/index.html> (23 July 2007).

This piece was innovative because it used cell phones as musical instruments. Although *Dialtones* is not a communication experience and it does not include voice, it can be regarded as a social and collective action happening in public space.[6] The distance from the mobile phone as a two-way voice communication device becomes even stronger when cell phones' power is used to create collective games.

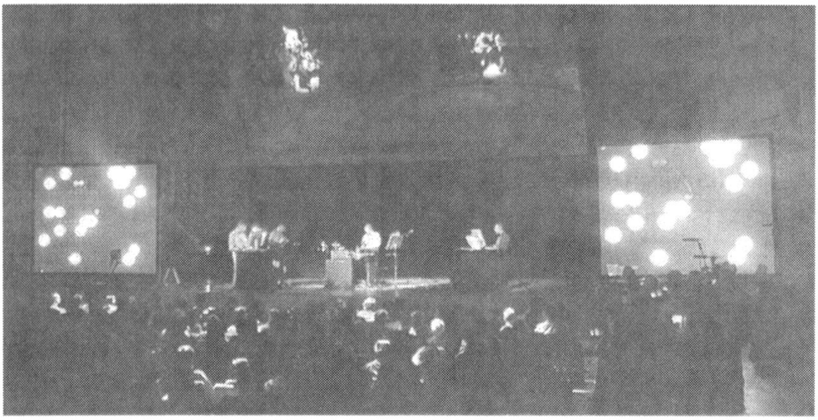

Figure 5: Dialtones in Austria 2001. The audience is able to see a mirror of itself on the ceiling, where the spotlights point to the actual ringing phones. The two projection screens placed on the side of the stage show the graphic interface used by the performers to trigger the audience's phones, constituted by the spotlights projected on the audience. © 2001 Golan Levin

3.2. *Blinkenlights*: the cell phone as a remote controller and game device

In 2001, the Chaos Computer Club transformed an eight-story building in Berlin's Alexanderplatz into the world's biggest interactive computer display. One hundred and forty-four lamps were arranged behind the building's front windows, which were

---

[6] A past work combining telephones and music was *Telefonmusik, Wiencouver IV* (1983) in Heidi Grundman (ed.), *Art Telecommunication*, Vancouver, Canada: A Western Front Publication, pp. 112-125. (1984). However, whereas this project was mostly concerned with the idea of transmitting and receiving music over the telephone, *Dialtones* transformed the cell phone in the music instrument itself. In 1983, one of the major characteristics of the project was the limited frequency bands which the telephone could provide for music broadcast. In 2001, Midi (Musical Instrument Digital Interface) ringtones enables the creation of polyphonic musical compositions simulating an orchestra on the handset.

independently controlled by a computer to produce a monochrome matrix of 18 x 8 pixels. Users could "control the building's façade" either via their cell phones or Internet, creating animations, playing Pong, or sending love letters.

Participants could use their mobile phones to call a specific number and play Pong against the computer. At first they heard instructions like "use the 5 to move the paddle up and 8 to move it down." If a second person called the system simultaneously, one caller played against the other. The difference between *Blinkenlights* and an ordinary computer game was the size of the "screen": a whole building in the middle of "Alex."

Blinkenlights transformed cell phones into game devices, and brought the game board/screen outside closed spaces into social urban spaces. The enlargement of the game board to the size of a building façade immersed large numbers of players and passersby into the game, transforming physical space into the game board. The possibility of carrying around the game control (that is, the cell phones) allowed people to interact with the screen and with each other in an open space. *Blinkenlights* explored the cell phone's potential to engage large groups of urban users and viewers in a social space that is both digital and physical.

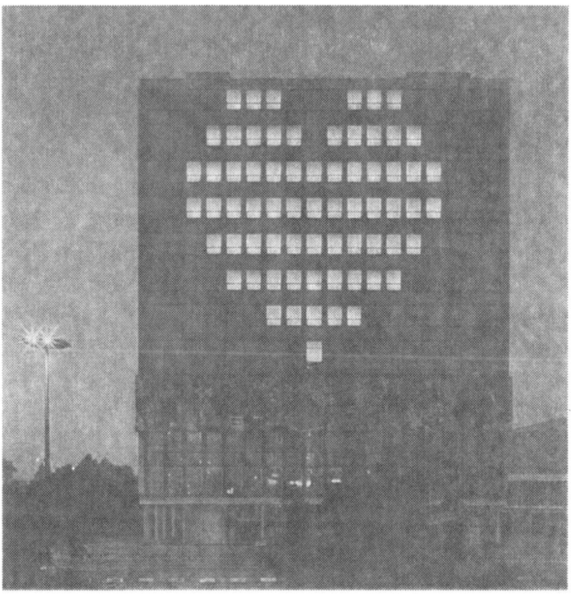

Figure 6: Blinkenlights at night showing an animated love letter.
© 2001 Chaos Computer Club.

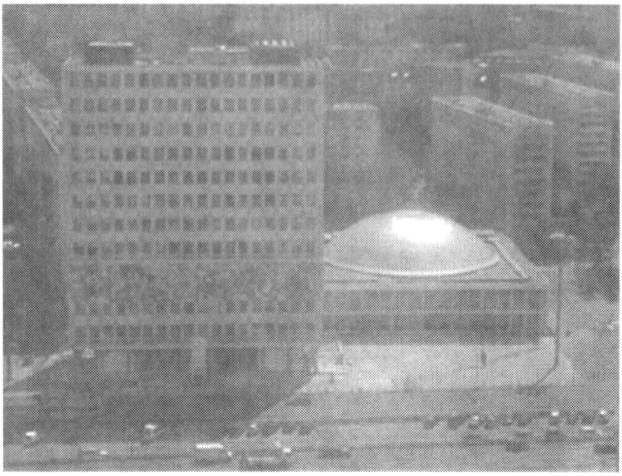

Figure 7: The Blinkenlights' building during the day on Alexanderplatz. © 2004 Adriana de Souza e Silva.

3.3. Broadening the concept of mobile technologies: Blast Theory

Although the British group Blast Theory did not work with cell phones initially, their projects foresee new ways mobile communication devices may function in the near future. In conjunction with the Mixed Reality Lab at the University of Nottingham, England, Blast Theory employs handheld computers and wireless devices to mix physical and virtual spaces, transforming the city into a playful multiuser experience. Their work focuses on developing games that happen simultaneously in physical and digital spaces, integrating and forming communities between players who walk on the street and online players. In their games, an action in the physical space might influence a decision in digital space and vice versa.

Their first collaboration, *Can You See Me Now?,*[7] resembles a traditional Pac-Man game played in hybrid space. Players from anywhere in the world can play online against the members of Blast Theory. Tracked by satellites, Blast Theory's runners appear on a digital map of the city next to online players. On the streets, handheld computers showing the position of online players guide the runners in tracking online players down. Street runners are equipped with handheld computers connected wirelessly to the Internet, GPS receivers, and walkie-talkies to communicate with other users. Up to 20 people can be online simultaneously. Online

---

[7] Blast Theory <www.blasttheory.co.uk/cysmn/> (23 July 2007)

players run away from street players in order to elude capture. If a street runner catches a virtual player, she is supposed to take a picture at the place where the chase ends, which is obviously an empty space. Street runners catch an online player if they are within 5 meters of each other. The game has been played since 2001 on specific days in many cities in the world, such as Sheffield (UK) , Rotterdam (Holland), in Oldenburg (Germany), and Tokyo (Japan).[8]

Similarly, their collaboration *Uncle Roy All Around You* [9] sets online players alongside players on the streets. Street players search for Uncle Roy with the aid of handheld computers. On the other hand, online players search for the street players and also for Uncle Roy in a virtual model of the same physical area where the street players are running. Online and street players must work together, and they have 60 minutes to complete the task. Street players can see online players on the map of their handheld computers and online players also see street players in the virtual modeled city. During the gameplay, online and street players can communicate through walkie-talkies and ask each other for help. The game was played in 2003/2004 in Westminster, Manchester, and West Bromwich (UK).

With the increasing availability of 3G phones[10] that incorporate all of the above-mentioned features, Blast Theory started using cell phones as their primary interface. Their project *I Like Frank*[11] is a similar experience that uses 3G cell phones to connect virtual and physical players in Adelaide, Australia (2004). Blast Theory looks to establish cultural spaces for mobile devices via games. Future games might allow the public to play on the streets using their own cell phones.

The rapid worldwide spread of 'smart' phones may increase the potential for this type of games and ludic experiences with cell phones to bring together users in different and distant places in the world. Commercially available location-based and hybrid reality games can be experienced in Japan (Mogi Mogi),[12] Finland, Sweden and Russia (Botfighters), and Brazil (Alien Revolt),[13] among other countries.

Within this context, it is important to understand that when mobility was added to telephones they became more than mobile phones. Mobility brought new artistic

---

[8] *Can You See Me Now?* was nominated for an Interactive Arts BAFTA in 2002 and won the 2003 Prix Ars Electronica Golden Nica for Interactive Arts.

[9] Blast Theory <http://www.uncleroyallaroundyou.co.uk/> (23 July 2007)

[10] Third Generation Cellular System. "Third Generation Cellular Systems include the possibility to offer data services without the need of establishing a connection (permanent connection) and speeds up to 2 Mbps. The main systems are WCDMA and CDMA2000 1xEV. The ITU refers to 3G as IMT-2000." <http://www.teleco.com.br/glossario.asp?termo=3G> (23 July 2007)

[11] Blast Theory <http://www.ilikefrank.com> (23 July 2007)

[12] Newt Games <http://www.mogimogi.com/>(23 July 2007)

[13] M1nd Corporation <http://www.alienrevolt.com/> (23 July 2007).

meanings to the telephone interface: bringing phones into the city space, releasing them from a fixed place, and transforming them into collective/social mediums and ludic devices. Henry Jenkins suggested,

> "Games have been to the PC what NASA was to the mainframe—the thing that pushes forward innovation and experimentation."[14]

Location-based activities will play the same role for cell phones, differentiating them from fixed phones, and increasing their power for communication and community formation.

Mobile and pervasive technologies help us to be aware of the physical space in which we live. Digital technologies in the 1990s have been mostly criticized for creating sociability in a virtual space, which was disconnected from our reality, placing users in a simulated and "unreal" world. Mobile technologies bring these multi-user and playful experiences to physical spaces, encouraging users to go out on the streets, and bringing new meanings to familiar spaces.

As art always foresees new uses for technologies, it is wise to look at these artistic experiments and try to picture the future, imagining how contemporary society will incorporate mobile devices into its everyday activities. Mobile phones are no longer just telephones.

Bibliography

Castells, Manuel, et al. *Mobile Communication and Society: A Global Perspective*. Cambridge, MA: The MIT Press (2007)

de Souza e Silva, Adriana. "Cell Phones and Places: The Use of Mobile Technologies in Brazil." *Societies and Cities in the Age of Instant Access*. Harvey Miller. Dortdrecht, The Netherlands: Springer, 2007.

Grundman, Heidi, ed. *Art Telecommunication*. Vancouver, Canada: Western Front, 1984.

International Telecommunication Union (ITU), 2007. <http://www.itu.int/ITU-D/ict/statistics/

Jenkins, Henry. "Games, the New Lively Art." <http://web.mit.edu/cms/People/henry3/GamesNewLively.html> (23 July 2007).

---

[14] Henry Jenkins, "Games, the New Lively Art." <http://web.mit.edu/cms/People/henry3/GamesNewLively.html> (23 July 2007)

Kac, Eduardo. “Aspects of the Aesthetics of Telecommunications.” *Siggraph Visual Proceedings.* Ed. John Grimes and Gray Lorig. New York: ACM, pp. 47-57 (1992). <http://www.ekac.org/Telecom. Paper.Siggrap.html> (23 July 2007).

Popper, Frank. *Art of the Electronic Age.* New York: Harry N. Abrams, Inc., Publishers, 1993.

Wilson, Stephen. “Chapter 6: Telecommunications.” *Information Arts: Intersections of Art, Science, and Technology.* Cambridge: MIT Press, 2002. 457-601.

Umble, Diane Zimmerman. “Sinful Network or Divine Service.” *New Media—1740-1915.* Ed. Lisa Gitelman and Geoffrey B. Pingree. Cambridge, Massachusetts: The MIT Press, 2003.

Artworks:
Blast Theory, *Can You See Me Now?* (2001) <http://www.blasttheory.co.uk/cysmn/ > (23 July 2007).

Blast Theory, Uncle Roy All Around You (2003/2004) <http://www.uncleroyallaroundyou.co.uk/> (23 July 2007).

Blast Theory, *I Like Frank* (2004) <http://www.ilikefrank.com> (23 July 2007).

Chaos Computer Club, *Blinkenlights* (2002) <http://www.blinkenlights.de>(23 July 2007).

Heath Bunting, *Kings Cross Phone-in* <http://www.irational.org/cybercafe/xrel.html> (23 July 2007).

Disembodied Art Gallery, *The Answering Machine Solution CD* (1996). <http://www.dismbody.demon.co.uk/stcd.html> (23 July 2007).

Disembodied Art Gallery, *Babble* (1993). <http://www.dismbody.demon.co.uk/home.html> (23 July 2007).

Disembodied Art Gallery, *Temporary Line* (1993/94). <http://www.dismbody.demon.co.uk/home.html> (23 July 2007).

Golan Levin & The Ars Electronica Festival, *Dialtones* (2001) <http://www.flong.com/telesymphony/> (23 July 2007).

Stephen Wilson, *Is Anyone There?* (1992). <http://userwww.sfsu.edu/~swilson/art/isany/isany.des.html> (23 July 2007)

Section 2: Critical Issues in Mobile Art

2.1 Critical Contexts and Definitions

Mobile/Audience: Thinking the Contradictions

Mary Griffiths and Sean Cubitt

1. The legacy of screen audience typologies; audience / users / producers

Screens 'audiences' are part of a legacy handed on to theorisers of mobile technologies, having always been considered to be a phenomenon with a taken-for-granted, yet prodigious, societal impact. In the case of older technologies, with audiences fixed in time and space (such as, traditionally, theatrical film and domestic television screens), what took an early theoretical hold was an exploration of the characteristics of an audience situated in those shared viewing locations. The focus in screen studies was initially on text-context interactions, and it led to formative and prevailing ideas of reception, viewer positioning, and of generalised audience attributes, characteristics and subjectivities. In film theory, work on the spectator's tendencies and attributes (such as gendered voyeurism, and fetishism), time and narrative, and the apparatus of cinema itself regulated the theoretical debates. Binary categories such as classic/experimental and realism/modernism and the coding and conventions of genre studies gave rise to certainties about shared audience expectations, and literacies. A 'language of audiences' developed which has not altered much with the arrival of VHS, DVD and web downloads, except that it now routinely includes the sociological notion of 'actual' as opposed to a textually defined 'implied' reader.

Twenty years after television screens began to proliferate in the living rooms of the suburban family home, cultural critic Raymond Williams' early concept of the 'mobile privatisation' of situated domestic audiences initiated theorising about the wider social meaning of sharing (across the nation) an invasive and public communication technology, capable of delivering 'education, information and entertainment', along with powerful ways of thinking about oneself and one's neighbour.

His work moved debates about 'screen uses' by audiences along a trajectory exploring the anchorage of media forms in social norms, and thus towards an investigation of the work of technology uses in reproducing culture, in its widest sense as 'the whole way of life of a people.'

The everyday, material and culturally specific field of mediated socialisation was at the core of this work, and it side-stepped the frequent essentialism of earlier aesthetic approaches to 'the viewer' and the 'text,' and provided sociologists and cultural theorists with more material questions about the social practices of media use.

This sociological, cultural approach to the theoretical debates about screens gained traction in television studies, and then the concept of 'audience' started to change with the arrival of fixed computer screens, and the simultaneous fragmentation and convergence of screen uses after the arrival of the internet: developments which impacted on, and gave complexity to, the theorising of screens, content, and user interactions. The questions of what constituted public and private spaces online, the individual user's power over communication practices, and the empowerment and limitations of code, make it increasingly difficult to theorise what is going on under the sign of 'the internet user,' except in micro-studies of user groups or in issues-based research such as the kind undertaken by Pew Research in the U.S. A large part of any user's day is spent travelling to virtual communities and imagined worlds, formed momentarily, in fragments of time, when an individual takes on the attributes of the 'cyborg' explorer: surfing their own sometimes unrepeated, and often unrepeatable journeys. User subjectivities become unfixed, unstable, privately formed, and often discarded. This can happen with only the traces of the online journey recorded in the retrievable data. 'Texts' and 'contexts' theory does not adequately provide the framework for understanding what is going on. The rise of 'usability labs' accounts for the micro-level of research about technology-human interactions, and relates back to the functionality and aesthetic concerns of earlier paradigms, with the renewed focus on 'content design' for the web.

Art-based discourses have been slow to follow the audience turn of media studies, with the notable exception of Pierre Bourdieu's [1986] analysis of class-based attitudes to art institutions. The 'implicit reader' model still dominates catalogue essays and art reviews. The flourishing fora for discussion of digital and interactive arts, meanwhile, stress the semiotic productivity of audiences [often on models deriving from Roland Barthes: see for example Landow 1994, Lanham 1993]. The concept of the (v)user elaborated by Bill Seaman, Joseph Nechvatal and Moroslav Rogala in the 1980s formalises another, more systems-theoretic, conceptualisation of the interactive audience as a group charged with taking up responsibility for completing the artwork, to the extent that the artist hands over that task to her audience. While discussion continues as to the extent of audience freedom in interactive media (do they merely explore options pre-programmed for them? Can they genuinely create something new?), the (v)user concept shares with the active audience theory a sense that, however thoughtful, the viewer of traditional artworks is always their subject, and that a more active participation in producing the artwork is in itself a good thing. Deriving from pre-digital forms like happenings, mail art and collaborative works like David Medlla's *Eskimo Carver*, but inflected by the aesthetics of networks and of computer games, digital art aesthetics privileges the construction by the artist of spaces in which audiences can use the work in their own fashion. The degree to which they are limited by the design of the code remains both a challenge to digital artists, and the last bastion of the ideology of authorship, which is so prevalent in artistic circles and the art economy.

The move from 'audience' to 'user' took another turn during the mid- to late 1990s. Laptops and other portable communication screens start to become extensions of the private world of home and work in public spaces, outside of institutional settings. An important extension of the customisation and privatisation of virtual and offline environments occurs. By 2005, technologies are smaller, wearable, and are both fashion items and extensions of the body's mobile and communicative power, while converging technologies such as internet-enhanced mobiles and iPods provide the user with even more capacity to create and share places of virtual assembly and content.

The figures for new media and mobile uptake and uses are, in themselves, a compelling argument against safely predicting mobile futures. There is always excessive market hype, but the industries, the content designers, the content aggregators, and the amateur uses are multiplying and developing exponentially. To take just one example: those who used their mobiles to upload 'amateur' witness footage to the internet and mainstream news agencies of the London bombings on July 7, 2005 are citizen newsmakers, as those holidaymakers were before them are, who recorded tsunami effects around the Pacific. Everyone, as Rheingold argues, can now be a distributor. The 'language' of mobiles and their users is being invented and re-invented daily: 'citizen journalism', 'mobilistas', 'smart mobs.'

The theoretical focus on users of these small screens has shifted slightly to meet the new paradigm of convergence of content, modes of delivery and publics. Sheller's work, for example, harks back to Williams's with the notion of the ubiquitous screen's 'mobile publics' and 'social gels' [Sheller, 2004], although it reverses the nomenclature to address an individual's personal mobility: wearable communication devices mean 'private mobilisation' has arrived. But mobile socialisation is at a number of removes. Individuation through technology has increased; privacy has both increased (people can make calls outside of normal social surveillance, and organise their social life unobserved and at the last minute) and yet decreased (mobile etiquette allows private conversations to take place in public places, and cell-phone users are trackable, as a London bombing's suspect discovered). The points of assembly are still privately organised through personal address books (or, at the higher end of the market and in business, through GP3 conference functionality). The convergence, content and regulation protocols are still being worked out (for example: the Viacom Outdoor digital bus-side signs in London may need offline protocols to be arranged, to avoid driver distraction (BBC). And there is, despite the hype, a lack of shared access to mobile technologies, and sometimes a gap in mobile literacies. These gaps may divide along classed, gendered, work-determined, or generational fault lines.

There is a point to this brief restaging of routine thinking about audiences: screen theory has always also been the theory of audience use and formation. Mobiles present a few challenges to conceptualising the 'audience' for aesthetic practices. In addition, what is at stake in theorising small communication screens is the question of what each new information technology throws out as a challenge to democracy, to the artist's position in society, and to ideas of shared public space. Aesthetic practice, and theory, has not faced such a challenge since the age of mechanical reproduction arrived.

2. Reach out: communication, interactivity

All disciplines create their own knowledge frameworks. British sociologist Jim McGuigan argues that the research methodologies on mobiles can be contained (for argument, because other priorities can be set) by demography, political economy, discourse and ethnography [McGuigan, 50]. He notes,

> "The mobile phone is not reducible only to a material object, a commodity circulating in the global economy of transnational operations… it is an obscure object of desire and a sign of the times." [McGuigan, 46]

They are not just desired; they are now part of creative practice. According to Peter Lunenfeld [following Mitchell et al, 2003], any discussion about the impact of computer-inflected media on the design disciples over the past five years should be situated within the move from information technologies (IT) to information technologies and creative practices (ITCP).

As comments to consider, in any attempt to theorise mobile + audiences, the first questions one must ask of both are: where is the other 'c' word—communication—and where is the other 'i' word—interactivity? How should this insight into new kinds of design be expanded to include person-to-person or many-to-many technological communication processes, and thus enable the thinking through of a theory of aesthetic practices?

Two concerns are at issue: the use of the mobility of the viewer in unexpected art display settings, and the deployment of mobile immersion techniques and the expectation of viewer interactivity. Both aesthetic practices provoke a re-conceptualisation of the material stability of artworks, of the places of viewing or participation, the spaces, which the technologies help audiences imagine, and the ensemble of attributes these acts help form. Though they reference the use of ephemeral media such as performance art, the history of mobile audiences or 'users' is different from almost every other kind of screen audience of a fixed technology.

So, where does one begin a descriptive analysis and a history of 'mobile' audiences for art? Mobile audiences are individuated, and yet they also are a set of overlapping and forming and reforming publics. These publics are not only mobile in the sense that they are hailed fleetingly by artworks embedded in urban space and practices, and no longer isolated in the bourgeois, avant-garde or radical spaces of public or private galleries. They are also mobile in the sense that an increasing number of urban dwellers wear the means to be contacted 'anytime, anyplace,' not just by their family and friends but also by marketing organisations, local councils, and news services as well as artists.

There is potentially an excess of forms of mobile communication. In addition, mobile technologies are empowering. They provide the capacity for users to become increasingly sophisticated producers and distributors of their own and others' creative content: a technological capacity which extends to being able to be co-producers of art, albeit within, as later examples below will show, rulemaking and protocols not entirely originating from the audience itself. These factors shape the mobile audience for art, and give a contour to the collaborative nature of mobile aesthetics. Mobile technology has the potential to help construct specific dispositions in users, and also to assist the user to create, and make unexpected relations between themselves and others.

3. Northern lights, postcolonial space, mobile art producers

> *"All around us inanimate things are slowly coming to life, veneered with layers of digital information…[mobile tools] will stop being purely at the receiving end of end of data streams and become conduits, mustering bits from objects and infusing them into other objects"*
> [Fabio Sergio, mobile design interaction strategist]

"Art lovers can 'text' sculpture," a BBC news website reported in April 2004. The story concerns the 'opening' or 'launch' (if one can use those words) of *Spectra.txt*, a light sculpture by Peter Freeman, commissioned by Middlesbrough Council and situated at the crossroads of the city centre pedestrian mall. Freeman's work shows an increasing interest in light and with interactivity. A photograph of the two-tonne light sculpture made of steel and fibre-optics also illuminates the homepage of the local City website. It gives a 21st Century, edgy, 'new media,' 'mobile' feel to the city's online branding which, like most municipal branding, now functions as a point of citizen identification with local place, and a form of civic GPS for outsiders.

This city, a decaying industrial town in North East England's rust belt, has a history of waiting for the trickle-down effects of globalisation and neo-liberalism: for Middlesbrough, decades of labour market change and the loss of traditional industries have not been easy. Mayor Ray Mallon, a former police officer, was elected on a zero-tolerance-of-crime ticket. On a quick visit taken there in June 2005, it was noticeable that the city had an uneasy pulled-up-by-its-new-byte-straps feel: the ubiquitous new chain department stores in the mall, the imposing Town Hall standing as a memory of another, more self-important time, new luxury 'enclave' hotels for business travellers, extensive Sunday afternoon displays of youthful drunkenness, boarded-up traditional shop fronts, and a patrolling CCTV-enabled police van, all confirm the impression formed on the journey from Darlington. In the dusty carriages of the unrenovated Northern line train, no passenger mobiles rang. The language of mobiles—for example, 'user-driven personalisation', 'augmented reality check', 'software interfaces', 'key technology enablers'—seems quite out of place.

Spectra.txt is another part of the unsettling urban mix. It is interactive, an elegant 10 metre metal tower which responds in six pre-programmed ways to text messages sent globally. It is iconic and referential, not just for Middlesbrough, which invested 80,000 pounds sterling in the commission as part of its urban revitalisation. In 2006, in another response to Richard Florida's doctrine of 'invest in creativity, release economic well-being,' Middlesbrough is also opening the Middlesbrough Institute of Modern Art, Mima, a stylish gallery built to house the city's collections, and to reach out into the local community and beyond. As in so many instances, contemporary culture is required to encourage economic development, to replace lost social welfare structures, and to provide moral cohesion in a period when neo-liberal policies have resulted in the loss of community values [see Yudicé 2003]. This setting for art is increasingly familiar, and calls for an almost ritualistic response.

Already there have been pre-opening events. One of note has been the 2004 pre-Christmas projection by John Wood and Paul Harrison of short films from the department store chain building, Debenhams, onto the side of the BHS building. Judith Winter, the council officer responsible for 'the alternative to the traditional Christmas decorations explains the impetus for the screening in terms of accessibility and curiosity value, arguing that 'We don't just expect people to walk through the doors of the Mima. ' Showing what is usually seen inside in the dark, outside in public arenas, where there may already be a celebratory mood of imagined belonging and more openness to seasonal 'sights,' belongs to a curatorial move to reach out to new publics, engaging them with new ideas about the 'civic.'

Before turning to the questions this raises about mobiles, expectations of art and its audiences, a few more thoughts on the earlier local example. In many-layered ways, not least because the sculpture involves mobiles and urban renewal, it also speaks to the historical reflexivity and the connectivity impulses of public art aesthetics. Freeman himself addresses his sources of inspiration directly, both in his BBC interview, and on his website. He makes explicit the locally relevant reference to 'placing' Middlesbrough in a global and historical context through claiming Captain Cook's journeys to Tahiti. When he mentions Ridley's Scott's use of the alien landscape of chemical complexes on Teeside for Blade-Runner, his art is engaging with media-savvy but fragmented global publics everywhere, publics which form, even if they do not see each other because they share categorising knowledges, familiar sensibilities and a disposition towards enjoying complex intertextuality. The cinematic reference probably reverberates with audiences of many other screens, and with theorists of screen audience formation, and of spectacle's part in shaping active responses in actual viewers. The world of hi-tech, the information-age wireless techniques, the sparse, clean-lined functionality of industrial architecture are unavoidable points of reference for theorists of new media contexts and also strike a cord with information age and other post-industrial workers. Today, the connections between post-industrial-Middlesbrough viewers of *Spectra.txt* and Tahiti are not the ships which carried local peoples to places as far away as they could go before returning home, but are made through wearable information and communications technology. Mobile phone users are, in this context, always at 'home', because 'home', as in 'homepage', is a portable and ubiquitous property.

So, *Spectra.txt* represents the ways in which the forms of public art are changing, recognisably, becoming part of global flows of perpetual connectedness [Katz and Aarhus], and addressing and addressed by the fleeting look. This mobile-enhanced public artwork, by trading on the desire for immediate gratification, the filling up of every minute, becomes a trackable node in the flow of points of global identification, local belonging and an avant-garde aesthetic. Mobile connectivity, coupled with the desire or need for contemporary artists to transgress the boundaries of 'installation' in confined gallery space, makes this sculpture also representative of the new audiences for contemporary art. Urban 'audiences' are now marked by immersive contexts, and individuals and groups who crisscross each others' paths. They flow around public art in the streets, seeing it (or not registering it) as part of a changing mediascape of visual noise: signs, windows, advertising, computer screens, CCTV, large television screens, teletexts and billboards.

Both Freeman's historical annotations can be taken as light but significant references to the contemporary remix of the city's postcolonial memories and post-industrial landscapes. Public art can offer this remix in the very heart of historicised, real and transcendental spaces. For a multicultural city implicated in imperial history, the voyages of discovery still resonate: both in the present ethnic composition of the population, and for local viewers of any ethnicity watching themselves and the commercial buildings nearby mirrored in its shiny, luminescent surfaces. Possibly they are speculating about who and where the messengers are who are txting the program instruction resulting in the latest show of coloured light emitting diodes (LEDS). There are no webcams recording the activity as there are with the Swiss *Hello World* Project of Johannes Gees. But, txters may take a pxt of themselves in front of the display they have helped produce. This is, after all, a world 'where there are 150 devices on the market', where 'viral is a good thing, where immediacy trumps quality, and there are so many protocols it makes the VHS/Beta Debate look like rock/paper/scissors' [Mobile Media Blog]. And where there is a desire for perpetual connection and an increasing familiarity with personalised broadcasting, as is shown in the following exchange, taken from a *mo: life* interview with a vice-president of a leading global producer and distributor of mobile content, The Mobile Media Company.

> *Mo: life*
> "Is the mobile content you are seeing pushing new ground by utilizing the distinctiveness of mobility, or, because you deal with branded content, is the majority of fare extending into mobile from television, film and the web?"
>
> *Steiner Svalsen*
> "We are seeing that there is a growing commitment to 'made for mobile' or using mobile as a way to promote TV, film or web content/programmes/films. Mobisodes are a good example with both Vodaphone and Verizon running those from Fox. All of the big content owners are integrating mobile interactivity into their strategic plans for the US. "

Spectra.txt is indisputably 'quality'—it is 'art' and not 'content,' but it is 'immediate,' and changing, a visual display of the desire for connection and, in those fleeting connections, a guarantee of presence for users in their demonstration of shared and visible mobile literacies.

In both cases—*Spectra.txt* and Mima's pre-opening events—audiences and users of such inter-textual experiences are invited to escape traditional confining notions of artistic intent, and even the authority of critique and commentary. The physical placing of *Spectra.txt* in a centre of commerce and exchange, and the interactivity it offers, helps construct new, almost mutually exclusive spaces and subjectivities. Consumer and citizen, the mobile co-producer can call into being other imagined spaces and collectives (one txt program responds with a colour display of the local football club colours). These places and identifications are glimpsed in the process of txting, in the transitory flows of walkers, pausers, glancers and watchers in the city-space.

There are different ways one can theorise this kind of activity. The sociologist Sadie Plant, in her much-cited study of mobiles for Motorola [2002] (also discussed in McGuigan), categorises mobile user-behaviour in terms of categories of birds. The 'activity theory' of psychologists of interactivity like Bonnie Nardi would break down and explain the mediations of objects and technologies like *Spectra.txt*, as a series of acts of heuristic comprehension and user learning, using Vygotskyian 'scaffolding.' But, for those interested in mobile aesthetics, the issues might be the ease and openness of address, value systems, access and desire to participate and connect. *Spectra.txt* requires its mobile interlocutors, wherever they are. Democratically, the sleeping structure has a metonymic argument to make about the wholesale civic participation needed if Middlesbrough's renewal is to work. Of course, as Lessig and others have argued, code is law: only six txt message programs are possible at the moment, the interactivity thus fixes possible actions within its material and temporal circumscriptions.

What more can be said about these art works and events which invite co-production, either in meaning making as 'active audiences,' or through the active intervention by members of publics moving through civic space? Can we throw out the inventory of audience attributes shaped through fixed screen uses, until something like an 'audience,' recognizable in 'audience theory,' forms? Does a mobile aesthetic require a 'public,' in the sense, which Habermasian theory would conceive of such a thing? Does such a thing exist anymore?

4. Composing 'things public': the crowned giant, and networks

It would not be wrong in this context to remember that one of the most influential treatises on the philosophy of the body politic, namely Thomas Hobbes' Leviathan [1651], did not actually start out as a political essay but as an attempt at an aesthetics...For Hobbes himself brings together—just as we do—the three types of representation customarily kept separate: the science that represents things; the politics that represents people; and the art, which represents the meeting of humans and things [Exhibition catalogue. Latour and Weibel, 1].

At ZKM in 2005 the sociologist of science Bruno Latour co-curated with Peter Weibel an interactive exhibition called *'Making Things Public: Atmospheres of Democracy.'* Its aim was to chart the Dinge which make up democratic spaces, (the taken-for-granted Foucauldian surface technologies of democratic practice), through a collection of works by a hundred artists. There are many exciting things to look at, but the most interesting and innovative part of this exhibition is the tangible creation of a 'public sphere,' 'a thing' which can be shared and which is responsive at a micro-level to one's movements.

To enter the clean, postmodernist exhibition space, situated in one of the spacious ground-floor ZKM galleries, visitors have first to accept a barcode identifier to wear around their necks, which tracks their journeys around the exhibits and which collects data for further research on visitors. Wearing the coder, one walks first over a backlit interactive map of the exhibition, a reference to familiar European and US ID card and surveillance debates about citizen rights and responsibilities. Conceptual mapping of social connections, and the task of making visible the unseen networks and chains of meanings is a familiar trope of Latour's, and of Actor Network Theory. One is instantly subjectified by the spatial and temporal democratic ideas of 'horizonality', 'participation' 'inclusion', 'limits,' 'freedom', 'permission,' and 'consent.'

These ideas—call them traces—lie at the heart of the many exhibits, which collectively represent a Leviathan of small and fluctuating differences, not a unified aesthetic of mobility, or of mobile users. The concept is grounded conceptually in multiple enmeshed 'mobilities' because the invisible exhibit, the 'phantom public,' is activated by the movements of the viewers. The viewer-participant, by walking through a number of historically and geographically specific exhibits, experiences the 'atmospheres of democracy.' He or she senses, if obedient to the command to 'be attentive' that individual movement is impacting on the exhibits, and on other viewers. Designed by Michel Jaffrennou, Theirry Coduys et al, the features and techniques of 'The Phantom Public' [2005] are: 'interactive, virtual, variable scenography based on the behaviour of visitors to the exhibition, 382 digital objects: sensors, RFID-stations, microphones, loudspeakers, spotlights, screens, video projectors, displays, dimensions variable, development systems software' [Exhibition catalogue, 29].

Laid out in circuitous designs, the touchable exhibits differ in their use of multi-media modes, and in the rhetorics of address deployed. 'The Puzzle of Composite Bodies' represents issues of assembly through pixel portraits of 'Joe, War President' and 'Bin Laden's Shadow Over the World.' GPS is used in 'Milk,' the visual tracking of milk production in France, accompanied by footage of milk workers. Large wall screens display axioms about democracy on a video loop, responsive to viewers. One watches repetitive film of a woman practising privately, with a personal trainer, the techniques of public speaking; and then, footage of her giving a talk on what she has learned ('that anyone can do it') to audiences flowing around her soapbox at Speakers' Corner in Hyde Park. This short film ironically juxtaposes the neo-liberal discourses of public relations and self-improvement, with those of horizontality and participation.

Hand-drawn pictures and text generated by the issues of privacy and consent cover the curved walls of an exhibit dedicated to recovering but immobilised victims of trauma. In 'Pindices' or 'Personal Political Indices', a sociologist and an artist collaborated 'to make political activity visible.' Coloured badges with legends such as 'I kept myself informed' and 'I protested' were piled up in supermarket-style sweet dispensers waiting to be taken by the visitors, worn, and thus become part of the world outside the exhibit, changing the atmosphere of democracy inhabited by the badge wearer. This is an 'adbusting' kind of activity, a dynamic mobile redistribution of modes and meanings in which the viewer chooses the civic pins which best reflect beliefs: a practical demonstration, but outside gallery space, of the formation of democratic, material publics.

5. Towards a theory of mobile + audiences

The discussion so far shows that mobile technologies (dispersed, personalised, consumption-inflected), and aesthetic practices in which mobile behaviours are deployed and /or questioned, are underpinned by the artists' taking up of 'postmodern rationalities.'

To adapt McKee's list [116-17], these are contextual, plural, multiple norms, subjective, rhetorical, performative, and allow for positioning and culturally specific logics. In both the cases described, the technology which provides mobile connectivity does not work to 'reinforce the system' as Lyotard warned in *The Postmodern Condition*. The 'Cultures of Mobility' pages of Nokia's website bravely puts the alternative case for mobile technology:

> "In a world where restricted access and limited mobility have crumbled under the latitude of global connectivity [...] Mobility is having a profound effect on the design industry as it responds to the needs of a time-starved, information hungry society."

The market necessarily provides the technological conditions of interaction and 'co-production,' but only limited mobile activity is possible, primarily because of access and code restraints. Aesthetic choices, which buy fully into postmodernist play and the subjectivities of mobile-me media individuation, may lead to an anchoring of the market-driven and technologically determinist consumption of new gadgets. The attribute of 'personalisation', which this assists into being, left unchecked, stands to become undemocratic. Baumann, and then Deleuze and Guattari, have used the metaphor of transcendent 'nomadism' to describe the characteristics of global culture. But in Middlesbrough and Karlsruhe, mobility is experienced by viewers , visitors, audiences within specific social contexts, and as contextualised, coded, delimited by affect, attention and imagination, and it is plural.

Mobile aesthetics will need to address three categories: the subject, the object, and the audience. The aesthetic attitude, essential for the construction of aesthetic objects, has a history running through the 18th century aesthetics of taste as a shared judgement, which, however, the connoisseur can only acquire by subjecting herself to prevailing norms; through the Romantic and early modernist thesis of the 'aesthetic emotion', an innate sensitivity to the beautiful, towards a crisis in which, on the one hand, beauty is constructed from the shards of a once-shared but now fragmentary culture (postmodernism) and on the other the ineffable sublime (and its opposite, the vile and abject) struggle over the artwork's core tasks: to explore the conceptual limits of art and the semantic limits of meaning. The art object meanwhile has traced a trajectory from the organicist paradigm (a whole greater than its parts linked in morphogenic unity), through variations on montage, to the immaterial forms taken by minimalist, conceptual and Fluxus objects and actions. Digital and networked art add to this history by emphasising the co-production of the object (and even its concept), and the possibility of an art whose content is exclusively made up of its connectivity (as in Alexei Shulgin's *Refresh*). The price of such connectivity is that the links will die, and the work be abandoned, a risk which is integrated into many net.art projects.

Finally the audience for art now contains the artist him or herself as first spectator, for example in a-life projects. More significantly, the audience is neither a recognisable mass audience, as for theatre or film; nor an idealised domestic unit; nor yet an equally idealised individual spectator, as proposed in Panofsky's analysis of perspective and reinvigorated by apparatus theory in film studies. Instead the audience / participants in mobile practices are networked. That is to say, they are both individuated and simultaneously linked to one another. The primary content of any network is its members and the number and intensity of their connections. For a networked event to occur as event this dispersed but interconnected spectatorship must witness it. Without such witnessing, the event has no more significance than flicking a light switch. While the one-to-one relation of love poems and words of sympathy has not disappeared, it is the networked work of witnessing that draws what otherwise is only an anonymous databank of individuals into becoming an audience; and it is only through that act, freely undertaken by disparate individuals, that the mobile artwork becomes itself.

A mobile aesthetic or—better—an aesthetic of mobilities, is best conceptualised as 'social.' There is a global 'compulsion to mobility' [Urry, 1] and an anxiety about what can still be shared in public spaces. Mobile art in a 'post-secular' world does not need to become one of the secular false prophets of which Matustik writes [15]. It cannot be technology's servant by reinforcing consumer habits of use, which can be market system anchors. But an aesthetic of mobilities which might give reasons for small 'impossible' hopes about a 'democracy-to-come' would be ones which, like *Spectra.txt* and 'The Phantom Public' make attention lengthen and become productive, and the glance rest questioningly longer on objects, oneself and others.

Bibliography

Australian Network for Art and Technology. 2005. <http://www.anat.org.au/> Downloaded 12/8/05.

Balnaves, M., T. O'Regan, and J. Sternberg, eds. *Mobilising the Audience*. St. Lucia: U of Queensland P, 2002.

Barry, A., and L. Kimbell. (2005) 'Pindices: Personal Political Indices.' ZKM, Atmospheres of Democracy. March 20-August 7.

BBC News. 2004. 'Art lovers can 'text' sculpture.'
<http://news.bbc.co.uk/go/pr/fr/-/2/hi/uk_news/england/tees/3627277.stm>
Downloaded 12/12/04.

BBC News. 2005. 'Buses to display moving adverts.'
<http://newsvote.bbc.co.uk/mpapps/pagetools/print/news.bbc.co.uk/2/hi/uk_news/england..
Downloaded 1/9/05.

Beaton, J., and J. Wadjman. 2004, 'The impact of the mobile phone in Australia : social science research opportunities.' Report. The Australian Mobile Telecommunications Association Conference.

Benjamin, W. The Work of Art in the Age of Mechanical Reproduction.' *Illuminations*. London, 1970. 223-227.

Bourdieu, P. *Distinction: A Social Critique of the Judgement of Taste*. Trans. R. Nice. Cambridge MA: Harvard University Press, 1986.

Brown, B., N. Green and R. Harper, eds. *Wireless World: Social and Interactional Aspects of the Mobile Age*. London: Springer-Verlag, 2002.

Castells, M. *The Rise of network Society*. Oxford: Blackwell, 1996.

CitizenShift. http://citizen.nfb.ca/onf/info Downloaded 2/9/05.

Clarke-Dickson, P. "UK cellcos, broadcasters step up initiatives in mobile and iTV." Mobile Media Analyst 6.10 (2005), 5-6.

Freeman, Peter. <http://www.peterfreeman.co.uk/home1.htm> Downloaded 12/1/05.

Gees, J. (2005) *Hello World Project.* http://www.helloworldproject.com/ Downloaded 3/8/05.

Grether, R. 2005. *Directory of Mobile Art and Locative Media.* <http://www.netzwissenschft.de/mob.htm> Downloaded 5/9/05.

Katz, J.E. and M. Aakhus, eds. *Perpetual contact: mobile communication, private talk, public performance.* Cambridge: Cambridge UP, 2002.

Immersive Media. <http://www.immersivemedia.com/commercial_medai.php> Downloaded 31 August, 2005.

Landow, George P., ed. *Hyper/Text/Theory.* Baltimore MD: John Hopkins University Press, 1994.

Lanham, Richard A. *The Electric Word: Democracy, Technology and the Arts.* Chicago: University of Chicago Press, 1993.

Latour, B. and P. Weibel, 2005 (curated). "Atmospheres of democracy." ZKM, Karlsruhe. <http://on1.zkm.de/zkm/e/> Downloaded 30/5/05.

Ling, R. *The mobile connection: the cell phone's impact on society.* Amsterdam: Morgan Kaufmann, 2004.

Lunenfeld, P. "Media design: new and improved without the new." *New Media and Society* 16.1 (2004), 65-70.

Matustik, M. B. "Between hope and terror: Habermas and Derrida plead for the im/possible." *Epoché* 9.1 (2004), 1-18.

McGuigan, J. and the Agora Center, University of Jyvaskyla. "Towards a sociology of the mobile phone." <http://www.humantechnology.jyu.fi/articles/volume1/2005/mcguigan.pdf> Downloaded 5/9/05.

McKee, A. *The Public Sphere: An Introduction.* Cambridge: Cambridge UP, 2005.

Middlesbrough Council Website. <http://www.middlesbrough.gov.uk> Downloaded 12/8/05.

Mobile Journeys. <http://dlux.org.au/mobilejourneys/profiles.html> Downloaded 12/08/05.

Moores, S. *Interpreting Audiences: The Ethnography of Media Consumption.* London and Thousand Oaks, California: Sage, 1995.

Nardi, B. <http://www.darrouzet-nardi.net/bonnie/> Downloaded 3/9/05.

Nokia. 2005. 'Culture of Mobility'.
<http://www.culture.nokia.com/hasflash.jsp?section=0&division=0> Downloaded 7/9/05.

Oskman, V. and J. Turtianen. "Mobile communication as a social stage New Media and Society." 6.3 (2004), 319-39.

Plant, S. "On the mobile: the effects of the mobile telephone on social and individual life."
<http://www.motorola.com/mot/doc/0/234_MotDoc.pdf> Downloaded 5/9/05.

Svalsen, S. Interview. *mo:life* 5 September 2005. <molife@lists.s7digital.com> Downloaded 5/9/05.

The Technology Log. <http://www.technologylog.com/archives/miscellaneous/index.php> Downloaded 5/9/05.

Urry, J. 1999, 2003. "Mobile Cultures." <http://comp.lancs.ac.uk/sociology/030ju.html> Downloaded 1/7/05.

Williams, R. *Communications*. Harmondsworth: Penguin, 1968.

---.Television : technology and cultural form. London: Fontana, 1974.

Yudicé, G. *The Expediency of Culture: Uses of Culture in the Global Era*. Durham NC: Duke University Press, 2003.

Towards a Language of Mobile Media
Jon Dovey and Constance Fleuriot

Background:

In November and December 2004, Constance Fleuriot and Jon Dovey organised a series of five seminar discussions aimed at refining the language we use to describe and understand our experience of pervasive, mobile and located media applications based on a selection of experimental work by the Mobile Bristol research unit.[1]

The aim of the authors was to contribute to the development of a common discursive framework for description, understanding and production of mobile and locative media. This was prompted by the perception that interdisciplinary mediascape authoring teams lacked an 'agreed set of terms' upon which to proceed. This definitional difficulty was additionally compounded by the blurring of previously distinct categories of media and reception, which mobile media bring about e.g. spaces of media reception lose their 'borders' as media becomes more and more pervasive, and media genres similarly start to blur as reception practices change in response.

The authors' own backgrounds and experience in computing/information systems and media studies research reflect our perception that practitioners of mobile media are drawn from very many different fields, (eg Human Computer Interaction, Media production, Performance, Experience design) each with their own discursive formations that might be useful starting points for the development of a vocabulary that enables people to discriminate and choose between different applications. Participants in the seminars were chosen to represent a wide set of disciplines: Media and Cultural Studies, Experience Design, Human Computer Interaction, Psychology, Media Production, Performance, Information Systems and Computer Science. Invitation to participate was extended to practitioners and theorists working in the field, Constance Fleuriot, Phil Stenton, Jo Reid and Richard Hull from Mobile Bristol (University of Bristol Dept of Computer Science, Hewlett Packard Research Labs), Jon Dovey (Bristol University Dept of Drama, Film, Theatre and Television), Martin Rieser (Bath Spa University), Teresa Dillon (NESTA Futurelab), Clodagh Miskelly (University of the West of England), and Mark Jacobs (BBC).

---

[1] See <http://www.mobilebristol.com/> (September 2007)

Methodology

All discussion sessions took place in Bristol. Each seminar started with discussion of a *Mobile Bristol* project as an example of a mediascape—a located media 'experience'. Seminar discussions were analysed so that vocabulary used could be presented back to the group for comment and amendment in the subsequent session. In this way, a 'taxonomy' of locative and mobile media was developed in successive iterations. Each discussion began with the participants (expert users of mediascapes) describing their experiences in terms that could be used to generate sets of appropriate key words and search terms. These generalist descriptions then led on to the development of a more specialist *discursive* vocabulary used by practitioners, both artists and technicians. This approach generated discussion of recurrent themes across different applications, which allowed us to generate a set of design dimensions that we hope will have direct applicability to designers working in the field. The applications used as examples were: *Riot, Moulinex, A Description Of This Place As If You Were Someone Else, CitiTag, Savannah and Node Boat tour.*[2] These applications all use the geographical/physical location of the user of the kit as the context to trigger the delivery of the media content, and as such may not be representative of all kinds of mobile/locative media. A final, plenary session was an attempt to draw together the vocabulary used in the sessions, and to identify the emergent research questions for wider discussion. Preliminary results were presented at the *PLAN:ICA* network session in February 2005[3] and on the New Media Knowledge website.[4]

Findings

When each piece was discussed the words generated were listed and then categorised as suitable for either a *general* or a *specialist* audience. There was discussion as to which category a particular word or phrase would fall into, eg was GPS a commonly understood term, a specialist term, or did it fall somewhere in between? One of the aims of the seminars was to generate a 'useful' description of each piece of work that could be understood by a general audience, and would for example be usable in a listings magazine. Key words were also noted that could be put into a search engine such as *Google*. If, as the authors envisage, there is to be a rich layer of digital media overlaying the physical environment, then people need to be able to search for these by location, genre, topic, personal preferences etc. The website http://www.mscapers.com encourages authors to include descriptive 'tags'

---

[2] See <http:www.mobilebristol.com> for more information on these projects

[3] See <http://www.open-plan.org/> (September 2007)

[4] See <http://www.nmk.co.uk/> (September 2007)

when they upload mediascapes to the site, so that users can search for the type of piece they want to experience. For example the tags for *Riot1831!* are: voice-play audio walk historical drama Bristol destination

Key Terms for non-specialists

We asked our expert users to describe each experience 'as if to a non-specialist audience'. A number of key terms were repeated across different applications, which became the core to our further investigations and suggest ways in which locative and mobile media experiences will find a place in the wider media landscape. Participants generated the idea of a *Parallel World* to describe the applications, which they further defined as a walk through a *Soundscape*, in which the experience was something like eavesdropping on invisible storyworlds. Users also phrased their experiences as: "A slightly surreal experience of being in a wall of sound in the open air, which dramatises ordinary space and tells you about specific places." (referring to *Moulinex*)

A common way to describe the pleasures of mobile media applications was by reference to the idea of *Magic Moments,* when particular convergences or overlaps between the media environment and the physical environment were generated. These moments needed to have the appearance of being spontaneous, 'discovered' by the individual user, and fortuitous, for them to hold what we identified as characteristic pleasure. These moments might be as simple as checking to see if the gravel scrunching footsteps were actually somebody behind you or part of the soundscape; or the serendipity of hearing a story about lovers' meeting on a park bench for lunch, whilst the user was watching (an entirely other) man and woman eating their sandwiches together at lunchtime. The magnifying of the significance of these moments to the user subjectively appeared as greater than the individual components in the real and virtual spaces.

These simple descriptions were not in fact the kind of terminology that we were seeking when thinking about how one might tag work for search engines. Nevertheless searching these phrases in *Google* did in fact throw up some interesting results. However these results did offer us an important clue as to how to develop the research process since they pointed to a concern to convey the *affective* quality of the experiences, rather than its technological or media form. Participants used these descriptions as key simple terms to convey what the applications felt like and what kinds of pleasures they held.

Short 'Everyday' Descriptions Of Each Piece

We then involved users in generating a short descriptive paragraph for each piece. For instance Dan Belasco's piece *A Description Of This Place As If You Were Someone Else* was written as:

> "This is a set of personal stories that refer to specific places in Queen Square. When you walk around there, alongside the stories, you get the real sounds that happened in that place. The small screen that you hold has pictures to help show you where to look and stand. You can't always tell if a sound you hear is real or virtual. Sometimes the virtual sound makes you look round. You get special "Magic Moments" where you hear someone talk about something and you see it at the same time, as if they're recording while you're there. Sometimes it's a bit spooky/scary, sometimes it is like eavesdropping on a real conversation. Its simple to use, easy to learn how to make it work, you just walk around."

When we compare the descriptions of all five pieces we were able to use them to generate a simple template that allows us to describe most of what it seems users might want to know about an application, to at least get some idea of what it will be like. This is the descriptive template, which addresses the following categories:

Activity	What your body does, 'you walk around there'.
Place	Where does it happen, 'in Queen Square'.
Equipment	The visible equipment eg headphones, backpack and 'The small screen that you hold'
Content	What is it about ? eg 'personal stories'
Media mode	Sound and still images
Genre	E.g. history, drama, documentary 'the real sounds that happened in that place'
Affect	How does it make you feel, 'special "Magic Moments," 'a bit spooky/scary', 'like eavesdropping'
Sociality	Is this something you do on your own or with other people?
Skills	Do you need any special abilities to do it ? 'Its simple to use, easy to learn how to make it work'

Table 1: Everyday Descriptors

Descriptive Dimensions for Mobile Media Design

During discussion it became clear that the parallels and differences between the different applications could be used to construct descriptive dimensions, which could begin to map the field of possibility opened up by mobile and locative media applications. We began to generate a 'design taxonomy' based on real, existing applications rather than on theoretical ideas.

Critical reflection on the mobile media sample suggested the idea of these 'dimensions' of mobile media being a useful way for authoring teams to conceptualise what kinds of mobile media application they may want to produce. Given that mobile media use is flexible, contingent and heterogeneous, the idea of a series of dimensional axes along which a team might 'place' its application may be a useful dynamic framework. In authoring mobile media applications teams could define their objectives by asking where on the following different axes they wish to place the user experience.

The Dimensions of Immersion

SURFACE <> DEPTH   INFORMATION <> EVOCATION

Our expert user group identified 'immersion' as that quality of the experience, which held them in an imaginary or imaginative world, and left them feeling removed from the everyday surroundings of the experience. This sense of being in a 'parallel world' was also complemented by a number of observations about the quality of the attention which the user brought to bear upon the experience, a particular kind of immersive concentration that relates closely to the idea of 'flow' [Csikszentmihalyi]: rather than the kind of concentration required to solve a puzzle.

Thus a *less immersive* experience will concentrate on the physical surroundings of the user, and may be more informative than evocative. A *more immersive* experience would use data that complemented or in some way went beyond the physical environment, e.g. by evoking more than its factual history or memories of it, or by using sound in creative or poetic ways. A different kind of immersion (closer to engagement) may also be created through the social interaction, which an experience necessitates.

Users described the sensation of being immersed in the mediated world of the application and being 'pulled back out' by real world stimuli, such as a siren or other ambient landscape sounds. This led to speculation that a key aspect of design success in the future would be the author's success in controlling the flow between different levels of user immersion—crucially this flow will depend upon a close reading of the users' relation to the 'actual' real world landscape of the application. That is to say: parts of a location-specific application may be related to the material surroundings in different ways at different points on the user's 'journey'.

The Dimensions Of User Control

NONE <>TOTAL

Participants felt that designers should be asked: "How much control do you want the user to have over their navigation of the data environment ?" For instance, in some applications, such as the *Node Boat Tour*, the user had no control—since the navigation of the data was subject to the speed and direction of the boat. At the opposite end of the axis is an experience like *'Riot'* in which the user navigates the data according to where they wander in a large open space. Screen-based applications offering a variety of levels of data may also offer the user a high degree of control, in so far as each of them may be able to create a unique experience even while following a prescribed route. It also seemed likely, therefore, that an application aiming to offer complete user control should also therefore expect to evidence a longer user-control learning curve.

CLEAR RULES<>UNCLEAR RULES

A second user control dimension can be brought to bear upon the design problem of making the user's role in the application clear. How does the application answer the question 'What am I supposed to do?' Are the 'rules of engagement' made transparent for the user at the start, or must they be experientially acquired in the application itself? Do users realize they are in control or even what they in control of? How explicit is this to be made? This is in fact a classic interaction design problem, where experiential learning might be contrasted with more didactic approaches to the user. For instance, in some applications it might actually be important that the user understands that returning to the same trigger-point three times will trigger three different sound inputs; in other applications it might be irrelevant whether or not the user knows this. The position of an application on this axis will also have a bearing on the ways in which the experience seeks to facilitate emergent behaviours. Rule-sets for instance can be established that prescribe certain behaviours in order to encourage others to emerge.

Dimensions of Space / Place

ARBITRARY MAPPING<>MEANINGFUL MAPPING

This dimension describes the nature of an application's relation to the physical/ social / cultural landscapes it pervades. *Arbitrary* mapping assumes no content related to a specific location—*meaningful* mapping describes applications where content is tied to specific locations. When an application is arbitrarily mapped onto a landscape there is no semiotic or meaning-making necessary between landscape and application for the application to make sense. That is not to say that meaning-making is not generated in the relation between the two—far from it. However the application could run equally well in any other location and generate entirely different meanings. *'Moulinex'* or instance ran in Bristol's Queen Square as a rich audio environment mixed from ambient sound and the sound tracks of two films ('The Matrix' and 'Moulin Rouge')—different sound files were triggered by moving round the square. Here the experience was 'arbitrarily mapped'—there was no semiotic connection between the Square and the sounds, although experiencing the application transformed the Square into a atmospheric performative space. Dan Belasco Roger's *'Description'* on the other hand was 'meaningfully mapped' in so far as the stories his narrators told stories all related to particular parts of the Square, it would have made no sense to run the application elsewhere. These degrees of site-specificity again had a major impact upon the quality of the users' experience.

A discussion over internal and external referencing revolved around the question of whether media files and their locations related to each other (e.g. in experiments such as *Savannah*) or to objects or places in the real world, such as a bench in a specific place in relation to a certain sound. For example in experiments in Queen Square, was it about the Square as a whole or about a file in a specific location in the Square, which may be not have even been a physical link, but an artistic reference?

Dimension Of Space
LINEARITY <> NON LINEARITY

At one end of this axis, a user has to follow a sequence along a prescribed route, and at the other is expected to wander about in a space in order to find the data in the application. This may be affected by the scale of the location, which in turn may either facilitate or make problematic the navigation of different densities of data at different points in the space.

In effect many designers will use a combination of both approaches—the 'Riot' application, for instance, made some assumptions about user-routing by offering an 'introductory' reading of the Riot Act as the user entered the space from a certain point, near to where the kit and headphones were handed out. Some information (a map of the historical locations) was given out in the form of a leaflet, which caused some users to walk towards those locations. All users would hear one of the dramatic high-points: the charge of the Dragoons across the Square, after 17 minutes of being in the *Riot* experience. All the other events that unfold in the Square, depended on where the user walked. As the user left the Square to return the kit, implying that their experience was complete, the final zone would trigger a rollcall of the dead and injured.

Dimension Of Time
FIXED RUNNING TIME <> OPEN RUNNING TIME

Should the user to be able to spend as much time as they like in the application or does it, at the opposite end of the axis, have a fixed running time? The *'Node Boat Tour'* for instance is time-limited by the length of the journey, the *'CitiTag'* game is limited by the length of each game's 'round'. 'Riot' on the other hand had no time constraint, users could take as long as they wanted-it would take a couple of hours before a user exhausted all sound files-at which point they would only hear background hubbub.

TIME SPECIFIC <> TIME UNSPECIFIC

Designers may also have to consider at what time of the year or day users are experiencing the work. In some cases. for instance. the expert users in our seminar group reported pleasure in the 'time slippage' aspect of the work, hearing a wintry location described in sunny summer glory for instance, or listening at night-time to day-time experiences.

PERMANENT INSTALLATION <> ONE OFF INSTALLATION

This last temporal aspect of a piece needs to take account of whether it is intended to become a permanent part of the environment or only exist there for a limited time period. This will effect the artist's design decisions. For example, some work may include 'registrational interactivity' where participants can add to the work over time. It is far easier to have higher levels of technical support for a one-off event than for a year-round installation.

Dimension Of Data Depth
ONE LEVEL OF DATA <> SEVERAL LEVELS OF DATA

Some mediascapes run as a single layer, whilst others may have further layers of data available to the user. When users navigate a mediascape other optional levels of information may also be offered. These options may affect the user's sense of immersion, for example by bringing them out of an evocative piece to discover factual information before re-immersion.

The Social Dimension
PRIVATE <> PUBLIC

The *Private* implies a personal experience, perhaps shared with family and/or friends, whereas *Public* implies an experience open to a wider audience, residents or tourists

SOLITARY <> SHARED <> COLLABORATIVE

Solitary implies a different route through a different, but related experience; whereas *Shared* implies that all users get the same media at the same time. *Collaborative* experiences such as *CitiTag*, depend on using the same media whether or not there is a team strategy.

Producer Expertise
PROFESSIONAL MEDIA PRODUCER<>'AMATEUR'

This is defined by the style of delivery and its seeming authenticity. The BBC's *Boat Tour* recordings and Belasco Roger's in-situ recordings, were both recorded by professionals, but had different production values. The quality of the recording does not necessarily detract from enjoyment of the content if it is compelling.

Relation To Existing Experience
AUGMENTATION <> NEW EXPERIENCE/STAND ALONE

Augmentation implies the enhancement of the existing eg *Node Boat Trip* or a Museum Guide. *New Experience* implies the substantive creation of a new environmental experience, which may or may not be specifically linked to the place, ie *Riot* and *Moulinex*

Specialist Descriptions: Interdisciplinarily generated language

It is important to note that these specialist definitions of locative media are derived from the descriptions and critical reflections produced by an interdisciplinary group and can therefore be seen as a first sketch of what kind of discursive terms might be used by interdisciplinary teams to produce and assess mobile media applications. We might observe that the nature of discussions across interdisciplinary boundaries led to what some specialists might understand as a somewhat superficial set of terms. We would argue that this is a useful reflection of the difficulty of generating common terms that allow 'in depth' discussion across disciplinary boundaries.

The following are a set of descriptions of the characteristics of the type of applications that were discussed in the seminars:

> Mobile or Locative media are situated or located in particular spaces and places. They are often location specific, containing location elements. They are usually device mediated. Mobile media may be meaningfully mapped onto space or arbitrarily mapped, they may create meaningful linkages with place or space. Mobile media content is controlled by movement through physical space i.e. not by gestural movement of the body.

The physical real world, the social real world and the data world are all overlaid in Mobile media. Mobile media may create solitary or group experiences. Mobile media may be linear or non-linear, making use of montage, collage, bricolage and fragmentation.

Mobile media may be immersive creating simultaneous parallel worlds that place the user in oscillating realities, that have something in common with going in and out of dream-like states. (like entering a daydream, rather than becoming unconscious of place as your mind is wandering)

Mobile media may take the user between *surface* experiences in which the media maps so closely to the physical environment that the user is focused on their surroundings and *deep* experiences in which the media evoke affective responses where the physical environment decreases in significance. Mobile media may also be ambient providing unstructured 'background' sound for the physical environment.

Mobile media need to have *hooks,* which draw the user into the experience in a pleasurable way. These may be purely informational or they may be 'magic moments'. Mobile media may dramatise space, making it into a performance environment. Mobile media may produce experiences of time slippage between different historical periods and the user's time. Mobile media have the potential to address the user in either the first, second or third person. Mobile media may have the appearance of apparent spontaneity, whilst in fact being carefully crafted.

Mobile media have the potential to prompt emergent behaviours and strategies in unpredictable ways, which may mutually effect and be affected by being either in clean (i.e. empty), or public places where other people outside the experience are also present. The quality of media richness may affect users' responses to the physical environment. Mobile media users will import their memories of similar classes of media experiences in order to help them understand; they will remediate the experience. Mobile media forms all have a user control learning curve. Mobile media may be informational or evocative. Mobile Media create invisible aural mapping for a place.

Some Preliminary Observations On Affect And Pleasure In Mobile Media Experiences

Deep immersion/concentration: loss of sense of physical surrounding seems important and pleasurable to users ('flow' in Csikszentmihalyi). The sensation of moving through different levels of immersion seems to be very pleasurable, of surfacing—of course the sensation of movement through levels is relative, only noticed as it changes.

Pleasurable discomfort and the excitement of immersive spatial navigation: Novelty and media experiences that do not have clearly defined boundaries produce uncertainty in users, which only becomes pleasure through experiential confidence.

'Magic moments': Very strong responses to certain moments in mobile media applications, in which the various elements of physical world and data world appear to 'line up' (e.g. hearing a story about lovers meeting on a bench for lunch and seeing some lovers eating lunch together on a bench in the same location, or the live-cam moment in the *Node Boat Tour*.) The pleasure in these moments correlated, in our sample, with a sense of unexpectedness.

Synaesthetic confusion is pleasurable. For example determining whether the sound is coming from the soundtrack or the actual world (e.g. Sirens or footsteps)

Response enhancement: Having one's response to a place manipulated and enhanced is valuable. This might be by the reinsertion of historical data, evocation or the transformation of a place with arbitrary data inputs; the end result is a pleasurable sense of having had a location 'enhanced' or enriched.

Other Research Questions, arising from the seminar discussion

The research seminars also raised a number of questions which are beyond the scope of this paper, but which could be the subject of further investigation, namely: how authoring teams can control and determine users levels of immersion. This could be further defined through an understanding of how the following elements determine the user's degree of immersion: seamless media mixes, levels of data (e.g. how many sub-menus are available); degrees of meaningful mapping; the quality of the sound/image; the protection of the media environment from actual environment 'bleed'; the balance of information and evocation; and the type and quality of narrative:

How can authors control the balance between pleasurable discomfort (eg excitement, anticipation) and hostile discomfort (eg this kit doesn't work! I don't know what I'm supposed to do)? Because mobile media are novel and also somewhat 'unbounded' compared to other media experiences, the conceptual skills of interface control, developed in HCI design, might need to be translated to mobile media arenas, for example: what is the mobile media equivalent of the 'help' function?

How can authors create environments that allow for 'magic moments' to occur with the appearance of spontaneity? What happens to the interface when the location is no longer the only context or is it not the context at all which triggers the delivery of media? (i.e. not a map-based interface.) The effect for authors of different sensor systems and their relation to interface design? As the development of delivery systems for augmented reality evolves, will the experience necessarily become more compelling? (e.g. degree of sensory immersion through head-up displays). How will a specialist descriptive language develop?

Bibliography

Csikszentmihalyi, Mihaly. *Optimal Experience: Psychological Studies of Flow in Consciousness*. Cambridge: Cambridge UP, 1988.

Snapshots from Curating Mobility
(If you build it, they won't necessarily come).
Beryl Graham

> *"Is curating mobile media like catching in still photography the moment the horse's feet leave the air? Or should the curator be on the horse?"* [Drew Hemment, 2004]

> *"... not many are on horse, but does that make the horse race less interesting?"* [Lize Black, 2004]

In an on-line discussion of curating locative and mobile media art, Drew Hemment captures several issues by referring to Eadweard Muybridge's early photographs of a galloping racehorse. How can curators deal with the time-lag in art institutions that are still debating if photography is art? How can curators 'make visible' the immaterial? The metaphor also pithily illustrates how the 'curatorial position' often sits uncomfortably in relation to locative and mobile media—the position may be face down in the dust, or trudging along behind with the bucket of conservation and preservation.

It could be argued that it is a responsibility of curating to catch up with the horse, and coax it into the bridle of a critical third space shared by an audience, without breaking the spirit of the artwork. In this struggle, it is obviously the artworks that form the beacons and lead the way. It is also often the experienced artists who manage to cut through the hyperbole of god-like technology and media-sexy jet-trash mobility. For example, Lucas Bambozzi's *Spio* (2004-) is an art installation celebrating the autonomous mobility and surveillance capability of a vacuum cleaner. Likewise, the Richair team have satirised the 'sexiness' of mobility by utilising both hi-tech and low-tech in their performance-oriented events, including wireless antennae, garlic, computers in plastic lunchboxes, roller-skates, and in the case of *Let's Do Lunch* (2005) the Route 26 bus in Hackney. Artists can also teach curators much about a critical awareness of the cultural meaning of the technology itself. *Inch* (2004-) is a work-in-progress by Spencer Roberts and Anneke Pettican, which uses GPS tracing to 'draw' a common graffiti tag on the landscape, a "territorial marker" that reflects the power relationship of both military and tourist uses of the technology.

"GPS technology has been lauded for introducing a new affinity with landscape and revitalising an appreciation of place. The dystopic side of this coin is that it is by virtue of this technology our landscapes are being mediated—along with our experiences of them. ... The politics of GPS prioritises stasis and fixity. The politics of experience prioritises movement and flux" [Spencer Roberts, 2005].

At this stage of the relatively new media, it is therefore less a time for curatorial cannons than for a series of snapshots (even if blurry ones of a horse's backside disappearing over the event horizon). This chapter takes opinions from media such as discussion lists, but seeks to relate them to the wider context of contemporary art, and to any useful histories. It focuses on some key issues, which are areas of both challenge and potential for curators: the issues of technology, audience, space, time, and history.

The Technology

"The tension between the technology's demands and those of its user as mobile agent tends to produce an equal awkwardness in us as we use these services." [Simon Pope, 2005]

"... poorly paid advertising executives for the R&D laboratories of technology companies ..." [Saul Albert, 2004]

"I wonder if with time the internet would become inconspicuous, or would it give rise to odd political reactions." [Yasir Husain, 2004]

These quotes reflect three key issues for the technology itself in relation to art and audience: Firstly the simple fact that the technology is emerging, and is prone to teething problems, unexplained tantrums and connectivity problems—as Simon Pope outlines, the user often falls into the gaps between networks, between desires for seamless mobility, and a patchy reality. Secondly, this 'beta-testing phase' means that artists often unintentionally catch the sexed-up rhetoric of technology companies, and thus present the curator with a challenge to avoid a 'trade show' context of display for the artwork.[1] Thirdly, there is the strong cultural meaning of the technology itself—Yasir Husain from Pakistan cogently points out that appearing in public places with a video camera provokes a strong cultural/political

---

[1] Peter Lunenfeld (1999) has critically examined the 'trade-show model' of exhibiting new media art, and Lev Manovich's 1996 bifurcation of art/technology into Duchamp-land and Turing-land has been analysed in the context of curation by Cook and Graham (2002) and Graham (forthcoming).

response, so how much stronger still is the visibility of newer technology, and the conspicuous spectacle of consumption or surveillance[2] that it creates, making it difficult to concentrate on the content of the artwork.

Thus, presenting art using emerging technologies is often framed in terms of the struggle between 'potential' and actual completed artworks:

> "Besides, locative game technologies might be the only way to make an immensely rich human heritage of otherwise invisible myths and history vitally interesting to ordinary people—in realtime/place. name one;" [Saul Albert, 21 April 2004]

> "... locative media as a kind of paradoxical extension of military control, as well as a potential means towards a liberatory "spatial practice": [<http://www.gpster.net/potentialmobiles.html> Mark Tuters, 2004]

However, curators also need to beware of any simplistic division between technology-led and content-led: As Naomi Spellman pointed out in 2004

> "... criticality for now at least all too easily becomes surface ornament, an easy way to equate a work with intelligence, hipness, or awareness."

Such an approach may lead to showing artworks which may be more familiar to a fine art audience, but which may not necessarily be using the particular characteristics of the technology in a sophisticated way.

Being able to identify these particular characteristics can help both the curator and the audience make sense of the artwork. Concentrating on characteristics or behaviours rather than media puts the emphasis on the user, and notably, it was Steve Dietz, a curator who has worked across new media, who suggested the useful characteristics of Interactivity, Connectivity, and Computability.[3] By identifying that

---

[2] The obvious histories of the media in military and surveillance were widely discussed on the New-Media-Curating discussion list (see also Graham, 2004). Mobile and locative media are not alone in this, as artists and curators have been struggling for some time with the dubious parentage of technologies—including the Internet (military and pornography), VR (military), video (surveillance and porn) and photography (porn, colonisation, military).

[3] Steve Dietz, 1999. These and other categories of new media behaviour are discussed in relation to curating new media art in: Cook and Graham, 2004; Graham, forthcoming 2005.

which is particular to new media, he helps to separate what is genuinely new from that which is merely novelty.[4]

If technology-led interests seem determined to present mobile media as the wild, free Arabian stallion in contrast to the lumpen Shetland pony of 'old media', then curators are perfectly able to defuse this by the context in which the work is presented: For example, the exhibition Art for *Networks* (2002) curated by Simon Pope showed no-tech work by Stephen Willats alongside contemporary internet work. Anna Best's artwork in this show, *Ice Cream Van Convention* (2002), also used the idea of national networks and location without using high-technology. In her case, she tracked relationships between ice-cream vans, territories and networks, a highly locative and mobile project, featuring free icecream at certain exhibition openings. Artists, who work new technologies into their practice from a strong history, and for good locative reasons, are also those who tend to effectively ground the hyperbole of the technology. Barbara Keating tracked fair trade and transport issues by tracking suppliers, and following shoppers and their goods home from an indoor market, with a tracked mobile video phone [Fountain, 2004]. Likewise, the *Migrantsoul* project [Shahidul Alam, 2004-] is starting to introduce locative media to a solid photo-documentary project, tracing migration routes from Dhaka to Europe.

Once the novelty of an emerging medium has worn off, it tends to be the value of the artwork that remains, and the context in which it was presented—a key responsibility of the curator. In terms of technology, art history and new media, the responsibility is quite an onerous one, as Saul Albert reminds us:

> "If the process is too messy or the critical discourse too weak, it will go the way of 'holographic art'." [Saul Albert, 2004]

The Audience

> *"I often find that public art projects can be documented in a certain way, which makes them look as if they were extremely effective, but the reality might have been completely different, or: different people's expectations or experiences of the project are so diverse: the artist might have thought it was terrible, but a participant might have thought it was fantastic."* [Anna Harding, 2003]

---

[4] Barbara London, Video and Media curator at The Museum of Modern Art, New York discusses the pressure to do the "Novelty Hustle" when curating new media art, in an interview with Sarah Cook, 2001.

"... until enough people get these 'locative' devices ... there will never be an accumulation of enough 'geo-annotations' to make accessing the 'geospatial web' a compelling experience." [Marc Tuters, 13 April 2004.]

"The term "locative art" itself already points to a state of closure—locative is location-based and site-specific and thus implies access limitations (there are many other constraints that could come into play here)." [Christiane Paul, 2005.]

As a curator of art including new media, Anna Harding is in an excellent position to comment on the strange gap between documentation and reality, and the occasional equally strange gap between artwork and audience. As the title of this chapter states, just because an artwork is designed for participation, this does not necessarily mean that it will be used, or that it will be used in the intended location, context or duration. As Marc Tuters states, there is a question of critical mass for participative projects, and in the case of locative media, this depends on factors of compatible equipment, and technical knowledge of how to use it, in additional to the usual factors of inhibition, art-world conventions, and knowing of the artwork's existence in the first place. 'Access' is a buzzword for both the 'potential' of mobile media, and for curatorial responsibility: if the artwork is not accessible to the audience, then what is the curator doing?

Some artists have deliberately played upon this, notably Heath Bunting's *BorderXing Guide* (2003), which concerns information on points to cross national boundaries, but is only available on the internet from certain 'authorised' computers, thereby confounding the 'international accessibility' rhetoric of the internet (and the simplistic access buzzword of arts funders). The audience may be narrow for certain artworks, but if it is a deeply engaged audience, then this can be justified in curatorial terms. The project *murmur* in Toronto, for example (Micallef, Roussel, and Sawhney, 2003-), is a located project where simple signs on street furniture inform the audience that they can call a number to hear memories of that particular place. The context and the audience are very specific, but the project appears to be a successful, continuing and growing one (at least from the documentation). These very local, very specific artworks often do gain a committed audience, and the very local approach of "everyday practices"[5] can also help to subvert the 'spectacular' nature of powerful technology.

Beneath the hyperbole, it is however, a notable characteristic of mobile technologies that they can indeed help to cross perceptual or cultural boundaries:

---

[5] Discussed by Simon Pope, 2005.

Examples can be found in the 'walks' of Janet Cardiff, which Lev Manovich described in 2003 as:

> "... the best realization of augmented space paradigm so far—even though Cardiff does not use any sophisticated computer, networking and projection technologies."

Her work *The Telephone Call* (2001), for example, was a walk with digital video camera and headphones, which guided the user around the public and not-so-public areas of San Francisco Museum of Modern Art. Wandering in the stairwells and office space of the institution was a dis-locating and revealing experience not only for the user, but for staff, and for other audience members, who were then drawn by curiosity into the experience itself, and the rest of the group exhibition. It physically crossed the boundaries of the institution, between library, office and front-of house, and, en route, created a questioning of these boundaries through an adventurous curatorial stance.

This creation of a located 'third space' between the audience and the artwork is one of the exciting potentials promised by mobile media—the sheer engagement offered by games and immersive tactics is an attractive factor for many artists and curators. The potentials, however, need substantial experience and skill to be fulfilled. Engagement is a complex area, attained mostly by the content of the artwork, but also by the characteristic of 'interaction'—one of Dietz's three categories, but also variously touched upon by Rafael Lozano Hemmer's "relational architecture" (1997) Nicolas Bourriaud's "relational aesthetics" (2002), and a longer history of participative works including live art, community art, art activism, and Situationism. Thus, in addition to the necessity of functioning technological systems, the systems of participation must also work. The skills of designing participative artworks often come from long experience, or from fields not necessarily within new media. For example, *Learning to Love You More* (2004-) by Harrell Fletcher and Miranda July, is a highly participative web site where anyone can send in responses to 'tasks' such as to take a photograph of two strangers holding hands, or record cover versions of pop songs. The project is remarkable for the quantity and quality of responses, which could be attributed to the artists' long experience in community or participatory arts projects including low-tech approaches. The artists also combined the project with physically located workshops and exhibitions, for example making portraits of a local taxi driver, and vouch for the fact that most responses come from individual contacts, located work, and word of mouth, rather than relying purely on the internet. Other artists working with mobile media, such as Patterson, Zurkow and Bleecker's PDA-based *PDPal* (2003-), can testify that curatorial approaches which allow for plenty of workshops, attendants and guided tours, help to gain real participation in more 'immaterial' contexts. Likewise Hector Canonge's Ciudad *Transmobil* project (2004-) managed to gain significant participation from immigrants New York City, using mobile phone cameras, via substantial outreach work so that even those without homes participated.

In many ways successful audience participation makes up for the essentially 'ersatz' nature of technology—the fact that it is often a poor imitation of real interaction, of real live experience, or real location. Those artists and curators who manage to get the technology to work as an effective means of enabling real interaction, are the clever ones, and often draw on knowledge from HCI, game-play, educational uses of technology [Graham 2002, and Cook and Graham 2001].[6] enabling practices, or live art, such as Rafael Lozano Hemmer. Such activities mean a handing over of some control to the audience, and this demands a good deal of trust given by the artists and the curator: As Rafael Lozano Hemmer said in 2001: "Successful pieces that feature "interactivity for groups" are usually out-of-control."

The Space

> *" ... you can't just as a curator think that you can work in local communities without any specialist knowledge. That is really problematic."* [Anna Harding, 2003]

> *"Site-specificity itself is also in need of a critical revision. Its re-use today is often poorly disguised recuperation and normalisation of highly politicised events and topics as melancholic, aestheticised cultural history."* [Saul Albert, 2004]

All the technology in the world will not make up for lack of experience of a "Knowable Community" [Pope 2005] or lack of curatorial knowledge of the history and debates of public art, site specificity, or activism [Kester, 2004; Lacy 1995; Raven, 1989].[7] Locative and mobile media are often caught between this, and the desire to use the potential characteristics of the media—engagement with context,

---

[6] There is a substantial amount of research on educational and uses of locative and mobile media, including PDAs used as interpretative devices at San Francisco Museum of Modern Art, and the Exploratorium. The Antenna-designed project at the Tate is briefly discussed on the New-Media-Curating discussion list. However, there is little crossover of knowledge from educational to artistic uses of the technology, as is described in Graham 2002, and Cook and Graham 2001.

[7] There are very many books which cover useful debate in public art in general—those that contain references to new media include: Kester, 2004; Lacy 1995; Raven, 1989.

the narrative of movement through space,[8] the non-art audiences, personal meanings of place, the no-strings wireless dream. Like other people's dreams however, the individual meanings of location simply might not be interesting to other people, and so a difficult balance must be struck between too much local knowledge, and not enough. Perhaps an early example of a located yet engaging approach are some early projects by Nina Pope and Karen Guthrie—*Homespun* in 1997, linked the parental homes of the two artists, and *A Hypertext Journal* (1996) which retraced a journey of Johnson and Boswell around the Highlands. The difference between potential use, and meaning of the works for locals or visitors is highlighted by various artworks: Darko Fritz has made a series of works called *Migrant Navigator* (2002-), concerning the meaning of 'home' immigrants, involving billboards, the internet, and flower beds. The group e-Xplo have made series of night-time coach trips around cities, using GPS to trigger audio soundtracks according to cultural meaning; both are good at using technology which adds to rather than competes with the urban environment, but how can the readings of visitors versus locals be resolved, however advanced the technology?

Ironically, the art world has worked very hard to make the traditional white cube space as much like 'no-place' as possible, and to make international arts festivals into an interchangeable no-mans-land. Another irony is that many places still lack venues in which to show certain media work. A positive example of curators using mobility imaginatively is *The Big M* by Isis Arts, a triangular inflatable black tent for showing video shorts, which is still touring to audiences that arts venues do not reach, in a modern equivalent of various past galleries on buses, in caravans, or on the backs of lorries. The major new venue offered by new media is of course the internet, that famous no-space which has formed an appropriate venue for curated exhibitions of locative media, such as Jillian Mcdonald's *Location is Everything* on the Rhizome site in 2005. Curators must understand the 'location' and context of both material and immaterial means of distribution, which often presents a basic challenge to those with an object-oriented history.[9]

Materiality is a key issue exhibiting something that, like the legs of racehorses "was not viewable" [Ivan Pope, 2004]. For the audience, not only might the works of mechanical reproduction lack the aura of the hand-crafted object, they also might lack objecthood altogether, yet, importantly, some see the locative as a means of reclaiming this:

---

[8] Manovich, 1996, refers to the way in which, in video games, narrative and time itself are equated with movement through space (i.e., going to new rooms or levels.)

[9] Sarah Cook (forthcoming) examines this aspect of immateriality in relation to curating new media art.

"... it seems that Benjamin's concept of the aura, which was obliterated by cinema and TV as well, is reconnected within locative media ..." [Naomi Spellman, 2004]

Curators, might, however, find useful precedents in the exhibition of another immaterial art form—sound:

"... visual art has so many preconceptions associated with it that sound doesn't. for instance, at the launch of Graeme Miller's project yesterday, the deputy mayor made a speech saying, when they were approached about this art project, they imagined a mural or a banner. That is their preconception of a public art work." [Anna Harding, 2003]

It is perhaps no coincidence that some of the most effectively exhibited locative and mobile media are sound-led: Harwood and Fuller's *TextFM* (2001) which broadcast text messages from mobile phones, or Akitsugu Maebayashi's *Sonic Interface* (1999) which subtly morphed ambient sounds of the city. Responses by curators on the CRUMB discussion list[10] suggest that sound curators meet challenges with art institutions, similar to those met by new media art curators, but also reap the same rewards of new audiences, and changes in perception of a public environment.

The Time

"*An atomised space and time is more susceptible to commodification and contestation. This typically manifests itself in relation to property (space) and work (time).*" [Spencer Roberts, 2004]

"*... so many artists working with PDAs and mobiles were in the development stages, so I allowed anything from people coming on 4-5 months after the beginning/opening of the show and the 'closing' of what I called the 'active cycle' of the show.*" [Patrick Lichty, 2004]

"*There must be a dialogue between the past and the present which acknowledges that the present is the more important space, but that the present requires the past in order to locate itself.*" [Clive Gillman, 2001]

---

[10] Curating Sound was the theme in March 2002.

As Spencer Roberts states, space and time are linked, and time relates to work, including the work of curators. Patrick Lichty has discussed this in relation to his show *(Re)distributions* (2001), and several other curators have remarked upon the different timescale of curating new media art, which can lead to an 'evolution' of the artworks during the exhibition, and can lead to anxiety for curators who are unused to the this tendency. Artists Nina Pope and Karen Guthrie stressed in 2001:

> "… one key point for those trying to commission and curate in the field of new or emergent media: … make sure you know where the work 'begins' and 'ends' for them …"

> The curator Tamas Banovich also stressed in 2001, that art institutions suffer from a huge "lag." Many are still struggling with the time-based aspects of video, never mind live, connected or located media, and as Patrick Lichty pointed out in an online panel in 2005:

> "There are certainly arts that operate on different temporalities, and Lyotard's assertion of art's timeframe beginning after fifty years is a cultural straitjacket under which New Media could not exist, nor could much Conceptualism or Performance Art (which I posit that New Media, as a fairly ephemeral, experiential time-based medium/genre/movement is closely related to)."

This lag means that curators are often ill-equipped and reluctant to act as 'beta-testers' for the bleeding edge of technology (or even technologies nearing 50 years old). Conversely, the speed of change can mean that important debates on media-specific characteristics such as interaction or embodiment that happened in the early 1990s, tend to be discarded before a critical mass is established, only to emerge later, in this case in connection to mobile or locative media.

Again, it is the inventiveness of the artists that pulls together the factors of time, space, and a sense of technological history. Germaine Koh's series of artworks often deal with location, mobility, and older technologies, including *Relay* (2005-), where the audience can text messages that are then translated into Morse code, and flashed by on/off lights from significant art buildings. Anne-Marie Schleiner's *Operation Urban Terrain (OUT)* (2004) was highly mobile project in New York which successfully satirised a variety of boys toys: Two women in strange techno-fetish clothes, one with a laptop and the other with a projector, pointed moving images onto building walls in key public locations. They were connected through a 'mobile wireless bicycle' to an online team of five game players located around the world who intervened on servers in a popular online military simulation game. Importantly, this was not only site-specific, but also time-specific, as it happened during the Republican National Convention in New York City.

Which History, What Future?

" *... the found object is no different from the found painting; the silence of Marcel Duchamp doesn't mean anything. If mobility is an issue relating to the use of a mobile phone for a work of art, it stops being different from, say, kinetic or other art, that just somehow doesn't sit on a plinth or in a frame, or is sited specifically in some space or other. One work sits in a phone, another in a frame—but that doesn't make it necessarily different.*" [Jorn Ebner, 2004]

"*A picture frame designates the status of that which it encloses as an art object, at the same time as it separates the art object from the world. A further act of distancing occurs by virtue of the critical distance of the perceiving subject and the physical distance from the perceiving body within the gallery context.*" [Drew Hemment, 25 Apr 2004]

" *The crucial dilemma of digital art, and software art in particular, may very well be that the understanding of its "backend" will always remain a fringe culture (closed system) that won't be integrated into the mainstream of (perception-oriented) art criticism.* " [Christiane Paul, 2005]

On the CRUMB discussion list theme concerning curating locative and mobile media, the posts include references to Marcel Duchamp, skateboarding, Eadweard Muybridge, Situationists, the Whitehouse web site, Lucy Lippard, Paul Virilio, geo-caching, Walter Benjamin, and how to make wireless antennas from noodle scoops. This reflects the complexity of the histories with which curators, critics and audience are faced.

Although it often seems like the theorists are ahead of the practice, way into the territories of "vapour theory," criticism is nevertheless important for new media art, if it is to be documented and historicised, and available to future researchers [Depocas, Ippolito, Jones, 2003; and Graham 2005].[11] Histories obviously inform curating, not only in the content of exhibitions, but in the way in which curating is done. Patrick Lichty has discussed this in relation to factors of time, but there are

---

[11] Issues for archiving, conserving and collecting new media artworks, including references to the Variable Media project (Depocas, Ippolito, Jones, 2003) are covered in Graham, 2005. The importance of press and criticism to new media art has also been discussed on the New-Media-Curating discussion list.

also different metaphors of curating new media art: is it more like producing films,[12] like public art, like research and development, like being an artist/curator, like the white cube of visual art or the black box of performance [Manovitch 2003], like riding a horse or standing back at a critical distance, like filter-feeding [Schleiner, 2003] or like the self-curation model of Rhizome?[13] With this range of possible models and skills, curators are taking a leaf from the model of new media artists, and sometimes working in small teams, or, as Joasia Krysa posits, "nodes in a network" [2005].[14]

Patrick Lichty usefully outlined in 2005 (July 7) how "transinstitutional" and other models kinds of curating have been influenced by the needs of the media themselves, but in 2001, Iliyana Nedkova very accurately described how the rhetoric of 'mobility' and the international, boundary-crossing nature of new media is much more ambivalent when examined in the light of reality. Her projects *Blind-Dating Technology* (2000) and *Crossing Over* (2000) were celebrations of the new media, but she is also unflinchingly honest about the difficulties of trans-national projects on the fringes of the military industrial complex, and the feeling of "curatorial inadequacy" that this can produce. The mobile, fast responsiveness of the freelance curator role, is perhaps not so glamorous if considered as "outsourced" labour.

Despite the challenges, the final snapshot considering the possible futures of locative and mobile media is a relatively sunny one: The openness of the curators cited here reflects a growing number of specialist curators with a firm critical grasp of the media, their own methods, and a willingness to be led by the artists. Curators are generally unwilling to be led solely by the technologies, but are making firm links with the wider art world. Inke Arns' exhibition *Dispersed Moments of Concentration* (2005) for example, concerns issues of locative media, but includes a wide range of media work, not necessarily using locative technology literally. If curators can integrate criticality and artwork, and ensure that the audience can access this third space, then Saul Albert's spectre of the dead-end model of holography, can hopefully be banished.

---

[12] The 'producer' model of new media curating is mentioned by Benjamin Weil, in an interview with Beryl Graham, 2002.

[13] On the rhizome.org web site, users can 'curate' their own selection of artworks from the ArtBase online collection of artworks.

[14] See Krysa, 2005. In particular, Franziska Nori mentioned the importance of working in curatorial teams.

Bibliography

Alam, Shahidul (2004-) Migrantsoul: an exploration of transitions across borders. <http://www.migrantsoul.net/>.

Albert, Saul. "Locative Literacy." *Metamute*, M28: Summer/Autumn 2004 (6.07.04). <http://www.metamute.com/look/article.tpl?IdLanguage=1&IdPublication=1&NrIssue=28&NrSection=10&NrArticle=1390&search=search&SearchKeywords=locative&SearchLevel=0>.

– – –. "Re: [Locative] Locative Gaming: Dawn of the Zombie Cyborgs." New-Media-Curating Discussion List. 21 Apr 2004. <http://www.jiscmail.ac.uk/lists/new-media-curating.html>.

Arns, Inke. "Dispersed Moments of Concentration." 2005. <http://www.hmkv.de/dyn/d/>.

Bambozzi, Lucas (2004-) *Spio.* <http://www.http.uk.net/docs/exhib4/press_release_spio.htm>.

Banovich, Tamas. "Tamas Banovich." Presentation at *Curating New Media*. Seminar, Gateshead, May 10-12, 2001. <http://www.newmedia.sunderland.ac.uk/balticseminar/banovic.htm>.

Best, Anna (2002-) *Ice cream van convention*. <http://www.annabest.info/cv_ice-cream-van-convention>.

Black, lize. "Re: April Thought For The Day: Exhibiting Locative Media." New-Media-Curating Discussion List. 13 April 2004. <http://www.jiscmail.ac.uk/lists/new-media-curating.html>.

Bourriaud, Nicolas. *Relational Aesthetics*. Paris: Les Presses du Reel, 2002.

Bunting, Heath. *BorderXing Guide*. 2003. <http://www.tate.org.uk/netart/borderxings/>.

Canonge, Hector (2004-) "Ciudad Transmobil: Creating Participatory Spheres through Communication Technologies." <http://www.ciudadtransmobil.net/>

Cardiff, Janet. *The Telephone Call*. 2001. <http://010101.sfmoma.org/>.

Cook, Sarah. "Multi-Multi-Media: An interview with Barbara London." 2001. <http://www.newmedia.sunderland.ac.uk/crumb/phase3/ilondon.html>.

– – –. "Immateriality and its discontents." *Curating New Media*. Ed. Christiane Paul. Berkeley: University of California Press, 2005.

– – –, and Beryl Graham. "A curatorial resource for upstart media bliss." *Museums and the Web 2001: Selected papers from an international conference*. Ed. David Bearman and Jennifer Trant. Pittsburgh: Archives & Museum Informatics, 2001. Also available from URL: <http://www.archimuse.com/mw2001/papers/graham/graham.html>. 197-208.

– – –. "Curating new media: Net and not net." Art Monthly (Nov 2002). London: Jack and Nell Wendler. 44-45.

– – –. "Curating new media art: Models and challenges." *New Media Art: Practice and Context in the UK 1994-2004.* Ed. Lucy Kimbell. London: Arts Council of England, 2004. 84-91.

CRUMB (2001-). New-Media-Curating Discussion List. <http://www.jiscmail.ac.uk/lists/new-media-curating.html>.

Depocas, Alain, Jon Ippolito, and Caitlin Jones, eds. *Permanence Through Change: The Variable Media Approach.* New York: Guggenheim Museum, 2003. Also available from URL: <http://www.variablemedia.net/e/preserving/html/var_pub_index.html>.

Dietz, Steve. "Why have there been no great net artists?" *Through the Looking Glass: Critical Texts.* 1999. <http://www.voyd.com/ttlg/textual/dietz.htm>.

Ebner, Jorn. "locative." New-Media-Curating Discussion List. 7 Apr 2004. <http://www.jiscmail.ac.uk/lists/new-media-curating.html>.

Fletcher, Harrell, and Miranda July (2004-) *Learning to Love You More.* <http://www.learningtoloveyoumore.com/>.

Fritz, Darko (2002-) *Migrant Navigator.* <http://darkofritz.net/>.

Gillman, Clive "Re: Too Interactive: Nov theme of the month." New-Media-Curating Discussion List. 19 Nov 2001. <http://www.jiscmail.ac.uk/lists/new-media-curating.html>.

Graham, Beryl. "An interview with Benjamin Weil." CRUMB. 2002. <http://www.newmedia.sunderland.ac.uk/crumb/phase3/iweil.htm>.

– – –. "Curating new media art: SFMOMA and 010101." 2002. <http://www.newmedia.sunderland.ac.uk/crumb/phase3/append/sfmoma.htm>.

– – –. "Webexclusive: Exhibiting Locative Media: CRUMB discussion postings." Metamute, 20 April 2004. <http://www.metamute.com/look/article.tpl?IdLanguage=1&IdPublication=1&NrIssue=24&Nr Section=5&NrArticle=1373&ST_max=0.>

– – –. "Taxonomies of new media art: Real world namings." *Museums and the Web 2005: Proceedings.* Ed. David Bearman and Jennifer Trant. Toronto: Archives & Museum Informatics, 2005. <http://www.archimuse.com/mw2005/papers/graham/graham.html>.

– – –. "The ironic heirs of serendipity: British new media art 1980s to now." *White Heat and Cold Logic: British computer arts 1960-1980. A historical and critical analysis.* Ed. Paul Brown et al. Cambridge, Mass.: MIT Press, 2008.

– – –. "Redefining digital art: Disrupting borders." *Theorizing Digital Cultural Heritage.* Ed. Sarah Kenderdine. Cambridge: MIT Press, 2007.

Harding, Anna. "Interview on Interference with Raimund Minichbauer." *Republicart,* July 2003. <http://republicart.net/art/concept/interview-harding_en.htm>.

Harwood, Graham, and Matt Fuller. *TextFM.* 2001. <http://scotoma.org/TextFm/>.

Hemment, Drew. "Fwd: April Thought For The Day: Exhibiting Locative Media." New-Media-Curating Discussion List. 13 April 2004. <http://www.jiscmail.ac.uk/lists/new-media-curating.html>.

– – –. "Re: locative." New-Media-Curating Discussion List. 25 Apr 2004. <http://www.jiscmail.ac.uk/lists/new-media-curating.html>.

Husain, Yasir. "Re: mobile media, education." New-Media-Curating Discussion List. 21 Apr 2004. <http://www.jiscmail.ac.uk/lists/new-media-curating.html>.

Isis Arts, (2000-) *The Big M.* <http://www.isisarts.org.uk/>.

Keating, Barbara. *Fountain.* 2004. <http://www.commissionsnorth.org/showcase/portfolio/223>.

Kester, Grant. *Conversation Pieces.* Berkeley: University of California Press, 2004.

Koh, Germaine. *Relay.* 2005. <http://www.balticmill.com/html/viekoh.html>.

Krysa, Joasia, in partnership with Tate Modern. *Curating, immateriality, systems: A conference on curating digital media*, 4th Jun 2005. London: Tate Modern.

Lacy, Suzanne. *Mapping the Terrain: New Genre Public Art.* Seattle: WA Bay Press, 1995.

Lichty, Patrick. "Curating Locative Media." New-Media-Curating Discussion List. 13 Apr 2004. <http://www.jiscmail.ac.uk/lists/new-media-curating.html>.

Lichty, Patrick. *[Re]distributions.* 2001. <http://www.voyd.com/ia/>.

Lichty, Patrick. "Re: Gere:Linked References and Open Systems." 4 July 2005. Tate Online Events: COdE Of practice: Online Panel Discussion June 13—July 18, 2005. <http://www.tate.org.uk/contact/forums/onlineevents/>.

– – –. "Participation/Curating/knowledge production ." 7 July 2005. Tate Online Events: COdE Of practice: Online Panel Discussion June 13—July 18, 2005. <http://www.tate.org.uk/contact/forums/onlineevents/>.

Lozano-Hemmer, Rafael. "Too interactive (belated)." New-Media-Curating Discussion List. 6 Dec 2001. <http://www.jiscmail.ac.uk/lists/new-media-curating.html>.

Lunenfeld, Peter, ed. *Snap to Grid: a user's guide to digital arts.* Cambridge: MIT Press, 1999.

Manovich, Lev. "The aesthetics of virtual worlds: Report from Los Angeles." Ctheory, Global Algorithm 1.3. 1996. <http://www.ctheory.com/e42.html>.

– – –. "The Death of Computer Art." 1996. <http://www.thenetnet.com/schmeb/schmeb12.html>.

– – –. "The poetics of augmented space." *New media: Practices of digitextuality*. Ed. John Caldwell and Anna Everett. New York: Routledge, 2003. 75-92.

Mcdonald, Jillian [curator] (2002-) *Location is Everything*. <http://www.rhizome.org/>.

Micallef, Shawn, James Roussel, and Gabe Sawhney (2003-) *murmur*. <http://www.murmurtoronto.ca>.

Nedkova, Iliyana. Presentation at Curating New Media Seminar, May 10-12, 2001. <http://www.newmedia.sunderland.ac.uk/balticseminar/nedkova.htm>.

– – –, and Maria Troy, eds. *Crossing Over*. Columbus, Ohio: Wexner Centre for the Arts, 2000.

Patterson, Scott, Marina Zurkow, and Julian Bleecker (2003-) *PDPal*. <http://www.pdpal.com/>.

Paul, Christiane. "Public Forum and Openness." 23 June 2005. Tate Online Events: COdE Of practice: Online Panel Discussion June 13—July 18, 2005. <http://www.tate.org.uk/contact/forums/onlineevents/>.

Pope, Ivan. "Re: April Thought For The Day: Exhibiting Locative Media." New-Media-Curating Discussion List. 14 Apr 2004. <http://www.jiscmail.ac.uk/lists/new-media-curating.html>.

Pope, Nina and Karen Guthrie. *A Hypertext Journal*. 1996. <http://www.somewhere.org.uk/hypertext/journal/home.html>.

– – –. (1997) *Homespun*. <http://www.somewhere.org.uk/homespun/index.htm>.

– – –. (2001) "On being curated—a critical response." Presentation at Curating New Media Seminar, May 10-12, 2001. <http://www.newmedia.sunderland.ac.uk/balticseminar/popgut.htm>.

Pope, Simon [curator] (2002-) *Art for Networks*. Cardiff: Chapter. <http://irational.org/artfornetworks/>.

– – –. (2005) "The Shape of Locative Media." Metamute, M29, 2 Feb 2005. <http://www.metamute.com/look/article.tpl?IdLanguage=1&IdPublication=1&NrIssue=29&NrSection=10&NrArticle=1477>.

Raven, Arlene. *Art in the Public Interest*. Ann Arbor: UMI Research Press, 1989.

Richair. *Let's Do Lunch*. 2005. <http://richair.waag.org/>.

Roberts, Spencer, and Anneke Pettican (2004-) *Inch*. <http://www.anthropo.org.uk/gpsweb/>.

Schleiner, Anne-Marie. "Curation fluidities and oppositions among curators, filter feeders, and future artists." Intelligent Agent, 3.1 (2003). <http://www.intelligentagent.com/archive/Vol3_No1_curation_schleiner.html>.

– – –. *Operation Urban Terrain (OUT)*. 2004. <http://www.opensorcery.net/OUT/>.

Spellman, Naomi "Re: locative." NEW-MEDIA-CURATING Discussion List. 26 Apr 2004. <http://www.jiscmail.ac.uk/lists/new-media-curating.html>.

Tuters, Marc. "locative forbearers." New-Media-Curating Discussion List. 13 Apr 2004. <http://www.jiscmail.ac.uk/lists/new-media-curating.html>.

– – –. "Re: locative? tracked!" New-Media-Curating Discussion List. 14 Apr 2004. <http://www.jiscmail.ac.uk/lists/new-media-curating.html>

2.2 Understanding Public Spatialisation

Beyond Mapping: New Strategies for Meaning in Locative Artworks

Martin Rieser

> *"... In the western Deserts, tattered Fragments of the Map are still to be found, Sheltering an occasional Beast or beggar; in the whole Nation, no other relic is left of the Discipline of Geography."* [Borges]

Introduction

Locative artworks, based on digital mobile technologies are a relatively new phenomenon. Yet art practice based on site-specific works and nomadic strategies is not just old, but ancient. Locative art, by its very nature, trespasses into the realm of Public Art, but by its interaction with the public transforms our notions of site-specific and ambulant practices, defined over the last three decades by artists such as Richard Long, Hamish Fulton, Vito Acconci and Sophie Calle. I pose here the question whether, by similarly rooting locative practice in profound cultural and psychological structures, locative work can gain greater artistic resonance. The exploration of the syntax of spatial language and its relevance to current practice is the subject of this essay.

A Gendered form

The political and economic shape of society ultimately forms contemporary modes of narrative. The contradictory pressures of neo-liberal economics, which drive the growth of personalised and peer-to-peer media and the interpenetration of workspace and private space, also seem to offer a unique opportunity to break Mulvey's determinist "male" control of narrative vision (which dominated in the 19th and 20th centuries) and to promote a more de-centered and subtle mapping [Mulvey, 12-15]. In this sense the hypothesis of the constructed self, raised by Lacan in his article "The Mirror Stage as Formative of the I" [Lacan] can be revisited in a newly emergent context. Certainly, the traditional passive (Hollywood) female role under the 'framing' of 'the male gaze' can be undercut by the freedom of viewing implied by an ambulant audience's experience. In this context, the intrusion of an ambient "real" environment can become a further interruption of the voyeurism of the viewer [Felluga]. In many ways the active participant appears as in a liminal state between worlds [Galloway], whose attention is constantly moving between absorption in diegisis and attention to an intrusive "real" world with all its immediate physicality.

Feminist critics have often raised alternative strategies to break the negatives of a culture of male control [Flanagan]. Not surprisingly, some interesting female locative practice explores precisely this area. Teri Rueb's *Drift*,[1] for example, ties a sound landscape to the movements of the tide on a north European beach. The installation covered a 2 km x 2 km region on the Wadden sea that is filled with areas of interactive sound. The piece creates a space of flows consisting of sounds and words that travel like particles on simulated air and water currents loosely based on actual oceanographic and meteorological data. The audience must either give itself up to the primal cycles of nature or risk terminal confusion and data loss.

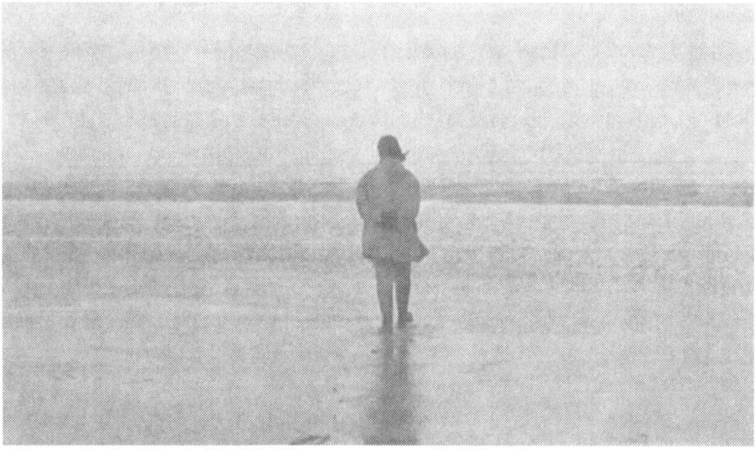

Figure 8: Teri Rueb "Drift" (Photo courtesy Teri Rueb)

Mapping as Critique

The GPS mapping practices of modern psychogeographers,[2] are seemingly related to the writings of Guy Debord and his practice of the 'Dérive' [Debord] but in reality

---

[1] Teri Rueb, Drift (April 2004) part of Ohne Schnur curated by Cuxhaven Kunstverein- the installation covered a 2 km x 2 km region on the Wadden sea that is filled with areas of interactive sound.The piece creates a space of flows consisting of sounds and words that travel like particles on simulated air and water currents loosely based on actual oceanographic and meteorological data. <http://www.terirueb.net/drift/> and <http://www.kunstverein-cuxhaven.de/ohne_schnur/englisch/index2.htm>

[2] See <http://www.gpsdrawing.com> and <http://socialfiction.org> (accessed 12/03/05)

seldom appear to achieve anything identifiably subversive. To quote one cultural critic:

> "Locative media is: Psychogeography without the critique. Algorithmic psychogeography, the term used by http://socialfiction.org to describe their rule-based derives through the city, is not just a development, but actually a fundamental reversal of the critical use of this Situationist tool."[3]

The "Dérive" or 'drift' was a method for subversion, of remapping the world with 'uncontrolled' clarity, for identifying the secret flows of money and power below the surface of the city. However, one strategy Debord does cite: "the introduction of alterations such as more or less arbitrarily transposing maps of two different regions,"[4] has been successfully adapted in several locative works. Jen Southern and Jen Hamilton in *Distance Made Good*,[5] used parallel mirrored journeys on two continents; in *Shadows from Another Place*,[6] Paula Levine creates a hybrid space between Baghdad and San Francisco composed of the superimposition of their city centres. A mapping of the initial US attack on Baghdad is superimposed upon downtown San Francisco. The longitude and latitude of each bombsite is marked in San Francisco using a GPS device. C5 in *The Other Path*[7] set out on a month long Great Wall trek, starting in the northwest desert of China and following the Wall eastward to where it runs to the edge of the Yellow Sea. GPS data collected during this trek was used to develop a pattern matching search procedure for locating the most similar data model in the most similar terrain in California.

---

[3] Saul Albert <www.twenteenthcentury.com> (message dated Tue Apr 27 2004) "Locative media is: Psychogeography without the critique. Algorithmic psychogeography, the term used by <http://socialfiction.org> to describe their rule-based derives through the city, is not just a development, but actually a fundamental reversal of the critical use of this Situationist tool."

[4] Guy Debord, Introduction to a Critique of Urban Geography , Published in Les Lèvres Nues #6, 1955

[5] Jen Southern and Jen Hamilton "Unfeasible Symmetry" Artists Newsletter Magazine (October 2003) article on Distance Made Good shown at The Gallery, Stratford-upon-Avon, UK. July 2002 <http://www.theportable.tv/dmg/index.html> (July 2005)

[6] Paula Levine, Shadows from Another Place (2003) Levine builds upon this link, creating Hybrid space between Baghdad and San Francisco composed of the transposition of Baghdad and San Francisco. A mapping of the first US attack on Baghdad is superimposed upon San Francisco. The longitude and latitude of each bombsite is marked in San Francisco using a GPS device.

[7] C5: The Other Path <http://www.c5corp.com/projects/otherpath/index.shtml> (July 2005). In April of 2004, C5 set out on a month long Great Wall trek, starting in the northwest desert of China and following the Wall eastward to where it runs to the edge of the Yellow Sea. GPS data collected during this trek is being used to develop a pattern matching search procedure for locating the most similar data model in the most similar terrain in California.

One of the most lauded recent locative artworks has been Blast Theory's *Uncle Roy All Around You.* Yet it appears an uneasy mix of performance and game, its full narrative only accessible to those who successfully complete their quest. The real and virtual sit in an uncomfortable relationship with the environment, which is only valued as a source of directional clues—and any casual bystander remains largely mystified and excluded. The charge levelled at Blast Theory at a recent conference was that of complacently and uncritically adapting new practice for the games industry, thus unwittingly acting as fashionable agents for intrusive and suspect technologies. For the flip side of ubiquitous communication and augmented location is the ability to track the audience: so a whiff of the totalitarian always haunts the liberating potential of the technologies. Matt Adams has rebutted this critique in interview, pointing to the collaborative co-dependency explored by the work.[8] William Mitchell identified early on this potential for pervasive surveillance as the era of "dataveillance" [Mitchell 157].[9]

Lev Manovitch has observed:

> "GPS, wireless location services, surveillance technologies, and other augmented space technologies ... want to make the map equal to the territory."[10]

There is nothing particularly surprising about this intentional (and impossible) trajectory of the technology: for neo-liberal economics extends its controlling force into every facet of our lives by voluntary technology adoption, blurring the boundaries between public and private, work time and personal time. But while the potential for monitoring and control grows exponentially, the map can never equal the territory and the opening quote of this essay forms a cautionary warning. The personalisation of space through information "tagging" through GPS systems and online tools has rapidly become an everyday occurrence. Unfortunately, the very same processes of neo-liberalism are simultaneously draining meaning (and Benjamin's definition of "aura') from the 'real' spaces of our lives. This process has also identified by Marc Augé as 'Supermodernity', a simultaneous culture of superfluity of places and of 'no place'-the mall, the motorway and the airport [Augé].

---

[8] Interview with Martin Rieser for "Mobile Audience" (January 2005)

[9] Mitchell rather uncritically applies the term to the tagging of net journeys—it is now superceded by the term "pervasive surveillance"

[10] Lev Manovitch, The Poetics of Augmented Space: Learning from Prada <http://www.manovich.net/DOCS/augmented_space.doc> (2003)

These issues seem to me to mark a step-change in the way technology governs our behaviours. But if the real and virtual can be so easily melded together by technology, with its potential for personal freedom and conversely, institutional control; where are the reciprocal responses from the artists? Very few mobile works have as yet tried to find a metaphoric critique of advancing ubiquity-I can only immediately think of Jonah Brucker-Cohen's *WiFi Hog*[11] (whose ironic intentions in 'hogging' a WiFi node were even misunderstood by the wireless community at the time) or Drew Hemment's *Loca*. Both these works are explored as cases studies elsewhere in this volume.

Maps and Subjectivity

From the Ptolomaeic to the Colonial, maps reflect cultural subjectivities. It is well known how Mercator projections exaggerated the relative sizes of the northern nations against their southern empires. The area-accurate Peter's projection still shocks by depicting the true area of the southern continents. Our mental representations are necessarily complex, and to me it seems problematic for artists to merely map literal representations back onto space using locative technologies, but this appears to have been the predominant practice of many early projects such as aspects of the Locative Media workshop[12] and *Urban Tapestries.*[13]

Research into spatial representation shows how mental maps create subjective distortion, describing not space, but the objects or nodes in it, and so these inner representations appear to be a direct contradiction to the continuous Euclidian 'space between' of a map.[14] If this perception of space resides in the mind of the beholder, that perception springs from a further complexity of sources. We construct our notions of space through the 'hardwiring' of the body. (The unconscious knowledge of space we carry is partially based on an invisible and forgotten sense:

---

[11] Jonah Brucker-Cohen, Wifi-Hog: From Reaction to Realization (2003). <http://locative.net> (acessed 2005)

[12] Locative Media Workshop. The international workshop entitled "Locative media" focusing on GPS, mapping and positioning technologies took place from July 16 - 26, 2003 at the K@2 Culture and Information Centre on an abandoned military installation in Liepaja on the coast of the Baltic Sea <http://locative.x-i.net/> (July 2005)

[13] Urban Tapestries <http://urbantapestries.net/>. Proboscis: Urban Tapestries (2002-4). The Urban Tapestries software platform allows people to author their own virtual annotations of the city, enabling a community's collective memory to grow organically, allowing ordinary citizens to embed social knowledge in the new wireless landscape of the city.

[14] Barbara Tversky, Joseph Kim, Andrew Cohen, Mental Models of Spatial Relations and Transformations from Language, Stanford University/Indiana University <http://www.psych.stanford.edu/~bt/space/papers/tverskykimcohen99.doc.pdf>(1999)

proprioception) and through culturally determined 'soft wiring', be it politics, gender, religion or class.

The advent of Locative Media means that the body could potentially reassume its primacy as measure and scale in the world. It gives us orientation through its physical structure—front, back, left, right, up, down and its relation to the force of gravity. The mobile body could offer continuous new perspectives on the world, allowing for a far richer and subtler perception by all the senses, than is possible in a static environment [Rodaway]. Acknowledging this physical potential might be the beginning of more engaged artistic practice for Locative Media. For example, Rueb has related how in her work Trace,[15] the narratives seemed to gain greater significance for an audience in proportion to the steepness of the terrain traversed at the time of hearing![16]

Building as Symbol

The symbolic qualities of architecture also supply valuable examples of such physical reinforcement of meaning. Locative art must absorb these lessons if it is going to embody more than a superficial mapping of experience. While referencing causality is not a precondition for art, it can augment its impact. To quote the architect Juhani Pallasmaa:

> "The technological environment all together weakens the causal understanding of the world and that is mentally harmful."[17]

This deracination of the image, also identified by Benjamin in his famous essay, as an adjunct of the age mechanical reproduction, can perhaps be redeemed through proper artistic alignment of the new technologies of mobility [Crary, 225].[18] Detachment from the causality of real space needs addressing by the artists who uses locative technology. My own experience of participating in 'serious' locative games such as *Uncle Roy* suggests that such games induce the same obsessional

---

[15] Teri Rueb: Trace site-specific interactive sound installation Yoho National Park, British Columbia (1999)

[16] See <http://www.terirueb.net/old_www/port_3.html>also described in interview with Martin Rieser (February 2005)

[17] Ruyter, Eystein Wilhelm AN INTERVIEW WITH JUHANI PALLASMAA : student architecture paper Issue 3 - (March 1999) <http://org.ntnu.no/a/mar99/juhani.html>

[18] Walter Benjamin, "The Work of Art in the Age of Mechanical Reproduction," in Illuminations, ed. Hannah Arendt, New York: Schochen Books, (1969), quoted in Jonathan Crary, Techniques of the Observer: On Vision and Modernity in the Nineteenth Century, Cambridge: The MIT Press (1990) p.225

tunnel vision of the video game player, rather than the subtle engagement required of a truly 'located' audience.

Architecture can already offer us an integrated and compelling model for embodied spatial narratives. One of the social functions of architecture is to create ritual spaces-a kind of social form in which a designed narrative can unfold harmoniously and simultaneously within the larger context of an interactive space. We are just beginning to understand this narrative potential of ancient architectural alignments, religious architecture's spatial organisation, and acoustic resonance. To illustrate my point, Chartres Cathedral, (which was built in France in 1235 AD) has the "Chemin de Jerusalem," a pavement maze in the nave. Pilgrims walked the maze as a substitute for a real pilgrimage to Jerusalem, shuffling along on their knees as penance; the physical act literally mirroring a spiritual journey.

We can also see the full integration of the physical and spiritual in the architecture of a Hindu Temple. Based on the form of the Mandala, the temple forms a microcosm of the Universe and also of 'Cosmic man'. The proportions of the central square corresponded to the proportions of humans, which were also the proportions of the Cosmic Man. Measurement and proportion were crucial to the proper construction of a Hindu temple. Like the Mandala, the Hindu preoccupation with mathematics originated with the Vedic sacrificial altar. Typical pilgrimages would involve walks around the various perimeters of the sacred spaces. [19]

Towards a New Narrative

Architecture In the West up until the 18th century, architecture preserved its ability to frame and support (ritual) actions that allowed the human subject to orientate to a Supernatural Being. In the late 18th century architects underwent a profound crisis of meaning in their discipline. Once cosmography and mythology disappeared as socially accepted realities, the external referents of architecture became problematic. In *Being and Time,* Heidegger identified the loss of this world given over to fundamental experience and its substitution by objects and routines [1927]. This disjuncture remains a universal problem for artists and architects alike, but Alberto Pérez-Gómez picks up the challenge raised by Heidegger and imagines a use of space, which revivifies the lessons of pre-enlightenment practice, citing Libeskind´s Jewish Museum in Berlin as a space capable of:

[19] Author unknown, Temples, a brief guide on mailerindia <http://mailerindia.com/temples/ history/> (July 2005)

"existing in a thick, vivid present, between the earth and the sky, as a unique place in the universe, always subject to forces larger than ourselves that in fact make us human."[20]

Libeskind's Jewish Museum takes the ritual narrative of architectural space to a new extreme. Corridor floors slope but appear level; ceilings lower from five to two metres on seemingly straight corridors, inducing a physical disorientation in the observer, which echoes the disorientation of the Jewish victims of Holocaust. Libeskind's work is also highly symbolic: its floor plan, based on a broken star of David and the void left by the 'disappeared', is echoed by a physical 'Void' running through the heart of the building. Locative media narratives have yet to match the engaging symbolic and physical experience of such architectural spaces. The holistic experience of such a space, makes one believe Pallasmaa's contention that:

"Architecture re-mythologies space and gives back its pantheistic and animistic essence." [Pallasmaa 2000]

Ancient Narrative Spaces

Research at Neolithic sites around the UK has also revealed striking similarities in their acoustic properties. Key examples, such as the huge passage tombs of New Grange and Maes Howe have a resonant frequency (at which sounds naturally echo and reverberate) of about 110hz. Standing waves, where sounds are reflected off walls and superimposed on to one another, have also been observed in these and other sites and evidence suggests that ancient architects consciously realigned stones to create these effects. Archaeologists have speculated that these frequencies would have produced reverberating echoes that may have induced a state of trance, adding to their potency as sites of spiritual importance.[21] I am not arguing here for a return to mysticism, but for a renewed understanding of past acoustic sensitivities to enrich current practice.

Other evidence of carefully aligned geometries can be frequently found in the organisation of space by earlier societies, for example in New Mexico the solar and lunar cosmology is encoded in Chacoan architecture: through the buildings'

---

[20] Alberto Pérez-Gómez Dwelling on Heidegger: Architecture as mimetic techno-poiesis(1998) <http://www.tu-cottbus.de/BTU/Fak2/TheoArch/Wolke/eng/Subjects/982/Perez-Gomez/perez-gomez_t.html>

[21] Newgrange Burial chamber reveals haunting sound of past, Sunday Times, Irish Edition, (15th July 2001) see also <http://www.orkneyjar.com/history/tombs/tombacoustics.htm> (July 2005)

orientations, internal geometry, and geographic relationships, which unified the Chacoan people with the major cosmic cycles [Sofaer,1997].

Locative narratives and present time

There are contemporary narratives resonant with the reinforcement of site and story. *Riot*[22] from Mobile Bristol depicted the Bristol Riots of 1831. This first GPS-enabled locative drama was an immersive and powerful experience, engaging with the immediate spaces of history, mapped onto a Georgian square where the original events took place.

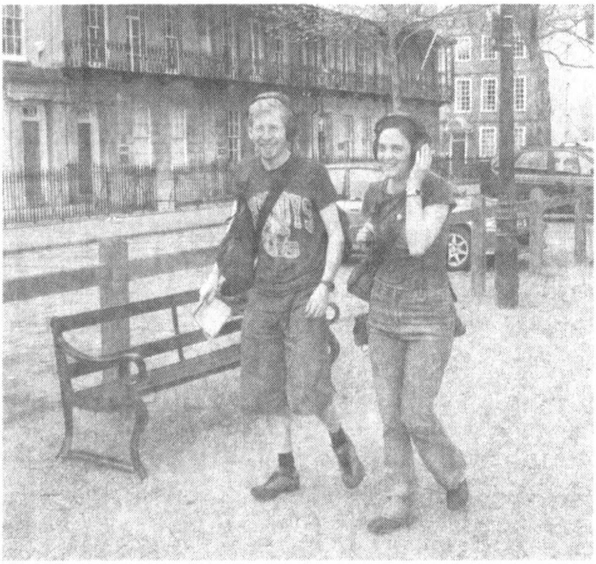

Figure 9: Mobile Bristol: Riot! Photo: Martin Rieser

Jeff Knowlton, Jeremy Hight and Naomi Spellman's *34n118w*[23] dealt with a similar historical hinterland in LA. To quote Hight:

> "The story world becomes one of juxtaposition, of overlap, of layers appearing and falling away. Place becomes a multi-tiered and malleable concept."[24]

---

[22] Mobile Bristol is a professionally co-ordinated team of business and academic researchers from Hewlett Packard, Bristol University, UWE and digital product experts The Appliance Studio. <http://www.mobilebristol.co.uk/QueenSq.html>

[23] "34 north 118 west" premiered November 15, 2002 at the Art in Motion Festival <http://www.xcp.bfn.org/hight.html> (July 2005)

At first sight it seems contradictory that such engaging locative works tend to deal with an historical past rather than the lived present. After all Paul Virilio identified new media as promoting the change from considered diegesis to continuous and automatic present, the user creating the narratives both as subject and object; the visual subject becoming transferred to a technical effect, which forms a sort of 'pan-cinema', turning our most ordinary acts into movie action.[25] However where these locative works succeed, they seem to overlap the user's enactment of the continuous present with the user's immediate perception of a contiguous past.

But even sophisticated site-specific locative work, which draws on, a resonant location can fall down on the issue of content. Rafael Lozano-Hemmer is one of its most famous exponents. His *Re: Positioning Fear* (Graz 1997)[26] was loosely based on the Cathedral's fresco "the Scourges of God," which depicted locust plague, the Black Death and the fall of the city to Turkish invaders. Halogen lights tracked people and created their sharp silhouettes. Internet Relay Chat Web Texts were sent by the public and included in the projections. However, as in all such cumulative public interaction, the included texts could often appear as trivial, and inappropriate to the visual impact of the work and the nature of the "fears" being addressed.

Physical Space as Diegetic Space

Bob Hughes, author of 'Dust and Magic', has postulated that landscape can be a model for new media narratives, with each track or journey mapping an individual trajectory through a story space.[27] But here we encounter a problem-if we can map a story space, don't we risk fixing the narrative as rigid terrain. What then happens to narrative subtlety and flow? In my own work I have attempted to use the ambiguity and resonance of language to ensure that the stories continue to vibrate and hold their numinous sense of lived experience.

---

[24] Hight, Jeremy, Narrative Archaeology, Streetnotes (Summer 2003) <http://www.xcp.bfn.org/hight.html>(July 2005)

[25] Lindroos, Dr. Kia, Non-Linear Narrative As a Form of Political Action: Viewing Chris Marker's Film Sans Soleil (2002) <www.essex.ac.uk/ecpr/events/jointsessions/paperarchive/mannheim/w22/lindroos.pdf>(July 2005)

[26] Re:Positioning Fear featured during the third international Film and Art Biennale in Graz, Austria, a Relational Architecture piece transformed the courtyard façade of one of Europe's largest military arsenals, the 350 year old Landeszeughaus. featured during the third international Film + Arc Biennale in Graz, Austria, a Relational Architecture piece transformed the courtyard façade of one of Europe's largest military arsenals, the 350 year old Landeszeughaus.

[27] Bob Hughes, Narrative as landscape <http://www.bobhughes.demon.co.uk/NasL.html> (2004)

Understanding Echo (2002)[28] was my attempt to root narrative in an interactive circular space where "memory, dreams and reflections" could rise to the surface of a central pool of water in the form of a projected figure. Based on the myth of Echo, trapped in a pool of water by her unrequited love of Narcissus, language played a central role, indexed directly onto the physical space. We are immersed in Echo's longings and become her sounding board: spatialised narrative and poetic monologues fuse together in the physical environment of the piece. Once an audience entered the installation, they became part of the diegesis. The closer a visitor approached the pool, the more intimate their knowledge of Echo.

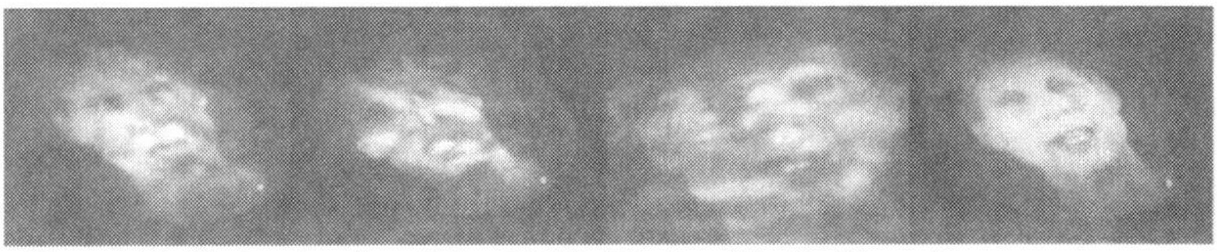

Figure 10: Understanding Echo 2002 (Photos: Martin Rieser)

This process of narratising space was extended in *Triple Echo* (2003)[29] where the sound tracked with the participant seamlessly from screen to screen, and from character to character without breaking the narrative thread. This spatial mapping culminated in *Hosts* (2006)[30], a mixed site-specific and locative work for Bath Abbey, similarly using "sticky video," where specific video characters tracked individuals from the audience across real space through a succession of screens, reacting to an individual's movement and attention. An audio landscape of *a capella* sound accompanied the visitor. If a visitor stood for a time in front of a particular screen, a video 'Host' would address the chosen individual with a series of ambiguous aphorisms. The piece was a reflection on human life and death, history and absence. The 'Hosts' could be taken to represent a variety of presences from the angelic to the historic. It was an attempt to map an invisible co-present onto a public space in a meaningful way.

---

[28] Martin Rieser, Understanding Echo , DA2 Commission, shown at Cheltenham Festival of Literature 2000, Watershed Media Centre, Bristol and ISEA2002 in Nagoya, Japan 2002 <http://vis.vision.ss.is.nagoya-u.ac.jp/isea/text/program/E/exhibitions.htmland> and <http://www.da2.org.uk/projects/echo.htm>

[29] Martin Rieser:Triple Echo (2003), research pilot, using a contemporary triple screen exploration of the Orpheus Legend, funded by the AHRB

[30] Martin Rieser: Hosts was funded by the AHRB and was shown in Bath Abbey (February 2006) <http://www.shipofools.org.uk/artists/martin6.htm> and <http://www.martinrieser.com/Hosts.htm>

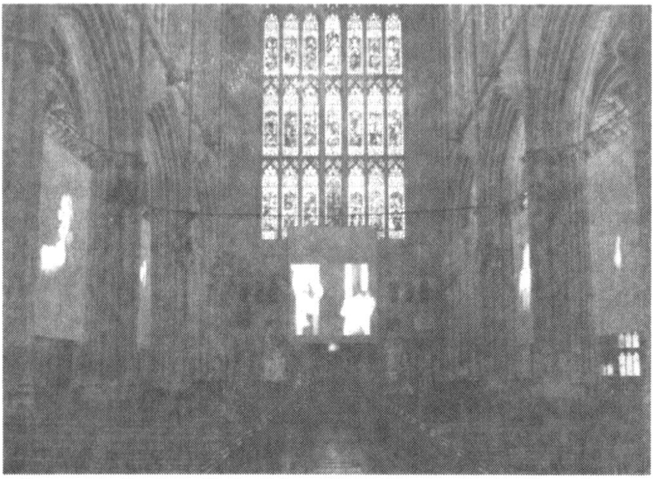

Figure 11: Hosts in Bath Abbey 2006 (Photo: Martin Rieser)

Conclusion

The emergent field of locative media art explores the convergence of computer data and physical location using portable media. I have argued that convergence should necessarily lead us to a radical reassessment of the nature of representation through the embodied form of the experience. It also seems to me that physically experienced locative art reveals a medium where magnified artistic resonance can be achieved through a deeper knowledge of spatial grammars. Whereas the history of cinema is one of the development of a visual language of increasing subtlety and expression; Locative Art is in its infancy and has yet to deliver work of comparable range and depth. It is too early to define as an artistic medium, but geographical space as extension of narrative is a concept as relevant to architects as it is to cultural theorists, filmmakers or media artists. It seems an essential first step in understanding the potential of a medium where the artist can allow the audience to also act as 'liminal' personae[31], to analyze how earlier artefacts have successfully embodied meaningful social and cultural transitions using spatial narrative. The role of the fully 'embodied" user in Locative artworks is all part of that understanding, allowing artists to forge new and necessary grammars of interaction.

---

[31] "Prophets and artists tend to be liminal and marginal people, 'edgemen,' who strive with a passionate sincerity to rid themselves of the clichés associated with status incumbency and role-playing and to enter into vital relations with other men in fact or imagination" [Turner 128].

Bibliography

Augé, Marc. *Non-Places: Introduction to an anthropology of supermodernity*. London: Verso, 1995.

Benjamin, Walter. "The Work of Art in the Age of Mechanical Reproduction." *Illuminations*. Ed. Hannah Arendt. New York: Schochen Books, 1969.

Crary, Jonathan. *Techniques of the Observer: On Vision and Modernity in the Nineteenth Century*. Cambridge: The MIT Press, 1990.

Debord, Guy. *Society of the Spectacle*. Detroit 1977.

Felluga, Dino. "Modules on Lacan: On the Gaze." *Introductory Guide to Critical Theory*. 2005. <http://www.purdue.edu/guidetotheory/psychoanalysis/lacangaze.html>

Flanagan, Mary. "Navigating the narrative in space: gender and spatiality in virtual worlds." *Art Journal*, Fall 2000.

Debord, Guy. "Introduction to a Critique of Urban Geography." *Les Lèvres Nues* #6, 1955.

Galloway, Anne. "Mobility as world-building/technologies at play." *Receiver* #10 *Connecting to the Future*. 2004. <www.receiver.vodafone.com>

Heidegger, Martin. *Being and Time*. Trans. John Macquarrie and Edward Robinson. Oxford: Blackwell, 1978.

Hight, Jeremy. "Narrative Archaeology." *Streetnotes* (Summer 2003) <http://www.xcp.bfn.org/hight.html>

Hughes, Bob. "Narrative as Landscape." 2004. <http://www.bobhughes.demon.co.uk/NasL.html>

Lacan, Jacques. "The Mirror-Stage as Formative of the I, as Revealed in Psychoanalytic Experience." Trans. Alan Sheridan. *Écrits: A Selection*. New York: W.W. Norton & Co., 1977.

Lindroos, Kia. "Non-Linear Narrative As a Form of Political Action: Viewing Chris Marker's Film Sans Soleil." 2002. <www.essex.ac.uk/ecpr/events/jointsessions/paperarchive/mannheim/w22/lindroos.pdf>

Manovitch, Lev. "The Poetics of Augmented Space: Learning from Prada." 2003. <http://www.manovich.net/DOCS/augmented_space.doc>

Mitchell, William J. *City of Bits*. Boston: MIT Press, 1995.

Mulvey, Laura. "Afterthoughts on 'Visual Pleasure and Narrative Cinema.'" *Framework*, Summer 1981.

Pallasmaa, Juhani. *The Architecture of Image: Existential Space in Cinema*, Helsinki: Rakennustieto Oy, 2000.

Pérez-Gómez, Alberto. “Dwelling on Heidegger: Architecture as mimetic techno-poiesis.” 1998. <http://www.tu-cottbus.de/BTU/Fak2/TheoArch/Wolke/eng/Subjects/982/Perez-Gomez/perez-gomez_t.html>

Rodaway, Paul. *Sensuous Geographies: Body Sense and Place*. London and New York: Routledge, 1994.

Ruyter, Eystein Wilhelm. An Interview with Juhani Pallasmaa. *student architecture paper* Issue 3 (March 1999) <http://org.ntnu.no/a/mar99/juhani.html>

Sofaer, Anna. “The Primary Architecture of the Chacoan Culture: A Cosmological Expression.” *Anasazi Architecture and American Design*. Ed. Baker H. Morrow and V. B. Price. Albuquerque, NM: University of New Mexico Press, 1997.

Turner, Victor Witter. *The ritual process: structure and anti-structure*. Chicago: Aldine, 1969.

Tversky, Barbara, Joseph Kim, and Andrew Cohen. “Mental Models of Spatial Relations and Transformations from Language.” Stanford University/Indiana University. 1999. <http://www.psych.stanford.edu/~bt/space/papers/tverskykimcohen99.doc.pdf>.

Digital Media and Architecture—An Observation

Anke Jacob

> "In the transient architectures of today's cities, spatial definition changes constantly as space is activated as much by electronic as by architectonic artefacts. It is in tribute to the ephemeral nature of these inherently unstable images that the questions of temporary exhibitions—whether explicitly staged or occurring fortuitously within architectural terrain—becomes important to the architecture of events." [Tschumi 556]

Introduction

Two important tendencies within Western culture—closely connected and influencing each other—increasingly captured my interest over the past recent years: the advent of digital media in the urban environment and the trend towards the resurgence of the decorative in design and architecture.

In 1966 the American architect Robert Venturi states in his book *Complexity and Contradiction in Architecture* that he prefers "elements which are hybrid rather than 'pure', compromising rather than 'clean', distorted rather than 'straightforward', ambiguous rather than 'articulated', … inconsistent and equivocal rather than 'direct and clear'" [16]. Here Venturi paints a rather precise picture of the social and cultural development towards the end of the twentieth century: a growing desire for playfulness, ambiguity and emotional expression, which results in an increasing revival of the ornament the decorative. This tendency goes along with the request for architecture to include more flexibility and multifunctionality in our ever growing and fast-paced cities. On the other hand, digital technology has been entering the built environment on multiple levels in the last ten to fifteen years. In recent years, new media became increasingly influential in particular for the aesthetics, the organisation and the use of urban and architectural space—thereby perfectly serving an architecture striving for flexibility and aesthetic surprise.

The desire and need for flexibility

It has been more than a decade since I moved to London. During that time I have watched London's urban landscape change. Looking at the skyline of central London at night, you see it glowing like a rainbow. The blue lights on top the Lloyd's of London building designed by Richard Rogers—the only unusual coloured light ten years ago—barely compete now with other impressive lights in the urban landscape. In the twenty first century city buildings are colourfully illuminated, almost vanishing behind the bright lights. Their surfaces change constantly with the light. Urban spaces and plazas are lit by huge media façades, transforming the night into day. Light installations play tricks with our perception of space, deceiving our eyes.

The technical and digital revolution has pushed the limits both of possibilities and imagination. When dusk is falling, the illuminated city—seductive and beautiful—becomes the essential place to be. Statistics reveal that since the year 2000, fifty percent of the world population now lives in cities.[1] The city changes our behaviour; our behaviour changes the city. Our urge towards independence and individuality leads to the wish to control the environment. We do not want to depend on the twenty-four hours rhythm dictated by the sun—modern technology ensures that we are not bound to determined places and times.

On the other hand, changing economic conditions, globalisation and centralisation result in an intensified density and continual transformation of urban life with extremely complex requirements. Long-term prediction and planning becomes almost impossible. Our lifestyle is less stable than it used to be thirty years ago. Society calls for flexibility in all areas and the need for changeable, multifunctional and responsive architecture is growing. Architecture needs to adapt to flexible programmes, which are based on openness and transformation considering the subject as generative contingency. Consequently a paradigm shift is happening towards time-based design, i.e. designing processes rather than objects.

London's light spectacle proves that aesthetics are at least as important as functional and programmatic performance. Our vision is considered the most important of our senses, as seventy percent of all information about the external world is perceived via our eyes. This might explain our constant visual thirst, our longing for the ultimate visual sensation and stimulation. In terms of visual experiences, nature itself is the classic example. It might be because of the lack of opportunities in modern city life to experience the magic of nature that we ceaselessly try to create magical moments within our visual surrounding to alleviate our hunger for aesthetic surprises. Thus emotional and the poetic needs to be addressed by contemporary architecture, as well as challenging functional and programmatic performance. Evolving technologies have always brought us a step nearer to controlling and perfecting our environment, changing its appearance and adapting it to our needs. So has digital media technology with its impact, directly or indirectly, onto architecture and urbanism.

---

[1] Source: Global Urban Observatory

The impact of new media and digital technology

In 1991, architect and theorist Peter Eisenman claimed that architecture was not only being confronted by electronic media, it had been swallowed up by it. With its emergence a reality shift took place and today there is no guarantee of what reality truly means [Eisenman 193-194]. There are multiple realities and multiple spatial experiences. Virtuality and virtual reality—the buzzword for the 1990s describing computer-generated environments—triggered the re-thinking of space. As Nicholas Mirzoeff states "Space can no longer be seen as simply an empty backdrop but as a dynamic entity with a history and characteristics that vary from period to period and place to place" [181,184]. So this rising instability influencing our lives, is it the result of the emergence of new technology? Or do less stable living circumstances cause or allow the evolution of new technology?

Within the contemporary architectural discourse the debate focuses on the problem of events and event structure because the place in traditional terms has ceased to exist. Bernard Tschumi explains "architecture is as much about the events that take place in spaces as about the spaces themselves" [13]. In his theoretical and practical work, Tschumi was one of the first architects to question the static and permanent nature of the built environment and therefore welcomed the arrival of digital technology, foreseeing the fulfilment of his vision. He argues that the "appearance of permanence ... is ... challenged by the immaterial representation of abstract systems (television and electronic images)."[2] And further:

> "Rather than questioning the technology of construction, architects will be involved increasingly in the construction of technology, including the computerised processes already altering building and design." [559]

Digital media technology not only affects the thinking of architects in terms of changing spatial conditions. Architects start to adapt principles, structures and the terminology of digital media within their design proposals, as reflected, for example, in one of Tschumi's recent projects, the master plan for the EXPO 2004 in Paris with the overall theme *Image*. Tschumi treats the exhibition area as a picture, which only comes to life through the actions and interactions of people. Each visitor represents a 'pixel' and is part of the flow, manipulated and animated by spatial interventions

---

[2] When Bernard Tschumi designed the fireworks for the opening ceremony of Parc La Villette (designed by himself in Paris in 1992), he created the ephemeral representation of the park's architectural concept in the sky. It was his expression of the idea of transient architecture, architecture as event, as installation, time-based, existent only for the moment.

such as gates, bridges, pavilions, and kiosks. The spatial situation is in constant flux and is constantly being reconfigured to give diversified readings of the place.[3]

Zaha Hadid's proposal for the new Guggenheim Museum in Tokyo is based on a similar concept. The building is understood as a picture dissolved into pixels. Thus its structure follows the pattern of an image rather than the law of construction and it is this that determines the appearance of the building. The snakeskin-like pixelation of the shell, which incorporates various digitally programmed surface performances, provides a flexible exhibition space "receptive to constant internal redefinition."[4]

Certainly a crucial impact on architectural designing and planning came from the introduction of CAD/CAM and better, more complex computer software is constantly evolving. The application of computers and three-dimensional modelling software as a design tool has influenced the design and manufacturing processes within architecture and therefore consequently the shape and appearance of buildings and their surfaces. Today, thanks to computer technology, it is almost as cheap to produce a curved surface as a straight one. Softer, more organic, more sensual forms are possible, tailored to special environments. So towards the end of the twentieth century, architecture has been freed from its restriction to linear forms and hard edges. Furthermore, advanced architectural theories can now be visualised. Experimenting with forms and shapes has reached another level, as there is a tool to test and communicate ideas with no limits to the imagination. These conditions create and satisfy a changed aesthetic attitude, which has been growing with the beginning of the postmodern era. However, in the built environment itself digital media technology entered the scene through a back door. It was the advertising industry that quickly realised the potential of LED technology (explained in more detail in the section *New display technologies—the fusion of digital media and material*).

Digital media screens replaced the traditional billboard animating whole building façades, for example the Lehman Brother building or NASDAQ building in Times Square, New York. Before architects and urban planners even thought of digital media as a device for façade design it was here already as part the urban environment. Huge moving images become integrated within the architecture and urban space, changing and dissolving the shape and structure of buildings and places.

---

[3] B. Tschumi, "Expo 2004," *Innovation*, pp. 85-87

[4] Z. Hadid, "Temporary Guggenheim Museum in Tokyo," Innovation, pp. 43-46

In an interview in *The Guardian* in November 2006, media guru William J. Mitchell says that, if those billboards could be programmed with "a coherent artistic vision," they could be used to give "an aesthetic consistency and a unity to the area."[5] So since the beginning of the new century it seems that an unstoppable process has unfolded: digital media technology has been embraced by architecture on a remarkable scale. As previously stated, digital technology is not something new to the architectural and design world. But now we see the digital visualising within buildings. "Those in the know call it *dynamic architecture*" [Mitchell].[6] In effect, architectural surfaces—inside and outside—become ambiguous, mutable and hybrid. Spaces become defined or distorted through the application of digital media technology. Programmable and interactive devices create conditions that allow openness, flexibility and individuality. Physical boundaries become blurred.

For example, the architect Jean Nouvel introduced digital media technology for the Concert Hall of the Danish Broadcasting Corporation in Copenhagen, Denmark (the winning competition entry in 2004). In his proposal, interior and exterior surfaces function as large projection screens immersing the architecture and visitors in constantly changing imagery and light. "It will be a volume, a mysterious parallelepiped that changes under the light of day and night, whose interior can only be guessed at" [Nouvel].[7] The technology, in combination with transparent material, adds an ambiguity to the architecture that expresses the chosen context of mystery and uncertainty. The concept for the concert hall building underlines Nouvel's aim of dematerialising architecture, making it a matter of light and surface effects. He also used projection technology for re-designing *The Hotel* in Lucerne, Switzerland, which was finished in 2000.[8] Sophisticated digital media technology is applied to project film sequences onto the ceiling of hotel rooms giving the place a sense of poetic ephemerality and each room an individual identity.

Resurgence of decoration—play and illusion

It seems no coincidence to me that architects and designers have been enthusiastically starting to exploit the potential of digital media technology. Its emergence overlaps with a rising tendency towards favouring the decorative in our

---

[5] J. Harkin, "Cyborg City," *The Guardian*, 26 November 2006

[6] William J. Mitchell is Professor of Architecture and Media Arts and Sciences at MIT and author of the book *Me++: The Cyborg Self and the Networked City*.

[7] See <www.jeannouvel.com> (10 June 2007)

[8] E. Franzoia, <www. floornature.com/worldaround/articolo.php/art13/10/en/text/text/html> (3 April 2005)

environment. Ornament and colour has been gaining ground after a period of minimalism. Some even call it the *decorative revolution*.

I believe that this trend is not just a fashion that will disappear after a few years. As a result of globalisation, our Western culture is confronted with cultures from different ethnic roots, with their aesthetic languages, their arsenal of ornaments and ways of decoration. The pleasure we gain from this influence reflects on our attitude towards decoration. At the same time today's society in the West is geared towards differentiation, which is often achieved by visual means. Our culture is moreover extremely image-orientated and our relationship with the visual, with imagery is almost obsessive. It seems today that the growing complexity of the world and our culture is most easily understood and imagined visually. We live in a visual culture and imagery is used to attract attention, to inform, to please or shock. Furthermore, visual culture is also defined through the visualisation of the non-visual existence. Quoting Nicholas Mirzoeff: "This visualizing makes the modern period radically different from the ancient and medieval world in which the world was understood as a book ... This visualization, always common, has now become compulsory" [6]. And after all, there are the technologies offering an endlessly growing variety for designing our environment that we cannot resist playing with them, driven by our longing for visual stimulation and experience.

The increasing desire for the playful, the ambiguous and the illusionary might be a way of unconsciously adapting to and dealing with the fast changing circumstances of life, with uncertainty and instability. Rigidness and static belong to the past. Softness and ambiguity are the preferred features of today. We want to experience softness in the visual, in the tactile and in the comfortable use of machineries (for example touch screens, the design of mobile phones and MP3 players and so on). More and more we want to give up our striving for eternity and durability in exchange of the short-lived experience of the moment.

Modern technologies, especially the use of animated light effects and moving imagery make spaces and surfaces less tangible and increase illusionary feelings. Images and sequences radically shape the architecture, distorting and redefining space. Walls come to life and spaces start to float, losing their physical markers and replacing them with virtual definitions. Virtual movements and movements in reality fuse together and for a moment one is not able to distinguish between them. When video sequences are applied on the outside of buildings, the façade gains depth adding the dimension of the space shown in the image to the building or street. The virtual becomes part of the urban space.

Transformable surfaces

Softness, playfulness, the desire for decoration—these features are certainly reflected in contemporary architecture. Today's façade design aims at a maximum of transparency and openness and tries to overcome the rigour of the grid and strict rhythm. As a result we can sometimes see the ornamental element developing almost naturally. However, the Swiss architect duo Herzog & deMeuron utilise the ornament intentionally as a means to dissolve and destabilise the classic façade. They understand the ornament in functional terms and recognise it as a form of experience. For designing the Eberswalde Library in Germany they invited artist Thomas Ruff to create sequences of images completely covering the façade like a tattoo. Accordingly our perception of the building changes completely. It is perceived as a picture and the walls are firstly image, secondly walls. The actual building dematerialises behind the image. In a similar way, the façade defines the expression of the previously mentioned Laban Dance Centre in South London. Translucent, coloured polycarbonate panels are used to give the building's skin different looks in different light conditions. Our visual perception is tricked by the mutation of material and colour and the shape of the building visually changes.

In another project, the new football stadium Allianz Arena in Munich, Germany, Herzog & deMeuron introduce digital media technology as a new form of ornament and decoration that is open to specific manipulation. The stadium's façade is made of translucent EFTE pillows, which are illuminated and can change colour throughout the whole spectrum. This flexible display system enables the outer skin of the building to display animated patterns or team colours. It allows the building to take on different identities and to transmit different information. Other recent projects, where the design of the façade is of particular significance, are the new library building of Brandenburg Technical University in Cottbus, Germany and the new extension of the Walker Art Center in Minneapolis, Minnesota, USA.

Visual disintegration and mutation, dissolving the building and façade into patterns and experimenting with various appearances in different light situations is also dominant in the architecture of Japanese architect Toyo Ito. At the flagship store of Italian accessory brand Tod's in Tokyo, the whole façade is a single continuous pattern, a construction made of concrete and glass wrapped around the entire building. The pattern was generated from the stylised two-dimensional images of the silhouettes of Zelvoka trees, which line the pavements in that area. At night the whole building turns into a huge paper lantern. The philosophy behind Ito's work is to eliminate materiality, to make architecture 'disappear'. Ito understands the building as a stage where transformation and play is self-evident. No surprise that he was one of the first architects who started to work with electronic media technology and interactivity. His *Tower of Winds* in Yokohama, an opaque silver object during the day, comes to life during the night when it looks like a 'dancer' in the city illuminated by an ever-changing pattern of lights on its surface. Controlled by computers, the display reacts to ambient sounds, wind forces, time of day and season and transforms the tower into an interactive landmark.[9]

Because of the new possibilities digital media technology now offers, architects have to rethink their approach towards façade design, as ornamentation can be flexible, changeable and animated. Tim Edler of *realities:united*, a Berlin-based architectural practice and creator of *BIX*, says, "architects need to develop a culture of the changing façade" [Bullivant 83-85]. *BIX* is the name of the animated electronic façade of the Kunsthaus in Graz, Austria. The way the media skin works is fairly simple yet very innovative. Circular fluorescent tubes, placed behind the translucent acrylic membrane covering the building function as pixels forming the low-res, grey-scale screen, forming a symbiotic part of the whole building. The light matrix affects the impression of the building's shape as it blurs and softens its edges. Recognisable from a distance, at close range the displayed content—images, text or sequences—becomes a totally abstract pattern. At night the light even determines the architectural form as the only thing that can be seen is the displayed pattern glowing from inside. The actual shape of the architecture disappears in the darkness of the sky. *BIX* integrates and synchronises architecture, information technology and communication content, "used as an instrument and platform for artistic production in a new level of mediation" [Bullivant, 83-85]. Since the opening of the Kunsthaus in Graz in 2003, *BIX* has also been employed on the façade of an eleven-storey high building in the centre of Berlin, forming a temporary exhibition space and displaying works of artists Carsten Nicolei and Rafael Lozano-Hemmer.

Buildings today often have to be multifunctional. Complex programmes define their diverse and flexible use and communication between the building and the user is crucial. With digital media technology, architects have a tool for responding to this, and today's communication can be much more individualised as well as more connected to the global and local context. Buildings can transport complex information. The digital display is the poster of the twenty first century, but we can read it like a clock tower—as for example the proposed building for the BBC White City Music Centre in London, designed by the London-based architectural studio FOA. Here the outer skin is

---

[9] See <www.designboom.com/eng/interview/ito.html> (10 June 2007)

"treated as a broadcasting device, ... allowing the building to literally reproduce the music performed inside through an audio scan that will turn pitch, rhythm and volume into changing digital patterns of colour and light ... The complex operates constantly in this double mode of relation between the public and the technology, between the real and the mediated." [Zaera-Polo & Moussavi, 620-622)

Using the façade as a device for complex communication can provoke a change in attitude towards a building or an institution challenging its conventional meaning. For the new Public Library building in Seattle, USA the Dutch architectural practice OMA initially planned billboard-sized screens showing various information, for example the latest books arriving in the building, event information, news and custom content introduced by the librarians. The library's image is transformed from a conventional and cut off place into a popular and open location, where through "the information flow, librarians become intellectual sports stars" [Ramus, 20].

Interactivity in architecture

As there is an increase of digital communication and new communication methods and tools are constantly evolving, inevitably digital interactivity is gaining relevance. Thus it is not a question of whether there will be interactivity. The emphasis is on what it will look like and how it will work. Searching for the right aesthetic is the actual challenge. Artists and architects have started to explore interactivity within architecture in many different ways. In 2002 Klein Dytham Architecture (KDa) and Toshio Iwai created *ICE* (*Interactive Communicative Experience*) for the Bloomberg headquarters in Tokyo. Digital information is visualised on both sides of a 5 x 3.5 metre wall, curved at the top forming a short 'roof'. Because of the arched shape the visitor becomes immersed in the displayed patterns. Data is experienced in a very physical way and the boundaries between data and body are challenged. The installation represents the personalisation of space because it responds to bodily movement in real time [Bullivant, 12-13].

The *D-Tower* in Doetinchem, the Netherlands, is an interactive urban sculpture, which reacts to data from a questionnaire (accessible via a website) to be completed by different groups of local people. Each evening the sculpture shows a certain colour expressing the emotional state of the questioned group of residents: red for love, blue for happiness, green for hate and yellow for fear. But it not only enables communication between subject and object, it also facilitates the exchange between subjects by sending out prewritten love letters and flowers from and to designated addresses. The creators of the *D-Tower* are artist Q.S. Serafijn and architect Lars Spuybroek of NOX, who say that the sculpture is "a coherent hybrid of different media, where architecture is part of a larger interactive system of relationships."[10]

The desire for playfulness and ambiguity in urban environments finds an aesthetic and very poetic expression in the project *Memory Wall* by architect/artist Jason Bruges. Bruge's piece of work is not about direct interaction where the personal influence is apparent. The user/visitor perceives the interaction over time, as the installation captures form and movement of individuals entering a certain space and moving around. The captured information is processed and displayed in a continuous loop resulting in unique images as the memories build up on the wall during the course of the day [Bullivant, 81].[11]

Interactivity does not necessarily always involve human activities as one of the earliest interactive façade installations by German artist/architect Christian Moeller demonstrates. In 1992 he set up the *Kinetic Light Sculpture* on the façade of a shopping mall in Frankfurt, the first light installation of this kind installed on a building's façade. By dusk, mobile blue-yellow clusters of light created changing patterns across the façade according to weather conditions, wind and temperature. Additionally a LED screen displayed oscillating graphic renderings of the ambient sound in the street. This and many other of Moeller's projects characterises his endeavour to understand the façade as an entity functioning as architectural skin and carrying images [Bullivant 65].

Increasingly within interactive installations, wireless technology is used, mainly utilising the mobile phone as a tool of participation. Within this technology lies a huge potential yet to be explored. Specifically in architecture, wireless networking can influence the way of organising spaces and the subject becomes more significant. "We're moving to a situation where you try to create very flexible spaces that can be appropriated to a lot of different uses depending on the functionality their users bring to them" [Mitchell].

New display technologies—the fusion of digital media and material

Within this process of evolving innovations, the development of the digital display and digitally managed lighting systems has been of particular importance and influence. Thereby the intervention of the digitally controlled, light emitting diode (LED) had been revolutionising. Because of its very small size, wide colour range and low maintenance, the application of the diode is extremely flexible and versatile. Gigantic, high-resolution electronic displays for outdoor and indoor use had been

---

[10] More information on <www.noxarch.com> (2007)

[11] *Memory Wall* is installed at the eighth-floor lobby of the Puerta America Hotel in Madrid, Spain, where the media display is integrated flush into the interior walls.

made possible by LEDs, and are employed for advertisement and, increasingly, for design purposes and displaying art.

Besides the application within such screens, we can observe an explosion of creativity in utilising LEDs for innovative light management, as well as in incorporating the diode into structures or material such as textiles, glass, metal mesh or lattice work—used for interior and exterior surfaces, even for whole façades. Several products and projects within architecture and design have already been implemented. For example the development of various compact, lightweight structural tiles such as *SmartSlab*™ and *iColor Tile FX*, which can be applied as modular elements within architectural façades or walls. Here the fusion of light, material, structure and digital technology creates a qualitatively new multimedia display system.[12] Architect Zaha Hadid, for example, incorporated *SmartSlab*™ in her winning design proposal for her competition entry for the new Guggenheim Museum in Tokyo, which features six hundred units built into the building's surface to display information and art from the museum.[13]

In 2004 the architects of Dutch UNstudio[14] and the lighting designers of Arup Lighting, UK set a new landmark by transforming the entire façade of the *Galleria* department store in South Korea into a single digital screen, which continually changes the appearance and shape of the building. The screen is made up of thousands of glass discs with LED lights placed behind them. The glass is backed with a dichroic-filter film diffusing the light of the LEDs into a homogenous glow. The result is a blurred display of the imagery, enhancing the decorative effect of the façade. Furthermore the material gives the façade an iridescent shimmer during daylight, altering the colour and visibility of the building according to the light situation. Because of the perfect interplay of material and display technology the architecture changes in appearance not only from day to night: the façade is rather in constant transformation, blurring the edges between the building, the sky and the surroundings [Bullivant, 42-3].[15]

Whilst investigating methods of visualising digital data it is particularly the fusion of material and media that can amplify the experience of the viewer by adding a physical dimension. Increasingly, this potential is explored. The results are devices,

---

[12] *Icon*, Issue 010, February'04, p. 38. For more information on *iColor Tile FX* go to <www.colorkinetics.com/ls/rgb/tile> (10 June 2007)

[13] A single *SmartSlab*™ unit measures 60 square centimetres and is 7.5 centimetres deep. The imagery shown is controlled by a standard PC system using custom *SmartSlab*™ software. The units can be combined for small installations, large display walls or can become part of façades of buildings. For more information go to <www.smartslab.co.uk>

[14] <www.unstudio.com/projects/year/2004/1/141> (10 June 2007)

[15] B. Nolan, "Dichroic Filter with LEDs," *FRAME*, Issue 42, Jan/Feb 2005, pp.154-155

which do not function solely as a display, rather serving as a decorative element addressing our senses and imagination on multisensory level. For example, London based designer Ron Arad[16] and his team used fibre-optics in combination with the material Corian® to create an ambiguous and tactile surface displaying digital imagery. The fibre-optics are embedded in the translucent material and the colours and shapes of the image shine elusively through, giving the impression of coming from the inside, flickering, vanishing and emerging again from the depths. When switched off, the screen has a tactile, white, stucco-like surface, which is beautiful in itself. The fusion of material and media results in an aesthetically surprising and intriguing, mutable device for presenting digital information.

Electroluminescent polymers have recently been discovered and explored by designers for innovative and decorative lighting devices, for example in the projects *Sonumbra*, *Digital Dawn* and *Blumen Wallpaper* by London based design company Loop.pH.[17] Electroluminescent polymers function as the basis for digitally controlled, monochrome displays. Applied onto flexible material, it can be cut, bent and twisted into almost any shape imaginable. Because of this flexibility, it can be utilised beyond the conventional display or light sources and made into patterned walls, knitted curtains and so on. However, the future, in terms of display technology, will probably belong to the OLED (Organic Light-emitting Diode/Display)[18] consisting of organic conductive polymers, which are intrinsically emissive under voltage.

I learned about light-emitting polymers for the first time in January 2004 when coming in contact with scientists in the Netherlands, who had been investigating into polymers. About half a year later companies such as PHILIPS and EPSON published their advanced research about OLEDs and its potential for industrialisation. Currently, this new technology is widely recognised as being the next generation of digital displays, as it has many advantages over LCD (Liquid Crystal Display) including low power consumption and high-contrasted image. But as a designer, I can see the revolutionary potential in its flexibility. Firstly, with OLED technology digital displays can be made invisible when not in use. Without voltage it can appear transparent, reflective or coloured. Secondly—almost more importantly—the substance can be printed via high-precision inkjet printers as an ultra-thin layer onto flexible materials such as foils. The scope of application might be limited for the moment, but the potential for future applications is huge. Fully wall-integrated

---

[16] See <http://www.corian.co.uk/Corian/en_GB/design/hi_tech.html>(10 June 2007)

[17] See <www.loop.ph> (10 June 2007)

[18] See <http://www.oled-info.com/introduction> (10 June 2007)

displays or transparent digital screens made from textiles, foils and other materials might become possible—merging the digital and the physical almost seamlessly.[19]

Conclusion

The architectural discourse about the application and practical implementation of digital media technology has just started. Even built projects still bear the characteristics of prototypes. Few examples are in the phase of being developed commercially. It is still a vast unexplored field. But it is certain that the fusion of architecture and media technology has defined a new benchmark in architectural practise. Such is demonstrated not only through the emerging projects, but also through the debate and exchange happening amongst academics and practitioners around the notion of *media architecture* and *urban screens*. Conferences and publications are dedicated to this subject. Architectural schools have begun to include it into their teaching programme.[20]

Architecture is not a static and silent entity anymore. Architectural surfaces are taking on the characteristics of textiles, are becoming flexible, decorative, soft and ephemeral. The steady representational façade of buildings dissolves into event surfaces. Constantly generated visual images invade the physical space, shaping and manipulating its content and spatial perceptions. Architectural space is transformed into a hybrid, where the real and virtual overlap.

Flexible structures allow us to set up our own preferences and parameters and space becomes increasingly customised. In the future we might be able to tailor our own specific spatial environment in a very flexible way. The notion of time within architecture is as important as ever before. The appearance of environments can change within seconds—and in the future possibly also its structure. The architectural facade offers the option to serve as a display for imagery and information is transmitted continuously to the surrounding. The built environment takes on a stage character, functioning as a performing act—with its own particular audience. In this context Ole Bouman acknowledges:

> "Architects have begun to understand their products as liquid, animating their bodies, hypersurfacing their walls, crossbreeding different locations experimenting with new geometries. ... We will see more and more architects realising spatialised

---

[19] Further information: <www.troika.uk.com/loresdolores.htm> (10 June 2007) and <www.mediaarchitecture.org> and <www.urbanscreens.org > (05 November 2007)

[20] For example at the Bauhaus University Weimar, Germany <www.uni-weimar.de/mediaarchitecture/english/index.html > (05 November 2007)

moments, through staging narratives, through event designing, working with effects and emotions" [Boumann, 22].

The future development is predictable and unpredictable at the same time since it depends on individual impact not only of creation, even more on the impact of participation and we shall be surprised, where it will go.

Bibliography

Bullivant, Lucy. "Architectual Design." *4Dspace: interactive architecture*, Vol. 75 No. 1, Wiley-Academy, London, Jan/Feb 2005.

Bouman, O., "Architecture, Liquid, Gas." *4Dspace: interactive architecture*, Vol. 75 No. 1, Wiley-Academy, London, Jan/Feb 2005.

Eisenman, Peter. *Aura und Exzeß: Zur Überwindung der Metaphysik der Architektur.* Wien: Passagen, 1995.

Frame. The international magazine for interior architecture and design, Amsterdam, NL Issue 42, Jan/Feb 2005

Icon architecture+design, Media Ltd, Epping, UK Issue 010, Feb 2004

The Guardian, Guardian News & Media, London, UK, 26th November 2006

Koolhaas, Rem, ed. *Phylogenesis: foa's ark.* Barcelona: Actar, 2003.

Content. Köln: Taschen, 2004.

Mirzoeff, Nicholas, ed. *Visual Culture Reader.* London: Routledge, 1998.

Mitchell, William J. *Me++: The Cyborg Self and the Networked City.* Cambridge: MIT Press, 2003.

Papadakis, Alexandra, and Andreas Papadakis, eds. *Innovation—From Experiment to Realisation.* NewArch 07, London, 2003.

Tschumi, Bernard. *Event Cities.* Cambridge: MIT Press, 1994.

Venturi, Robert. *Complexity and Contradiction in Architecture.* New York: Museum of Modern Art, 1977.

Urban Screens as the Visualization Zone of the City's Invisible Communication Sphere

Mirjam Struppek

> "Public space is the city's medium for communication with itself, with the new and unknown, with the history and with the contradictions and conflicts that arise from all those. Public space is urban planning's moderator in a city of free players." [Christ, 2000] [1]

Urban public space is a unique arena for exchange of rituals and communication. In a constant process of renewal this challenges the development of our urban society. How can Urban Screens contribute to this idea of a public space both as moderator and as communication medium?

Referring back to the old concept of the Greek *Agora*, urban public space is a unique arena for the exchange of rituals and communication. The constant process of renewal of these exchanges challenges the developmental direction of urban society. The architectural dimension of this space has played a changing importance in providing a stage for these interactions. Moreover, the architecture itself is a medium, telling narratives about the city, its people and the represented structure of society. Its audience can read the reoccurring social interactions and the way the space is inhabited as a participatory process. The whole urban structure is slowly becoming a crystallisation of the city's memory over time.

In the context of the rapidly evolving commercial information sphere of our cities, especially since the 90s, developers introduced new digital display technology into this memory landscape, which is more or less integrated in the architectural dimension.

> "The migration of electronic screens into the external cityscape has become one of the most visible tendencies of contemporary urbanism." [Scott Mcquire] [2]

Paul Virillio terms this new developing "pervasive architecture-style" of the screens covered high-rise facades as "Electronic Gothic." He refers to the narratives of Gothic church-windows, which always tried to influence people's moral behaviour. Yet such screens are trying to manage and control our consumer behaviour, also

---

[1] See page 2 of the Information Platform <http://www.interactionfield.de>

[2] Scott McQuire, *The Politics of Public Space in Theatre*, Urban Screens 2005

recently incorporating “interactive features supporting constant short-term stimulation” [Virilio 61-65].[3] The immersive effects on the audience will also be increased by further ‘perfect’ incorporation of the screens in the architecture of the urban landscape. The Media Architecture Group, based in Vienna extensively documents this development on their blog.[4]

What happens, if new art forms, using interactive tools such as the internet and mobile technologies, suddenly give access to visual streaming of these digital surfaces? Can it revitalise the urban information environment into a more human-oriented approach? In virtual space we have already seen various community experiments in the growing field of social computing, such as peer-to-peer networks, friend-of-a-friend communities and more recently mobile communities. Participatory experiments in content creation can be found in mailing list culture and more recently through wiki- and blogging-systems, serving an increased need for self-expression. Linda Wallace sees “the internet as a delivery mechanism to inhabit and or change actual urban spaces.”[5] Moreover, the air is already filled with electromagnetic flows of data exchange. Could large outdoor displays function as experimental visualization zones of this fusion of virtual public spaces and our real world? Can we localise the huge flows of information data through these screens and can these zones take a more active role. More active than simply being the canvas on which the digital world is rendered? Do digital screens, as a new modern mirror, reflect the public sphere as a medium of communication in the city, with its citizens as urban players? If true, this finally could bind the screens more to the communal context of the space, allowing them to contribute to the creation of local identity.

The first steps in broadening the commercial advertising content of large digital outdoor screens focused on the transfer of TV-features, slightly adjusted to the new circumstances. Soon we might have TV broadcast stations, specialised in urban public space and its local community. They will coordinate outdoor movie-screenings, the collective watching of soccer-games and special City-TV news channels. A citywide “networked web of spatial narratives”[6] with the fractured character of the mass media world will emerge. Imagine new soap operas, or Big Brother shows especially created for the urban space audience! An alternative, more

---

[3] Virilio, Paul. “We may be entering an electronic gothic era.” Architectural Design - Architects in Cyberspace II. vol.68, n. 11/12, nov-dec 1998, pp 61-65.

[4] See <www.mediaarchitecture.org>

[5] Wallace, Linda. “Screenworld” in: Material Media, artefacts from a digital age, 2003. URL: <http://www.machinehunger.com.au/phd>

[6] Ibid.

bottom-up approach truly would be to connect the rapidly developing *photosphere* to Urban Screens.

Offering already a more participation-oriented alternative, Superflex envisioned using a large outdoor screen to bring their collaborative Internet experiment called *Supercity*[7] out to the urban community. The aim was to connect the 3D virtual copy of Karlskrona with its original. While the virtual model allowed the citizens to easily play at home with a new vision of their hometown and meet as avatars to discuss the process of urban development, the screen should ultimately invite the urban audience to follow the negotiations sensitising them for their role as citizens. Similarly, in 2002 in Columbus, Ohio, Jeanne van Heeswijks project *Face Your World*[7] gave children, via special bus access to a multi-user computer game, the option to redesign their communities, as they liked. At three bus stops the creations were exposed on special screen sculptures, proudly presenting the results to the urban community.

> "It's about the way people look at the space around them. With everything being privatized now, people don't view the community as their own any more."[8]

> *Egoscópio* by Giselle Beiguelman invited participation through participants jointly creating the antibiography of a virtual character, by the submission of websites to screens in busy urban spaces.

> "A character living in the boundaries between art, advertising and information developed."[9]

"Intrusion" of the urban commercialized visual flow

What strategies do we encounter, by trying to capture attention, to draw people into the artwork? If artists are not creating specially designed screen sculptures, the effect of simply intruding alternative content into the commercial visual flow of outdoor screens is easily over-estimated. A questionnaire I undertook in the streets of Berlin in front of a large screen at a busy street crossing in a major shopping area, revealed the hard struggle for attention. When the Artist group *Strictly Public*[10]

---

[7] See <http://www.jeanneworks.net>

[8] See <www.thelantern.com/news/2002/06/27/Arts/Exhibit.To.Unite.Community-256824.shtml>

[9] Giselle Beiguelman is a new media artist and multimedia essayist who teaches Digital Culture at the raduation Program in Communication and Semiotics of PUC-SP (São Paulo, Brazil) Her Internet-streaming for electronic billboards was called "egoscópio" (2002) <http://www.desvirtual.com/egoscopio02/index.htm>

[10] See <http://www.strictlypublic.org/projekte/transmediale04/program-en.html >(September 2005)

had placed their video art in-between the normal advertisements shown on the large public screens, few people noticed that there was something different to the usual program on the screen (January 30th through to February 29th 2004).

Cultural centres have started to officially integrated screens in their Architectural Facade to function as an extension of their archives into public space. Examples are the Schaulager, the Federation Square, Melbourne or the new Screens at the Creative Industries Precinct and, in a more experimental style, at Austria's Kunsthaus Graz . Yet Sabine Gebhardt Fink (Zurich) questioned, in her contribution to the *Urban Screens 05* Conference, whether the screen of the Schaulager is suitably designed to catch public attention, claiming that the location can only be recognised by the people on passing buses.[11] The Creative Industries Precinct[12] (Australia's first site dedicated to creative experimentation and commercial development in the creative industries, located on the western fringe of Brisbane's Central Business District) intentionally integrated three different screens to addresses different audiences, as Peter Lavery explained at the conference. The three screens were orientated towards the passing cars at the main traffic route, or directed towards the new growing residential area or addressed the internal plaza of the square, clearly demanded different content strategies. The shape of the whole of Federation Square, the unique cultural precinct of Melbourne, is intentionally designed around a big public screen integrated into one of the Facades facing the Plaza. The screen and its content are centred around the idea of community building and are committed to representing Melbourne's multiculturalism.[13]

The orientation, location and size of a screen are in fact crucial to enabling suitable content to address its audience. Cities are engaged in the struggle against the "feeling of placelessness" caused by the spread of international architecture and branded shopping areas. In fact, screens can well look the same everywhere, so there is a need to consider the locality and sites-specificity of the content, to prevent further disconnectedness in the perception of urban space and the actual locality.

There are other 'easier' more flexible ways, for artists themselves to bring Video Art to a new audience into public space. In the tradition of graffiti and 'adjusting', they have used the context of the street and its facades to critically overwrite them temporarily with their mobile screening facilities. Troika for example developed a mobile *SMS Guerrilla Projector*.[14] It is a portable gun like looking machine, that can

---

[11] See <http://www.urbanscreens.org>(September 2005)

[12] See <http://www.ciprecinct.qut.edu.au/idea>(September 2005)

[13] See <http://www.federationsquare.com.au>(September 2005)

[14] See <http://www.troika.uk.com/sms-guerrilla-projector.htm>(September 2005)

be fed via SMS and screen them on the urban surfaces. The CECUT Project[15] in Tijuana aimed to give voice and visibility to the marginalised inhabitants. A wearable headset was used with an integrated camera and a microphone, connected to two projectors and loudspeakers to let people talk about their feelings. The projections on the 60-foot diameter facade of the Omnimax Theatre successfully transformed the existing space.

The project *Energie Passagen*[16] by Monika Fleischmann and Wolfgang Strauss, used projection directly onto the sidewalk to draw passers by into the installation and to connect them to the visualization zone, through interaction-generated news feeds on a larger screen via a text input device. *StalkShow*[17] by Karen Lancel played with the engagement of the audience through an interactive backpack touch screen carried by a performer in performance-oriented style. Passers-by could search in a large database of comments written by isolated people, while being observed by the connected big screen, showing their portrait combined with their inputs.

The new development of organic light-emitting diodes (OLED) will soon make it possible to create flexible plastic monitors that can be integrated into clothing. France Telecom envisions "a new means of expression," with their wearable screen displaying animated graphics on the wearer. Drawings and animations can be sent by MMS to another user with the same equipment.

> "Thanks to a dedicated embedded software application, the mobile can be used as a remote control to activate the screen's functionalities."[18]

Embodiment of Locative Media

Confrontation and created friction in the urban community needs to be more in focus of locative projects. It will encourage discussion and further development. Locative Media often uses the individual Screen of the personal, spatially aware mobile device, when visualising the located data attached to the Urban Environment. This way, the urban physical surrounding is reduced to a background to the individual experience of the active participants. It is important to consider a wide range of audiences to provide a clearer embodiment of the information environments, which can be brought to urban space. The providing of visible traces of this data needs to be conducted in an accessible way to also include the "non-

---

[15] See <http://web.mit.edu/idg/cecut.html >(Sept. 2007)
[16] See <http://www.energie-passagen.de/englisch.html>(Sept. 2007)
[17] See <http://framework.v2.nl/archive/archive/node/work/.xslt/nodenr-154563>(Sept. 2007)
[18] See <http://www.urbanscreens.org/> (Sept. 2007)

participants," who nevertheless are part of the environment. Projects like *Murmur*[19] and *Yellow Arrow*[20] have used the classic medium of stickers and street signs. Chalk would be another option; to point out the access points to the location based digital layers.

One other possible approach to this kind of more integrated locative interface is to eventually use digital billboards in a reactive manner. These elements need only be glimpses of the actual content, yet can lead to the deeper level, should the user decide to follow. The project *PDPal*[21] by Julian Bleeker simply used the big screen at Times Square as information sign, luring people to take part in the experiment to map the area and create a dynamic portrait of personal urban experience around Times Square, mapping the area. Mobile and networked platforms were used as a mediating and recording device.

In *CitySpeak*,[22] part of the Mobile Digital Commons network, used the 'Ile sans fil' hotspots in Montreal. Participants were able, via their SMS and web-enabled cell phones or wireless PDAs, to send text to a common server. The text was combined with real-time geo-encoded data from the location and processed using *NextText* text visualization software. The resulting "ephemeral graffiti" was layered back onto the locations. Both personal mobile screens and very public (large-scale projections) were combined to create an experience of giving a creative voice to the citizens, which is set visibly into the data context of the location.

The aim of *Biomapping*[23] was to share personal biometric information to construct maps that visualise where in an urban space we, as a community, feel stressed and excited. In his next version Christian Nold wants to visualize corporative data accumulation on a big outdoor screen. This kind of window, revealing invisible data also helps the passively engaged audience to "re-imagine the city around them" and of course invites them to participate.

Integration in the urban design elements and location

Finally. I want to point out a few examples, which try especially to integrate the existing urban infrastructure: installations using interfaces, already familiar as "tools" to the urban community. Combined with attractive content, I might suggest that these considerations are a good strategy for encountering a new urban

---

[19] See <http://www.murmurtoronto.ca> (Sept. 2007)
[20] See<http://yellowarrow.net/v3/>(Sept. 2007)
[21] See <http://www.techkwondo.com/projects>(Sept. 2007)
[22] See <http://www.cspeak.net/station>(Sept. 2007)
[23] See <http://biomapping.net/>(Sept. 2007)

audience through their daily routines, an audience that is not necessarily trained in the reception of 'New Media Art'.

Hole in the Earth [24] by Maki Ueda created a sculpture in the middle of a new plaza reminding one of the traditional European well. But instead of looking into your face mirrored in the water you can see other faces at the other side of the world. A screen, camera and microphone integrated into the sculpture connect a small square near Central Station Rotterdam, Netherlands to a small square of the Islamic mosque Daarut Tauhiid in Bandung Indonesia. This installation in a similar way to the old wells became a meeting point for exchange, yet this time with a remote community (a repeatedly formulated fascination for visually connecting remote places, following the first classical interactive experiment in public space: *Hole in Space,* connecting people in Los Angeles with people in New York via two large screens.[25]

Two projects, *Chat Stops*[26] and *Phonetic Faces*[27] envisioned using Bus stop windows for the integration of interactive screens. *Chat Stops* by Rude Architecture were equipped with interactive video technology. Thus communication between waiting people at different bus stops was made possible. If one liked to, one could start a videoconference with waiting people somewhere else. *Phonetic Faces* allowed people to contribute their image to a shared display and collaborate with others to create a collage of images using their mobile phones. Jason Lewis from the above mentioned project *City Speak* envisioned using old ATM machines as interfaces for a connection of independent text input and local screening.

If we venture with any medium whatever into a city's public space, we still need to face the responsibility for the sustainability of our urban society. I see the outer public space as the glue that holds urban society together. So be prepared! What will happen if the OLED technology makes it possible to cover buildings with large flexile planes of moving images? It is time for artists to urgently shape the future development of the emergent "screen world," in which moving billboards start to detect our behaviour and adjust to our consumer preferences.

---

[24]See <http://www.ueda.nl/earth>

[25] HOLE-IN-SPACE was a Public Communication Sculpture. On a November evening in 1980 the Lincoln Center for the Performing Arts in New York City, and "The Broadway" department store located in the open air Shopping Center in Century City (LA) were connected by giant video conference screens. See <http://www.ecafe.com/getty/HIS/>

[26] See <http://www.rude-architecture.de/chatstop>

[27] See <http://www.mee.tcd.ie/~bruckerj/projects/phoneticfaces.html>

Bibliography

McQuire, Scott. "The Politics of Public Space in Theatre." *Urban Screens* 2005. <http://www.urbanscreens.org/us05.html>.

Virilio, Paul. "Architectural Design." *Architects in Cyberspace* II. 68.11/12, nov-dec 1998.

Wallace, Linda. "Screenworld" *material media, artefacts from a digital age* 2003. <http://www.machinehunger.com.au/phd>.

2.3 The Creative User

Future Physical: The Creative User and theme of response-ABILITY

Debbi Lander

This critical essay explores the role of the public user/audience as a CREATIVE USER: someone who combines creativity with consumption. It adopts a user focus and draws on the research, observations, analysis and recommendations published in the report and DVD publication *The Creative User—the new market for interactive digital art*

Introduction

'To create today is to create with response ABILITY'

'Between stimulus and response is a space. In this space lies our freedom to choose our response' [Covey]

The explosion in digital technology has generated a multitude of digital artworks, and widespread use of digital technology has led to a critical appreciation of those artworks. Such artworks invite the consumer to engage in processes of interaction and creation, which shift the role of the consumer from an observer to a CREATIVE USER. A CREATIVE USER is required to act upon an environment and/or other people, determine right action, proactively choose a path and make decisions consciously and knowledgeably. But does interactive digital art respond to the creative interests and behaviour of the user? Does the user respond to the creative potentials of interactive digital art?

The concept of a Creative User is at the heart of development of multi-sited networked events and telemetric performances, responsive installations and environments, 3D interactive worlds and web/online games. At the centre of this practice is a focus on the user interface and the challenge of involving the public in creative and participatory processes. As Andrew Ward, Technology Director of *Future Physical* explains:

> "Art can be described as a process of communication within a medium. But by adding interactivity to art, you change the flow of that process where, originally it was asymmetrical: it now becomes symmetrical. At the same time the role of the user fundamentally changes, setting in motion a dynamic loop between the user and the end product. This concept is of central importance to all interaction designers."

With the development of ubiquitous computing, mobile technologies and wireless applications, attention has moved away from the technological tools themselves to the value of the experience they can offer. How do they enable us to create and interact? How can they enhance that intangible thing, our creativity?

Future Physical—placing the body at the centre of digital interaction

In 2001, Shinkansen was awarded a commission from the Arts Council of England East to design and produce the *Future Physical* initiative as part of a UK regional strategy to advance the development, profile and public awareness of arts and technology practice.[1]

The *Future Physical* focus on physical interactions with digital technologies provided a rich context in which to explore the role of the public/audience as a creative participant. The programme was designed as a human adventure for public and professionals alike and commissioned the development of new public installations, performances, exhibitions and online environments linked to four research areas: Wearable Computing/Smart Textiles, Bio-technology, Eco-Technology and Responsive Environments.

Future Physical presented some 70 events encompassing commissions, international co-productions and network exchange events were presented in a variety of different Locations and sites throughout the East of England, engaging over 59,000 people in a human adventure exploring future visions of the human body's interactivity with digital tools. Can technology give us a rich experience of life and human interaction? What is natural? What is live? What is authentic? Artistic director, Ghislaine Boddington explains the context behind the *Future Physical* focus on the body and digital interaction:

> "In arts and culture many experiments are taking place internationally to smooth out the physical and psychological awkwardness of the machine/human interface and to expand the creative possibilities. Software systems pre determine boundaries of usage and in artistic works, these boundaries are set by the artistic-creators, on stages and in installations for use by the performers or the public. The use of the word 'interactive' is to be questioned when linked to the intent of the artistic creators and the potential of digital technologies. Other terms such as relational and reactive are also in discussion. The immersion of the body in

---

[1] Future Physical was a cultural programme co-conceived by Shinkansen and Arts Council East, produced by shinkansen/Future Physical Ltd for Arts Council East. See www.futurephysical.org (archive website). shinkansen is an artist led research and production unit based in London and specialising in the performative use of digital technologies. See www.shinkansen.co.uk (archive website)

interactivity is a very different thing to the button pushing interaction of the last 30 years; the body needs to be touched by the experience in order that sensibilities, perceptions and responses are expanded. In staged space, the creator relies on the human intelligence of the performer to supply the surprise that we need from the interactive response. The focus is on that sense of liveness, which can be intensified by digital tools to give a kind of added excitement."[2]

Within this context, *Future Physical* wanted to involve the public/audience in new ways of behaving and learning through augmented play activities engaging the user in:

Evolved sensory perceptions
Extended physical and emotional interactions with digital tools
Expanded concepts of identity and consciousness
Virtual/physical connectivity and networking

As stated in the artistic brief for the *Future Physical* programme:

When the public takes on a creative and participatory role in the work, many new issues need to be addressed. Each member of the public comes to the experience with a unique history of gesture, movement, language and sensory experience. The public enters with an intent and expectation, which are solely their own, from internal predictions, which set certain expectations. They need to feel sufficiently confident to be creatively playful in the environment. Therefore, one-shop interaction and immediate gratification are expected; familiarity and recognisable navigation methods are helpful. Yet the expanded, unexpected experience is wanted. How advanced are artists at meeting these public needs? How do you cater for a wide range of people entering the environment with a wide range of skills? Can the user interface become the expressive arena of the work?[3]

Creative User Research—exploring the role of the public/audience as a Creative User

Future Physical integrated user research into the design and development of its programme to explore and examine the role of the public/audience as a Creative User: someone who combines creativity with consumption. Creative User Research

---

[2] Interactive/ Response?—Virtual Incarnations 2001 Chat Room: Briefing by Ghislaine Boddington, Artistic Director of *Future Physical*, Institute of Contemporary Arts, October 2001. Virtual Incarnations is a Dance Umbrella/shinkansen/Institute of Contemporary Arts/*Future Physical* focus on the evolution of dance and choreography in the context of developments in digital technologies 2000-2002.

[3] See <www.futurephysical.org>

focused on six case studies[4] within the *Future Physical* programme, which afforded the opportunity to research different types of interface and interaction designs and different types of users in different geographical Locations and venues in the UK.

Creative User Research was undertaken across 2001-2004 by a multidisciplinary team involving case study artists, specialist researchers from the fields of computer games, dance/movement research and audience development and, video documenters and photographers. The research had a focus on identifying commonalities found amongst users of the six case studies in relation to four main lines of enquiry: Role of the User, Interface and Interaction, Desires and Rewards and Information and Communication.[5]

Case Studies

The two case studies selected from the Creative User Research for this essay relate to the emergent use of mobile and wire-free technologies within interactive digital art and illustrate different approaches to creative user-participation within digital installation and exhibition environments. And, in common with all six case studies in the Creative User Research, the projects focus on the development of social, multi person activities using motion and movement as content and interface and are illustrative of extending the principle of collaboration in performance and visual based media practice to include the audience or end user. The goal is the engagement of a Creative User, an audience that actively participates not just in interpreting the work but also, in shaping and creating it.[6]

txOom—experiments with sensing technologies, smart materials and wearables

txOom is a responsive environment and PlaySpace which invites the user to actively take part in transforming the environment, shaping the sound, playing with visuals, tangling in textiles and thereby moulding their own experiences. Created by

---

[4] The six case studies involved in the Creative User Research were: *Cellbytes* – Telematic Performances – Ghislaine Boddington, Joseph Hyde, Armand Terruli, Leanne Bird, Vena Ramphal, Claire Keating, *Trajets* – Responsive Installation – Susan Kozel, Robb Lovell, Gretchen Schiller, Scott Wilson and Jonny Clark; txOom, Responsive Environment – *FoAM*, *Texterritory* – Interactive Performance – Sheron Wrey and Fleeta Siegal; Wear Me Magna – Exhibition (various exhibitors) and Fashion show – Ghislaine Boddington Tom Betts, Leanne Bird and Armand Terruli; *Coast Digital* – Public Walk and 3D Digital Environment – Masaki Fujihata and Takeshi Kawashima.

[5] See note 192 above.

[6] Creative User Research was commissioned by *Future Physical* for the *Future Physical* programme 2001-2004. Case Study fieldwork was financially supported by *Future Physical* (Shinkansen/Arts Council East), Arts Council England—New Audiences Programme (PlayGarden) and the Culture 2000 programme of the European Commission

an international group *f0AM* and comprised of media and intelligent system designers, bio/eco theorists, audio and visual artists, AI programmers and textile creators, tx0om utilises biomimetics, sensing/embedded technologies and smart materials to offer the user experiences in sensing and wearing media.

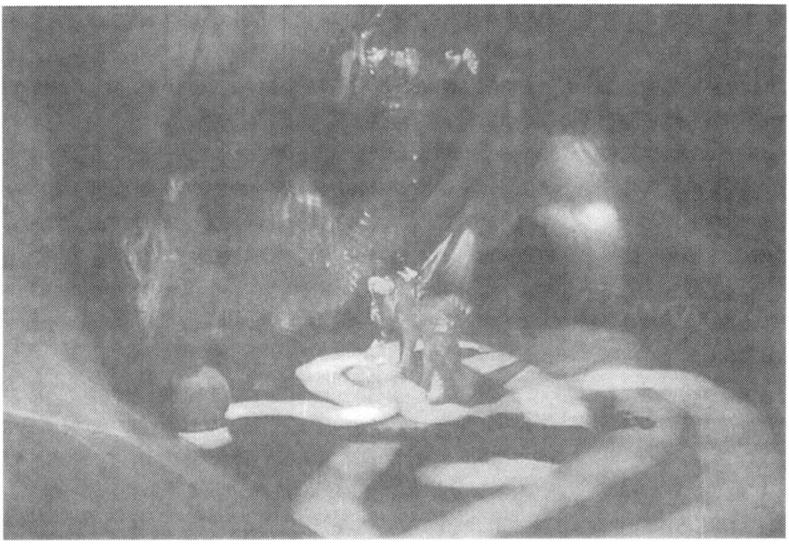

Figure 12: tx0om

As described by *f0AM*:

> "*tx0om* developed from T-Garden, with a tighter focus on materials, objects and spaces that opposed concepts of clothing and architecture as static and pre-defined structures. Within tx0om, we aimed to develop frameworks and prototypes of environments with an ability to adopt behaviours and properties similar to living organisms—spaces, objects, material and media that could be shaped by the activity happening within or around them."

The *tx0om* environment was sited in the historic and still active Hippodrome Circus Space behind the seafront in Great Yarmouth Town. The PlaySpace was composed of several wearable architectural elements, including wall shirts and swing dresses, functioning as shape able interfaces mediating experience between the user and the mixed reality of the PlaySpace. By wearing these garments, the participants literally wore the *tx0om* space, causing the physical environment to be reshaped by their collaborative activity within it. Each player was sensed by the PlaySpace using motion sensors and vision-tracking systems. The sound and visuals changed and evolved as the participants play together. *tx0om* PlaySpace catered for up to 5 or 6 users at hourly intervals. Users had their fortunes told while waiting and then were briefed and changed into white overalls before entering the PlaySpace.

tx0om was produced and presented as the main co-production for the WEAR ME!!! Network Exchange, linked to the wearable computing and smart textiles research strand of the *Future Physical* programme.[7] The *Wear Me* Network Exchange placed *tx0om* within the current context for wearable and sensor based media environment research taking place in the digital arts sector.

The development of the project involved local communities and national and international artists in six days of user testing of the system components and user experience. Specialist skills groups of movement, digital and textile artists of different levels of experience and educational groups of single mothers, adults with learning difficulties and young offenders participated. The aim was to enable a broad public to 'grow their own worlds' within the framework of the *tx0om's* action landscape.

Coast Digital—experiments with GPS, video cameras and stereoscopic projection

Coast Digital—Mersea Circles is a 3D digital artwork developed by Japanese Artist Masaki Fuijhata and Takeshi Kawashima. Its creation involved about 120 members of the public in the collection of media and data for the artwork using video cameras and GPS systems on group walks as public event (Coast Digital). The GPS systems told the users where they were on the earth's surface and the data collected was a series of half circles starting from East Mersea and ending in West Mersea. Video material of the shoreline, environment and people produced different perspectives on a particular place.

---

[7] *tx0om* was a co-production between FoAM VZW (Belgium), *Future Physical*/shinkansen (UK), Time's UP (Austria), Interactive Institute (Sweden) and KIBLA (Slovenia). Supported by: Culture 2000 framework of the European Commission, Flemish Ministry of Culture, Belgium, FOnds BKVB (The Netherlands), *Future Physical* and Arts Council of England (UK), CIEFFE srl, (Italy). The WEAR ME Network Exchange was a co-production between *Future Physical* (shinkansen/Arts Council East), Plan B, FoAM and Norwich Arts Centre in co-operation with Norwich School of Art and Design and Magna, with sponsorship from Xybernaught and funding from Culture 2000 Framework of the European Commission and Arts Council of England.

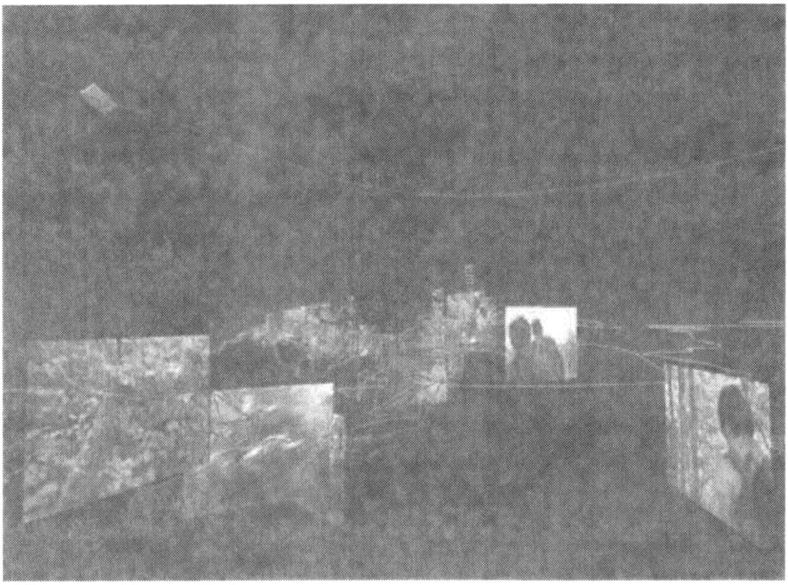

Figure 13: Mersea Circles

Masaki Fujihata explains his impetus and motivations for making *Coast Digital—Mersea Circles*:

> "One of the things that new media can do is allow us to reflect upon ourselves through processes of representation and retrieval. Depending on the particular structure of a medium, our perception or recognition of the world will be differentiated unconsciously. Mersea Circles was an experimental piece that attempted to share this process with the people of Colchester, drawing on their experiences: it is an archive of collective memory which can be shared."

The resulting artwork, entitled *Mersea Circles* presented an alternative map of the island, based on human exploration rather than physical features and true distance. The results interpreted the scale, speed and direction of human movement around Mersea Island. The artwork was presented in stereoscopic projection as an exhibition at FirstSite gallery in Colchester and explored the concept of stereoscopic perception. The user was required to wear 3D glasses to observe the projection, affording depth and distance in observation of media. Users could alter, by moving closer to and further away from the projection how they saw the observed images. Visitors to the exhibition could also interact with the documentation of the project in an adjacent process room through a mouse and a computer. A simple visual diagram gave visitors instructions on how to manipulate the documentation and invigilators were on hand to advise and demonstrate.

Coast Digital—Mersea Circles was produced and presented at Firstsite gallery in Colchester, UK as the main co-production for the ECO-TECH Network Exchange, linked to the eco-technology strand of the *Future Physical* programme within an exploration of the relationship between technology, nature and the environment.[8] The people involved in the collection of data on the public ramble spanned professional walkers, children and families, general public and invited video, digital and installation artists. The artists and production team briefed participants on how to use the equipment and what was required of them in terms of content.

User Analysis

Creative User analysis of the two case studies for this essay has a focus on one line of enquiry and explores and examines the role of the user in relation to participation in processes of creation and interaction.

Role of the User

Participation in *txOom* and *Coast Digital—Mersea Circles* involved the user in creative activity such as shifting images and shaping sounds, changing environments, manipulating media and textiles, playing with objects and gadgets, making shapes, creating physical movements, sensing space and time, triggering ideas and constructing mini performances and, interacting with people.

> ***TxOom:*** "It was like going in to another time and dimension…"; "The joy of pushing the dynamic door, understanding the play," "I could tell that us moving around was making a difference"; "… I wanted to play with all the objects to find out what they did and why they were there…"
>
> ***Coast Digital:*** 'When you have a camera, you look at things differently—it made us more playful"; "…I have been doing close ups of people—fantastic tales from Local area people, learning an awful lot…"; "'Used video and tried to film people

[8] COAST Digital was commissioned and produced by *Future Physical* (shinkansen/Arts Council East), firstsite and shinkansen for COAST, an Essex County Council initiative—Essex Visual and Media Arts with Commissions East, firstsite and *Future Physical* (shinkansen/Arts Council East) in association with Colchester, Rochford, Maldon and Tendring Councils and with Clacton County High School, Colbayns High School and Bishops Park College. The Eco-Tech Network Exchange was a co-production between *Future Physical* (shinkansen/Arts Council East), firstsite, Colchester Arts Centre and Signal

as they were walking along but was not adventurous with it" ;'I made solid camera position shots and some experimental shots.'

User participation can therefore be described as both a creative and social process, engaging the public/audience in activities, which are conceptual, practical and experiential. The means to undertake these activities is vested in the physical and the technological—in mental energy, physical activity, social interaction, media creation and in many cases, all types of input simultaneously. The engagement is therefore both a physical and an emotional one.

Creative User research undertaken amongst six case studies identified the role of the user as multifunctional and involving the participant in processes of observation, performance and creation. *txOom* and *Coast Digital* both give the user materials and tools to design, craft or make media at the primary content level. The type of content created is ultimately defined by the design decisions of developers: Users of txOom could make physical content and required using anything from small gestures to full body movements to activate audio and/or visual outputs. Participants in Coast Digital could make digital content in the form of audio and visual media and GPS data.

For participants of *txOom* it was clear to the user that the creation of content is at the heart of the process of participation and constitutes the 'art' experience.

txOom: '"I could not believe the responses I was creating. I felt part of the circus, like I was creating energy and putting it into the environment... you feel creative in the sense that you're making the sound and moving, or moving someone else. You ARE the interaction. You're making the atmosphere."

For participants of *Coast Digital*, the creation of content was seen as one part of the experience rather than as central.

Coast Digital: "The sense of your Location is an integral part of it ... "; "It was the walking I was interested in"; "... Feel a bit disjointed from what you are doing when using video camera ...It's more the idea that you are somehow, you are moving and GPS is recording your journey."

For participants of both artworks, the nature of the creation process was group based. Participation required sharing tools and collective control of media and its outputs. The group creation experience elicited both positive and negative user experiences:

txOom: "We nearly go there didn't we, when we all stood round in a circle, that was my favourite bit, suddenly you felt like you were together, you had figured something out' …

"It was not really easy to work as a group. You felt quite isolated and separated in this beautiful thing and that's really special" "…I saw the participants as part of the environment not a technology base trigger but as an environment playing on the participants" ; "…If you can clearly identify who is doing what you can interact more with others and play."

Coast Digital: "Thought we would have our own satellite devices and camera…. If we had our own GPS it would affect my behaviour… more of a group imprint than an individual imprint"; "Has pros and cons. Nice in the sense of group creation... Playful side is fine… immediate … sense of here and now is less that if you had your own GPS"; "We were all taking so much we lost track of where we were. Did not use video or GPS as left group behind"; "We got worried that we had left the gps behind, get into trouble. We felt we had rebelled against the system."

From the user viewpoint, *txOom* found more success with the group creation approach, largely because the environment was set up as playspace for groups to interact within. Walking along a coastline is a more individual experience as is holding hand-held devices.

In both artworks, the processes of observation and performance were also observable within the role of the User.

txOom: "I did not join in at first, I just watched people inside and I started to realise that the sound was changing' … 'Whilst being attached to the equipment, it felt like you were a performer'… I thought it looked silly if I was just hammering away at something. That wasn't exactly interactive. I supposed it would not have mattered if there wasn't anybody there."

Coast Digital: "I watched people filming each other and filming driftwood. Saw lots of interviews going on the walk, speaking to camera."; "…Did not behave differently from normal even knowing we were being mapped. We did try to be silly at some points"

Participation in both *txOom* and *Coast Digital* involved being watched by other users in the group and had a performative element where the user is required to 'wear' the media, be that sensors, GPS systems or video cameras. Observation remains an inherent part of the user experience, regardless of the level of interactivity on offer and serves as a form of adaption to the engagement or environment. Observation provides opportunities for the users to learn about the world they are in and/or the functionality of the interaction/product. The process of performance is very apparent when participation means engaging in an improvised human act, or at the very least, enabling elements of user improvisation.

The user interface employed by the *tx0om* development team required part or full body interaction on behalf of the user—the arm swing, the walk or wander, the jump and the swing. Here the immersion of the body in interactivity differs greatly from button pushing responsivity in terms of user input/output and experience. The type of interaction is active, requiring a high level of user input, reaction, response and output. Full or part body interaction was also required of the people who took part in the development phase of Coast Digital, with interaction facilitated by walking along a coastline with video and GPS devices.

Conversely, the user interface developed by Fujihata for the presentation of the 3D artwork *Mersea Circles* required a low mode of user input, reaction, response and output. The audience could observe the projection through 3D glasses and by moving closer to and away from the projection; they could alter how the images were seen. This type of interaction remains close to the traditional ways of perceiving and experiencing artistic creations. The user views an interaction product but does not have control of the media. The mechanics of stereoscopic projection afforded the viewer dynamic motion in observation and linked attention and observational behaviours of the user. More scope for user interaction was provided in the documentation room adjacent to the exhibition. Visitors could interact with the documentation through a mouse and computer which enabled the user to shape, move, alter and manipulate the digital content and form to varying degrees preset by design/order decisions through the use of simple finger work.

Involving the user in processes of creation and interaction elicits a diverse range of somewhat conflicting and contradictory responses from the public/audience. User feedback gained through vox-pop interviews and group discussions around *tx0om* and *Coast Digital: Mersea Circles* ranged across a wide spectrum from like to dislike, clarity to confusion, frustration to expression and unknowingness to new knowledge. Users described their experiences as 'magical', 'inspiring', 'soothing', 'energetic' and 'playful' and at the same time referred to their experiences as 'frustrating', 'disorientating'. 'unpredictable, 'unfamiliar' 'confusing' and even 'scary'.

*Future Physical*s analysis of the nature and scope of creative participation within tx0om and Coast Digital identified some of the key user issues and requirements associated with public participation in processes of creation and interaction.

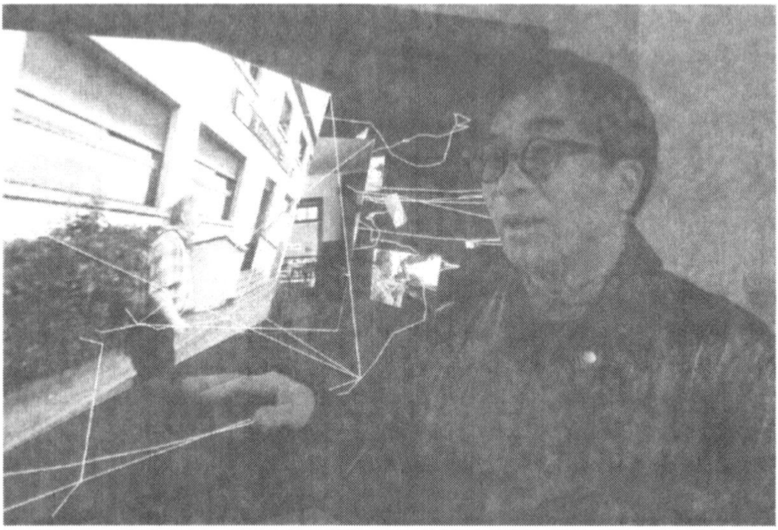

Figure 14: Masaki Fujihata and Mersea Circles

Key Role requirements

User skills and abilities: Participation requires the user to display and use specific abilities and skills which include imagination, intuition, confidence, risk taking, adventurous spirit, open mind, experimental attitude to experience, appropriate levels of intention and dedication, basis level improvisation skills and technical familiarity. One could distinguish users who venture towards exploration from users who convey to know the rules beforehand.

Time and duration within the media/environment

Time spent with the technology and in interaction is key to a rewarding experience and development of quality aesthetic output. More time spent widens creative scope through facilitating deeper levels of engagement. Rehearsal and returning to an interactive experience a second time aids understanding of role, use and impact.

> "You were trying to achieve all the layers, what happened though was that people were still having trouble with the first step, people needed more time. We did try to be silly at some points—it made us more playful but more time to be creative would have been nice..."

Safe Conditions:

The act of creating with media is an unfamiliar one to most users and issues of self-consciousness can function as a barrier to engagement. Users need to feel sufficiently safe and comfortable to participate fully in the role of a Creative User. Soothing fear and anxiety is a key requirement. The role of the creator can also be seen to carry ethnical concerns for users. The act of controlling media brings responsibilities and these are felt by the user and not always welcomed. Users also need permission to move and to create as a means to break down preconceptions as to how the public interacts with artistic work:

> "Found camera distracting. Worried a lot about the camera rather than the experience of being there." ; "I was so taken in by the experience that I forgot I was scared."

Key Design Requirements

Functionality: The inflexibility of ready-made technologies to realise artist concepts has led many practitioners down the route of creating unique products for their work. Innovations in this area seek to extend creative usage and scope and one can observe a preference for bespoke technological development in the field. Work is often released to market before system functionality is stabilised which leads to an element of unpredictability to interactivity. The unpredictable nature of interactivity leads to considerable variation sin user behaviour. Robustness and coherence are the base line requirements for functionality. Users want easiness of response/interaction, control of media and responsivity and to make decisions and be proactive.

Transparency: There is a general use view that interactive artworks have more depth that initially thought or than can be accessed during a first time experience. There is a common lack of understanding among users of all types as to why interactive systems fail to work. There is also a lack of awareness of appropriate navigation pathways when they are left for the user to fully figure out or explore. Users do not get it immediately. Users want something to happen, giving a sense of connection between themselves and the media/environment, noticeable changes and a response they can see/feel/know/understand and to see the personal impact of their actions.

> "The whole thing was quite confusing: constantly wondering what was going on the computer screens"; "My configuration range of interaction was not clear to me."

Media Choice: offering the users choices within interaction accommodates user diversity and provides the user with the means to make decisions. Catering to multiple viewpoints is a key challenge within the field of interface design. Users want a range of levels of interaction, a number of objects/toys to play with, a balance between too much and tool little to do and between abstract and concrete dimensions, the chance to interact with people as well as with media and the choice to observe or create. "It would have been nice to play with an image, play with a sound, change the environment, more precisely enjoy that manipulation"; "… If there is more things going on at once, you will be more encouraged to play."

Stimuli: something also needs to happen to initiate user participation in processes of interaction and creation. "If you are creating the narrative then you should really get that clear response back otherwise you lose interest"; "You want immediacy. You want to feel an immediate response." Varied stimuli, increasing the possibilities of interactions between different stimuli and thus enabling the creation of change encounters, creative routes and unseen options, people dynamics and group proximity. "If you want to go overboard, you should bring smell into the space to excite more senses, the sensory experience is very immersive. It's like stepping out into another reality." and "Audio enhances the experience."

Conclusions

tx0om was a very ambitious project in terms of design, implementation and situation and as with any innovative work it has both its plus and minus points. Sensor data was erratic and the sensor devices themselves not as robust as they should, possibility could not have been. This limited the number of users who were truly interacting with the system, an open source component named Oz. This particular, vital piece was delayed in its completion and was not truly tested or integrated within the overall system. This singular fact affected the ability of users to see the correlation between their actions and the changes/evolution in the sound and visuals. The *Future Physical* team felt that the full potential of the system was not reached artistically or technologically. Mediating natural and technological interaction is clearly a fruitful arena of experimentation although use of the physical body as an interface remains outside of the comfort zone of most people. User behaviour illustrated the need for public training in techniques for increasing body awareness, listening, undertaking and shifting presence into different states. However in terms of bringing new experience to the public it was a great success with Local community groups indicating the most positive reactions which extended to hugging the team from the joy of the experience. Users did express a desire for more creative input that was afforded, with many wishing they could bring their own visuals and audio to input into the base system but, on balance, tx0om offered more creative control than most artworks although developers still maintained authorship of aesthetics at the primary level.

Coast Digital—Mersea Circles on the other hand illustrated that interaction does not have to be complex or expensive and that simple user interaction/reaction is effective. The immediate relationship is a very rich experience. The depth of engagement in processes of creation was however limited by the budget constraints of the project. Groups of 5 or 6 had to share a video camera and GPS system and most users stated that they would have liked to have their own equipment. The ability to take on the role of a creator as an individual seemed not to be beyond the capacity of the participant but to conceive of this role, as part of a group in terms of co-authorship was a more difficult concept to grasp. The final artwork afforded relatively little opportunity for the user to affect the nature of the work. Whilst the user was fully engaged in creating content they did not continue to be involved in the selection and editing of the material for the exhibition. At the final stage of exhibition, aesthetic control remained in the hands of the artist. However, the allowance of different experimental objectives in the type and form of interaction/response plus different interpretations of interface and interaction is critical.

Both *tx0om* and *Coast* are examples of user experiences, which are 'open' in the sense that they enable the user to change the 'game' and leave it changed. Users were offered the opportunity to take an active role in a creative process. Experiences like *tx0om* and *Coast Digital* can be seen to offer a different form of engagement that the standard arts experience. The creative potential inherent in each is significant but certain user requirements need to be more fully understood and met if this potential is to be fully explored.

Summary

The Creative User—a broad definition

At the most basic level, the Creative User can be defined as anyone whose participation in an interactive artwork, installation or performance involves a level of interaction that results in them creating any form of content that did not previously exist without their input. It can be argued that once the participation moves from complete passivity to demanding an active role, interacting becomes creative use. As long as the user is extending the artwork, installation or performance away from its base state, that user is creative. This is a very general definition but given that different levels of creative use exist, a prototype definition is required. Different types of interactive works also enable different levels of creativity. Once we accept that an audience, faced with an interactive work, has become a creative user, we can then move on to deeper, more detailed observation of that participation.

Future Physical's research identified three basic modes of participation within the role of the user: the user an Observer, as Performer and as Creator. In typical non-interactive works, audiences could be thought of as receivers—mirrors, reflecting their own interpretation of an experience and absorbing experience that they could not directly affect. Even the most basis of interactive experiences as illustrated by Masaki Fuijhata's exhibition allow users to react, albeit in defined, rigid ways—to affect an outcome, but with little room for true creativity. It is debatable whether reaction can be defined as creative use. A higher level of interaction as illustrated by tx0om allows the setting up of a symbiosis between user and interactive work. If that feedback is at all sustained, and then creative use emerges. When the level of interactivity allows users—whether the raw materials are their own which they have introduced into the interactive system or pre-exist within the work—to create content that did not exist without their input, the creation of a unique experience and therefore creative use is taking place.

Are you response-ABLE?

The concept of user and product 'response-ABLITY' is key when planning, designing and improving the interactive art experience—which places new responsibilities on both user and artist-developer. The interface must be functional and transparent, plus offer the user choices, while the user must understand the rules of play in practical and conceptual terms.

Artist-developers have the most complex role in the creation of interactive artworks and they, more than any other members of the team, must bear in mind the needs, diversity and preferences of creative users. There is a conflict between artistic theory and the needs of the Creative User. But is vital that the needs and diversity of creative users are integrated from the outset at as early a stage as is possible. Whilst user needs in relationship to aesthetics, media and form must be born in mind, it is also the role of artists to extend, enhance and go beyond comfort zones and challenge people. User analysis of *Tx0om* and *Coast Digital* illustrate the user challenges inherent in-group control of media and delivering individual reward at the level of mastery and input/output within artworks, which require the user to create content in collaboration with others as a group art experience.

Future Physical's experience of co-producing several interactive artworks highlighted that overambitious design combined with lack of funding for prototype field testing could lead to levels of interactivity which were considerably more basis that those originally envisaged. Budget levels prevent the hosting of interactive works for a sufficient length of time to enable developers to test the marrying of different technologies and systems and for users to develop the necessary familiarity to engage creativity with the work. Lack of prior explanation of or instruction in the works also hampered user abilities and therefore satisfaction with interaction experiences.

Future Physical is fully aware of how difficult it is for development teams to fulfil visions completely, both in terms of receiving the necessary funding, exposure and staging. Despite these limitations it is clear that emergence of interactive digital art requires convergence between the worlds of art and marketing at a deeper level that is currently the practice.

Bibliography

Covey, Stephen R. *Daily Reflections for Highly Effective People.* New York: Simon and Schuster, 1994.

Future Physical—See www.futurephysical.org (archive website).

Shinkansen—See www.shinkansen.co.uk (archive website)

'A Fracture in Reality': Networked Narratives as Imaginary Fields of Action and Dislocation

Andrea Zapp

"... by connecting our similar yet distributed activities in physical space on a global scale, our methods of connection between ourselves and information become as important as the information contained within the transmission. Digital information has a meaning and networks are not only for data ..."[1]

Work in Progress

"Networks are not only for data...," and they create a relationship between their users. The subtext of the above quote brings up one of the Net's most current issues, namely: that it should be understood not only as an instrument for the transfer and distribution of information, but also as an 'open resource' for a participatory order. The Net is a comparatively unique cosmos of invented identities, partakers, and accomplices in joint forces, hidden in the endless labyrinth of home pages, chatrooms, and communities. (Or, as Steve Dixon formulates it: 'The World Wide Web is a site of therapeutic catharsis-overload, and constitutes the largest theatre in the world, offering everyone 15 megabytes of fame.')

These cinematic and performative elements of the Net serve as an intriguing aesthetic motif to explore as a dramatic setup for user participation, and as a testing zone for innovative 'actor' or 'character' constellations. Moving on from solely Web- or browser-based works, I am now interested in approaching these issues by combining physical installation with online components. This creates a more compelling and multi-faceted field of communication made up of several channels [Rieser & Zapp 77-89].[2] As these formations often only provide the frame for unforeseen content, they have to be regarded as works in progress, in which individual user contributions appear first as arbitrary anecdotal fragments and then turn into footage for evaluation and documentation of an open-ended dynamic chronicle (formerly known as "story"). We are now looking at the synchronized effects of immediate real-time exchange, as opposed to information channelled through a post-production filter and editing procedure.

---

[1] Brucker-Cohen, Jonah. "Report from ISEA (International Symposium on Electronic Art) 2002." <http://www.noemalab.com/sections/ideas/ideas_articles/pdf/brucker_cohen_isea.pdf> (07.08.03)

[2] For further information on these earlier works in relation to a discussion about character, role-play, and mimesis, please see my article: "net.drama://myth/mimesis/mind_mapping" in *New Screen Media.*

This produces short-lived episodic and transient content, which is originated by the textual and visual dialogue between participants of the online and remote physical environment.

In this article, my arguments reflect my own practice of using technical, creative, and media-philosophical keywords to define, in a more visual way, the aesthetic factors of a networked narrative environment as a central point in media art.

Imagination and Illusion

(imaginary, adj.: fantasy, make-believe, made-up, unreal, invented, pretend)

One of my recent projects, *The Imaginary Hotel* (2002/03), consists of an installation in a public gallery linked to a Website.[3] Gallery and Website visitors alike were invited to play with the idea of occupying their ideal room. The installation was set up as an open platform, concentrating on two adjoining walls to simulate a stage for (inter)activity and to reflect the room spanning between inner and outer areas, between the real room and the network. The installation architecture replicated a characteristic hotel room, with its universal furniture, carpet, and appliances. The usual text display on the room TV was replaced by a computer application, which was accessed via the remote-control mouse provided on a coffee tray on the bed. The software offered a wide range of image material with which users could modify the appearance of the room. Using a simple drag-and-drop interface, visitors could alter the interior on one of the walls and refurbish their room with different furniture, wallpaper, or pictures; on the other wall, they could also transport the hotel to another location by choosing an alternative window view, made up from photographic images. Their final room design was transferred via the Net to the projectors and displayed in life-size on the walls.

At the same time, Internet users could interfere and occupy the room virtually from a remote location. The TV interface also appeared on the Web site, enabling remote visitors to alter the projection and add their own material. A search interface collected personal images from the hard disk to be displayed on the window wall and in the (real) picture frame on the furniture wall. They were then inserted into the database choice on the TV and Web site. A webcam was constantly streaming real-time video from the 'hotel', documenting the visual changes and communication. The Website further reinforced this virtual architecture by providing a hotel lobby with a chat lounge. From the lobby users were able to ring up the telephone in the 'real hotel' installation. A special Web-telephone interface makes it possible to send text

[3] The premiere took place in October 2002 at the Chapman Gallery in Salford, Manchester and was linked to the Cornerhouse, Manchester, and the Folly Gallery, Lancaster. It was supported by the Leverhulme Trust London, The Arts Council England/North West, and the University of Salford.

messages from the Web site to the gallery telephone, where they were translated into voice messages.

Given that today, a wide range of collaborative projects are shown on the Web, *The Imaginary Hotel* was remarkably popular during its premiere in October 2002. The data and log file statistics over the duration of ten days show over 2000 different wall designs in which 10,790 objects were placed. 378 images were uploaded to the database. 507 users registered themselves into the chat lounge (not to mention the ones who checked in anonymously), and many users returned to the hotel site several times. Geographically the hotel guests came from places as diverse as Barcelona, Geneva, Marienbad, Dortmund, Tokyo, Rotterdam, San Francisco, Melbourne, and Montreal, to mention but a few.

The Narrative Environment

The hotel metaphor was chosen because of a hotel's structure as an empty shell in which neutral rooms turn into personal hideaways for a certain period of time. A hotel as such stands for an anonymous social melting pot in a constant state of flux, similar to the digital network. The vacant room was successively shaped by its short-term inhabitants of various identities, like a blank canvas,. Real and virtual guests arrive, meet, and disappear from out of nowhere and leave their personal traces, reflecting the seamless border between physical and suggestive places of being. Accordingly, the walls functioned as projection screens in a direct and more transferred meaning: the contributed material demonstrated that the walls had become a piece of scenery and backdrop for the personal narrative, referring through the photo-realistic image to the personal background. However, the theatrical space now added a fictitious meaning to the found footage from the individual desktop. So far I have experimented in most of my works with this particular type of individual—or user-originated—imagery streaming into an installation or context frame to fill it with episodic or sporadic prototypes of story. Peter Weibel outlines the general allegorical value of found footage in a media artwork as follows:

> "… then what we have are media-oriented observations of a second order, in which visual culture as a whole is exposed as a ready-made object for analysis. Consequently, observation of the world gives way to the observation of communication. The unconscious character of the visual code becomes evident in a kind of symptomatic reading …" [Rieser & Zapp 49].[4]

---

[4] Weibel, Peter. "Narrated Theory: Multiple Projection and Multiple Narration (Past and Future)" In Rieser and Zapp

The viewer's persona and reflection in front of the computer screen seemed to fuse with the virtual substitute beyond it, so that in this way the picture frame in the installation itself turned into a figurative object comparable to a mirror (and the majority of the images sent were made of portraits or similar personalised contents). In an equal emblematic technique, the projection wall was used as a window into the outer—but here digital—world. The public installation setting of the *Hotel* project, with its surrounding invisible wiring and obscure, circulating information, highlighted the theatrical backdrop as a crucial principle. It became a gateway to the Net as the main source material, and the boundaries between the individual and dramatic domain became less apparent. The networked platform then hinted at a cinematic scenario in which the former 'actor' or 'performer' gave way to the participant's own visual vocabulary filling the screens. To support this analogy, *The Imaginary Hotel* was explicitly designed as a wide-open stage, framed by the carpet and accentuated in the two fragile walls, curtains, and props reminiscent of a shoddy TV set. Thus it plays with an evident façade for the imaginary and illusionary.

I Spy with My Little Eye

The work as an imminent model of a circular *cause and effect,* between the remote physical and digital arenas, of interference and distortion of the installation setting, is epitomised in the webcam image.

If the Net and the 'mirror world' of cyberspace are spatially abstract, web-cameras can be interpreted as mediating devices: points of contact between the virtual and the real, or the spatial 'anchors' in a placeless sea. Web-cameras open digital windows onto real scenes within the far-flung geography of the Internet. The networked computer enables the exchange of text-based information with distant persons or machines; web-cameras add to that a degree of real-time visual knowledge. As Garnet Hertz put it, web-cameras constitute an attempt "to re-introduce a physical sense of actual sight into the disembodied digital self" [Goldberg 27].[5]

Webcams and related surveillance technologies have become widespread tools in Net artistic contexts, out of dramatic or critical motivation, and have further led to numerous related voyeuristic media formats (such as Reality TV, Big Brother, et al.). These technologies have further inspired filmmakers to explore the secret and anonymous bonds between observer and observed as an intriguing theme of storytelling, such as, for instance, *The Truman Show* (Peter Weir, 1998) or *EDtv* (Ron Howard, 1999).

---

[5] Campanella, Thomas J. "Eden by Wire: Webcameras and the Telepresent Landscape"in Goldberg

Next to the first *fascinosum* of a webcam (the covert observation of an incident through the spy eye), there is a second: the possibility of interfering with a location from a distance, without facing direct consequences, but with the ability to witness the results. This acts as an important catalyst for suspense and anticipation, similar to graffiti sprayed from the other side of the world. Yet the use of a web-cam also holds dangers when the authenticity of the images or the credibility of the cause-and-effect model is absolutely compulsory for the narrative method. (Ken Goldberg, whose instructive publication on the telerobotic and related interfaces has been quoted here already, analyses in his introduction the question of truth and belief in the webcam image. He notes that it can be easily troubled, worked upon, and presented in a falsified time frame) [Goldberg 13].[6]

In an earlier project I experimented precisely with this aesthetic model of the ambiguity of the 'live' and therefore apparently trustable authentic image:

Little Sister—A 24-Hr Online Surveillance Soap—has been running online since April 2000 at <www.azapp.de/littlesister>. The work transfers the conventional configuration of the TV soap opera to the Internet by linking the TV soap's typical locations and meeting spots to public and private webcams. Maintained by their original providers, these web-cams are chosen rather randomly from the WWW, and therefore they don't come under the author's influence.

Little Sister generates via its simple circular interface an arbitrary and parallel user narrative, in which the action takes place in its own spaceless and timeless micro-cosmos, so that the web-cam unfolds a separate fictitious meaning, outside of its actual intention, in the associated invisible home page. This plot is strongly based on the objective of secretly observing a remote activity in real-time, like a Peeping Tom. However, at the same, I distorted this effect by weaving a few fake QuickTime movies into the narrative, which more or less explicitly show the expected shoplifting or shootout. I edited them in accordance with the time-delayed and bleached-out aesthetics of a live webcam shot to blend them into the main action. Additionally, there were quite a few cameras based on a five-minute refreshes or even longer periods, dripping images onto the screen so slowly that, in general, the work was playing far more with 'the possibility of incidents' in the looming black void between the images than with the actual real-time experience. This somehow led as a dramatic side-effect to a much more 'imagined narrative', quite contradictory in itself to the real visuals, as a kind of wishful thinking, which actually reinforced the obvious parody of a rather well-known TV show.

---

[6] "Introduction: The Unique Phenomenon of a Distance."

In *The Imaginary Hotel* the act-and-react pattern was a reliable one, however, since the streaming video camera allowed an instantaneous look into the space, offering immediate proof of the action that had been taken. This often provoked a spontaneous non-verbal communication between the real and virtual visitors, the ones inside the installation addressing directly to camera, answered by the others outside on the Web with a few spontaneous text lines sent to the window to make themselves noticeable as observers from the outer space. Consequently the project met Goldberg's plea for an honest, thus-functioning tele(robotic) network of live visuals to fulfil its artistic purpose as a user-driven story space:

Does it matter whether a telerobotic site is real or not? Perhaps not to the majority of casual Net surfers, but to those who spent enough time to care, to patiently interact with a purported telerobotic site, discovering the site to be a forgery can be as traumatic as the discovery by a museum curator of a forgery among one of the Rembrandts in the museum's permanent collection [Goldberg 14].

Excursus: The Three Fs—Form Follows Function

When defining the term 'environment', such a mixed-media or mixed-location technique concentrating on a physical stage design shows further interesting and current overlaps between the fields of architecture, design, and media arts. If architecture can be defined as 'structural design', then media art might be termed 'designed structures', leading to a simple formula in which 'a networked installation equals a corporeal room in a hybrid space leading to generic content.'

The installation architecture then serves as a screen for personalised visuals as described already; the conceptual idea providing the frame for a subjective lining of the space, this final outcome being wholly dependent on the user's subjective interpretation of the offered space. It leads to an individual definition of the function over the given form or, in other words, the architecture of the site then defines the later composition and dynamics of the narrative.

Modernist architects such as Walter Gropius, Mies van der Rohe and Le Corbusier embodied in their philosophy the architectural concept—and later catchphrase—of 'Form Follows Function' as a guideline for a strongly purpose-oriented, experimental design:

> "The formula 'form follows function' hides a remarkable claim, namely that function is something that precedes form, that it exists independently of form, that it is there before form emerges. Forms, in other words, can be said to follow functions only if we consider functions to be entities that precede, and predate, forms. Only then does it make sense to urge the designer to make functions his starting

points, to make forms follow functions. But is there really such a thing as a function that exists prior to form? No contemporary science, natural or social, uses the notion of function in the sense that functionalist designers and architects did. No matter whether these sciences deal with material objects or immaterial phenomena, the scientific notion of function always refers to what an existing object or phenomenon *does* within a certain context. Whether we wonder about the function of the heart in human physiology or the function of façades in a townscape, hearts and façades have to exist before anybody starts wondering about their functions. In both natural and social sciences, form predates function: the notion of function is born from observing existing forms or phenomena."[7]

But critics of the functionalist design philosophy argued that:

"In the functionalist architect's eyes, the forms he 'brought forth' had no addressee, and they were not to have one either: they were not aimed at any particular individuals, or any particular public—just as the forms of leaves of grass, or of a snail's house (a functionalist would argue) were not meant for any public. The visual articulation of buildings and products was in principle never meant to appeal, in any sense of the word, to those who were to use them, or to see them: functionalists conceived of their buildings and objects as natural expressions of 'functions', as a function of 'functions', so to speak."[8]

This rather radical approach to the form as the result of a purpose-ridden, growing function has been criticised as an egocentric strategy that neglects the user/inhabitant. However, the rigid concentration on an 'anti-form' concept as a starting point remains very interesting in this context. Especially in retrospect, the antecedents of the movement correspond with the current idea of a networked narrative architecture as an adaptable room of subjective interpretation and use (*The Imaginary Hotel* touches, slightly ironically, another modernist principle 'that the walls come first and the wallpaper and deco comes second'), stating that a reinforced objectivity of the space—or the *tabula rasa* of the vacant hotel room—leads to a more subjective comprehension of the actual piece:

---

[7] Michl, Jan. "Form Follows WHAT? The Modernist Notion of Function As a *Carte Blanche*." *1:50—Magazine of the Faculty of Architecture & Town Planning*, nr. 10. Technion, Israel Institute of Technology, Haifa, Israel, Winter 1995, pp. 31-20. <http://www.geocities.com/Athens/2360/jm-eng.fff-hai.html> (05.08.2003).
[8] Ibid.

The dictum 'form follows function' was coined by the American architect Louis Sullivan in his 1896 article: 'The Tall Office Building Artistically Considered'. In the article Sullivan presented his approach to the emerging building type he referred to as the 'tall office building', soon to be called the 'skyscraper'. In connection with arguing for his tripartite concept of skyscraper design and for the upward character of the structure, Sullivan claimed that his design was a 'natural' result of an all-pervading law.[9]

A Hole in Reality

As the 'networked narrative environment' serves as a processor for unpredicted action and readings through exchanges between physical and digital spheres, the space and time continuum as it is defined in media history is questioned as a surreal and timeless experience. The possible narrative in this has moved on from being a mode of message, documentation, and linear story axis to being a matter of circular context and matrix-like structure, accessibility, and open variables. Genre-specific media art theory has been laying claim to quite a few new terms throughout these debates.

In the well-known concept of the 'spatial montage', by which Lev Manovich opposes the 'sequential narrative' as a tradition of film language to the spatial split-screen arrangement of story segments, displayed in parallel and simultaneously on one visual surface. (Compare here once again the *Little Sister* interface or Mike Figgis' 2000 film *Timecode*, which shows a similar principle.) But the montage techniques, which I have been discussing so far as dispersed, spatial-geographical parameters to launch meaning and associations, become more apparent at the point where Manovich relates this model of synchronised action and perception to the network, while referring to an early vision by Michel Foucault:

> "We are now in the epoch of simultaneity: we are in the epoch of juxtaposition, the epoch of near and far, of the side by side, of the dispersed ... our experience of the world is less of a long life developing through time than that of a network that connects points and intersects with its own skein..." [Rieser & Zapp 72].[10]

The montage strategy in a telepresent setting is then classified by a sequence of networked geographical locations and different parallel fields of activity to produce a

---

[9] Ibid.

[10] Foucault, Michel. *Dits et Ecrits: Selections, vol.1.* New York: New Press, 1997. Quoted in: Manovich, Lev. "Spatial Computerisation and Film Language," M. Rieser, A. Zapp, 2002

variety of allegories, as described in the practical samples above. Or, in a more direct way, it also:

> "Allows the subject to control not just the situation but reality itself. Telepresence provides the ability to remotely manipulate physical reality in real time through its image. The body of a teleoperator is linked in real time to another location where it can act on the operator's behalf: repairing a space station, doing underwater excavation, or bombing a military base in Baghdad or Yugoslavia" [Goldberg 175].[11]

Slavoj Zizek identifies the virtual space as a topological structure, and so as a 'hole (or fracture) in reality', becoming the opening to the 'phantasmatic room' and supernatural universe in a kind of topological inverse or 'in itself turn' of reality:

> "No longer at home in the material world, in search of the other scene, however, it will stay forever virtual, a promise in itself, a floating anamorphotic shimmer, only accessible with a glance over the corner of one's eye." [109-10]

In addition to this, Söke Dinkla, in her debate of the viewer representation in an imaginary construct, specifies the 'floating' work of art:

"Part of the authorship transfers from the artist to the user in the floating work of art through the imaginative act. … the user feels the effects of his own behaviour, which is determined by social norms and rules, and which he is reluctant to put aside. ... In the cybernetic circle his own gaze, which is determined by social conventions, is thrown back at him and makes him realise that it is he who generates reality with his gaze. In the floating work of art the user becomes conscious that he is an accomplice in a fundamental sense. However, he only seemingly occupies an omnipotent position that allows him to control events, since he is always victim and perpetrator at the same time. In a web of relations he is only one of many controllers. This becomes particularly clear when the floating work of art is extended directly into the Internet. In this case it is possible to try out 'shifting personalities', which are capable of continually changing their appearances and adapting them to new conditions" [Rieser & Zapp 37-8].[12]

---

[11] Manovich, Lev. "To Lie and to Act: Potemkin's Villages, Cinema and Telepresence" in Goldberg

[12] Dinkla, Söke. "The Art of Narrative—Towards the Floating Work of Art." M. Rieser, A. Zapp

Everything Flows

Söke Dinkla curated the media art exhibition 'Connected Cities—Processes of Art in the Urban Network' (1999) which focussed on the former German coalmining region known as the 'Ruhrgebiet'. The commissions were spread out in a metropolis of urban sprawls and leftover industrial landscapes, the artistic views and experiments trying to discuss the radical social, cultural and architectural changes forced by the move from the industrial to the digital age.

Consequently my own project for this venue—*A Body of Water, Waschkaue Ewald/Schlägel und Eisen II*—created a very site-specific networked narrative. As the work was produced in collaboration with the artist Paul Sermon, we used the telematic interface technique he developed for most of his installations to date. These allowed the projection of a whole body of remote viewers through a network and video conferencing system into one interactive space of fictive-documentary imagery.

The installation was located simultaneously in the Wilhelm Lehmbruck Museum in Duisburg and the 'Waschkaue' (a local term for the miners' changing halls) at the disused colliery Ewald in Herten, a small neighbouring town. A video camera captured images of the audience in Duisburg, standing in front of a chroma-key blue backdrop, and the images were sent to Herten and mixed with a camera image of the audience there. The joint image was then projected onto a fine wall of water, sprayed from high-pressure showerheads, in the Waschkaue shower room. A camera captured an image of this water curtain and fed it to monitors in the other participating venues, so that all the networked participants were interacting with each other in the virtual water wall.

The water screen was located in the centre of the shower room and had two different image sequences projected onto it simultaneously from either side. Floating independently on the water, the two images did not overlap and appeared as completely different scenarios. The audience was able to walk around the water screen and experience the images changing from the colourful telematic link and interaction with Duisburg to the black-and-white documentary footage of former miners showering after work in the original Waschkaue.

The flowing medium formed the heart of the installation and metaphor; it transported the public interaction and at the same time it reflected the area, the Ruhrgebiet, as a pulsating web of rivers and waterways. All the visual and conceptual layers met here. On the one hand, the viewers were confronted by the new era, the interactive platform of networked communication, as a possible future. Yet on the other, they discovered the ghost-like shadows of the past miners showering in the water—a flashback to the abandoned space and its former working culture.

The Ruhrgebiet and its inhabitants have ever been characterised by a strong local profile, individualism, and pride in local traditions resulting from the working culture. The coalmine Ewald/Schlägel und Eisen II and its impressive shower and changing rooms have been closed down since 1997. Over a thousand miners in one shift were using the Waschkaue each day and it was once one of the largest mines in Europe, employing over 7,000 miners.

Conclusion

The local context plays an integral part in the 'networked narrative environment' in general. It raises the essential question about our representation in between the societies, the digital and the (corpo)real one. Linking the past or present to the future, the site-specific network not only connects mutual (historical) backgrounds in a shared environment, but even more so speculates about particular and definite effects of the growing data space on the local community. The disappearance of the individual and its 're-emergence' in the network, as experienced in *A Body of Water*, puts the ordering parameters and established patterns of societal identity into question.

Likewise, *The Imaginary Hotel* examines the global village itself and reviews the prototype of an already digitised, nomadic identity. Its network emphasizes the insignificance of location within a supposedly borderless existence, in which the individual room is populated while a personal history can be invented. The above models recapture a creative approach that presents networked technology as crucial artistic equipment: the open-ended configurations of a multi-channelled 'story' space implying different readings. Finally, they equally explore their surrounding terrain with the intent to fuse the existences, the digital and the real, into an ephemeral silhouette within a momentary illusion.

Epilogue

The crucial point is not only based on the fact that the human being is a 'zoon technikon' that puts artificial technical environments, its second nature, between itself and its raw natural environments. The concern is that the status of this 'second nature' is virtual in an irreducible way. To return to the example of the interface: 'virtual' is the room which we can see on the screen of the interface, this universe made out of signs and shiny images, through which we are free to surf. It's the universe that is projected onto that screen and that evokes the wrong impression of depth. In the moment, where we pass its threshold and glance on what really lies beyond the screen, we see nothing apart from useless digital devices. This phantasmatic scenery and the symbolic order are strictly correlative: there is no symbolic order without the phantasmatic room, no ideal order of the logos with the pseudo-material, 'virtual' other scenery, in which the phantasmagoric phenomena can appear. Or, as F.W.J. Schelling, a key figure of German Idealism, expressed it:

> "There is no spirit without ghosts, no truly spiritual universe of ideas without the obscene, ethereal, phantasmatic physicality of "spirits" (the undead, ghosts, vampires…)." [Zizek 110][13]

Bibliography

Campanella, Thomas J. "Eden by Wire: Webcameras and the Telepresent Landscape."

Dinkla, Söke. "The Art of Narrative—Towards the Floating Work of Art." *New Screen Media: Cinema/Art/Narrative*. Ed. M. Rieser and A. Zapp. London: BFI Publishing/Karlsruhe: ZKM, 2002.

Foucault, Michel. *Dits et Ecrits: Selections, vol. 1.* New York: New Press, 1997.

Goldberg, Ken, ed. *The Robot in the Garden: Telerobotics and Telepistemology in the Age of the Internet*. Cambridge/MA/London: The MIT Press, 2001.

Manovich, Lev. "Spatial Computerisation and Film Language." Rieser and Zapp, 2002.

– – –. "To Lie and to Act: Potemkin's Villages, Cinema and Telepresence." Goldberg, 2001.

Michl, Jan. "Form Follows WHAT? The Modernist Notion of Function As a *Carte Blanche*." *1:50—Magazine of the Faculty of Architecture & Town Planning*, nr. 10. Technion, Israel Institute of Technology, Haifa (Winter 1995).

Weibel, Peter. "Narrated Theory: Multiple Projection and Multiple Narration (Past and Future)." Rieser and Zapp, 2002.

Zapp, Andrea, "net.drama://myth/mimesis/mind_mapping." Rieser and Zapp, eds.

Zizek, Slavoj. "Die Virtualisierung des Herrn." *Wunschmaschine Welterfindung: Eine Geschichte der Technikvisionen seit dem 18. Jahrhundert*. Ed. Brigitte Felderer. Wien/New York: Springer Verlag, 1996.

---

[13] Author's translation

Section 3: Case Studies
3.1 Locative Art

What makes mediascapes compelling?
Insights from the Riot! 1831 case-study.

Josephine Reid and Richard Hull
Hewlett-Packard Research Laboratories

Introduction
The availability of affordable personal mobile devices with excellent media capabilities, combined with pervasive computing technologies and infrastructure such as GPS and wireless networks has enabled a new medium for creative expression to emerge around locative media. Mobile Bristol is a programme investigating how mobile devices and pervasive information technology can be used to enhance the ways in which residents and visitors experience and interact with their physical environment and with each other in urban and public spaces. The programme has created an authoring environment to enable the easy creation of mediascapes by various authors.[1] Mediascapes are context-sensitive combinations of digital media linked to the physical world and typically triggered by interaction through movement or gesture.

The *Mobile Bristol* authoring environment has been used to develop a range of location triggered mediascapes such as a *Ferry boat tour*, an educational game called *Savannah*[2] situated digital stories, a nature walk, *CitiTag*[3] and many others. What these mediascapes share is the same simple interaction paradigm of using movement to trigger different media so that "where you are is what you get." In this chapter we will discuss the role of the mobile audience in these locative mediascapes by drawing on analysis of public reaction to a rich mediascape called "Riot! 1831" deployed in a public field trial lasting three weeks.

The purpose of the field trial was to use the Mobile Bristol authoring tool to create a professional mediascape experience so that we can understand the capabilities of

---

[1] Mobile Bristol: <http://www.mobilebristol.com>

[2] Benford, S., Rowland, D., Flintham, M., Drozd, A., Hull, R., Reid, J., Morrison, J., Facer, K. "Life on the edge: supporting collaboration in location-based experiences" CHI 2005

[3] Dovey, J. and Fleuriot, C. Experiencing Mobile Media: Bristol Seminars Presented at PLAN:ICA (open-plan.org) Feb 2005

the new medium, what people liked and disliked and how we should shape the underlying technology to best deliver those values. We commissioned two writers, Liz Crow and Ralph Hoyte, to develop an interactive narrative for Queen's Square in Bristol. We chose Queens Square in Bristol because of its physical setting, grandeur and wealth of history. It is an open pedestrian square recently returned to its eighteenth century magnificence. The square is relatively quiet but open, light and safe. The brief to the writers was to develop a mediascape that was strongly associated with the square that would utilise the affordances of the new medium such as the ability for people to choose where they walk and thus the non-linear nature of the content that they hear. We will concentrate on the insights that we have, based on extensive analysis of the data from the field trial, on what makes a mediascape compelling.

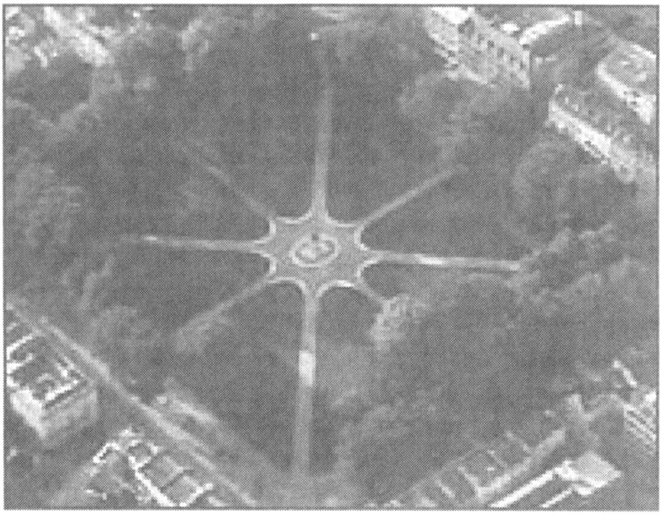

Figure 15: Queens Square

The role of the mobile audience in locative mediascapes

Riot! 1831 was set in Queens Square, a large open public square in Bristol. The mediascape was in the form of an interactive audio drama based on the actual riots that took place in 1831. During those riots thousands of protestors filled the square to protest against the defeat of the reform act. The final day was a massacre: mounted dragoons killed hundreds, and much of the Centre was destroyed. The mediascape experience was designed to take you back to 1831 and make you feel as if you were walking through the rioters and eavesdropping on their conversations. The participant or "audience member" wore headphones and a small backpack containing an iPAQ and a GPS unit and they walked around the square to trigger the sounds of the Riot. Thirty-four trigger regions covered the 150m wide Square, and associated with each region were three different sound files.

In these locative mediascapes, the audience were not simply passive consumers of the media; they interacted with the media through their actions and choices. They could choose how long to keep listening, where to walk and whether to interact with others in the environment or not. In this way they became both participants and actors to others in the environment. In a shared public environment, other people, plants, trees, street furniture, buildings and natural events—such as passing cars or barking dogs—all become part of the experience. The audience became part of a hybrid, parallel world that existed in both the physical and virtual environment.

Non-participants of the mediascape, who are also in the square, also form part of the backdrop to the experience and become an audience of the mediascape participants. They can choose to engage, ignore or rebuff the people who are participating in the mediascape. The social acceptance of mediascapes will have much to do with how they are designed and how they impact on the non-participant. Designs that do not factor in awareness of other activities in the physical space are likely to cause resentment by non-participants. Time-based challenges and games need to be especially aware of this danger, as the excitement of the challenge can reduce social and environmental awareness.

We will now report the headline results from the *Riot! 1831* field trial, which were very positive and then, go on to describe the reasons why people found it an enjoyable experience.

Results from the *Riot! 1831* Field Trial

Over 700 people tried out the experience. At the end of the trial we had 563 usable questionnaires, 531 trace file recordings of people's movements around the square, 30 semi-structured interviews and four in-depth ethnographic case studies. On returning their equipment, visitors were asked to rate six questions in a graphic rating scale format. Three questions asked about their experience of the mediascape. Scores from visitors were high for enjoyment, immersion and history coming alive with mean ratings of 74.5, 73.7 and 73.3 respectively on a scale ranging from 0 (not at all) to 100 (very much). There were also highly significant inter-correlations between the three questions making them powerful determinants of the overall experience (Table 2).

The strong interrelationship shows that immersion is strongly associated with enjoyment (and vice versa). We also infer that history coming alive is a special form of immersion that is related to the nature of the play and the historical setting. For example interviewees reported empathy with the people involved in the riots and a sense of walking in their footsteps:

"You can actually relate to this area and in some instances imagine that the people who were talking are in front of you because that was actually happening to people in 1831. You have to have that."

	Enjoyment	History
Immersion	$r = .65$, $df = 561$ $p < .001$	$r = .65$, $df = 561$ $p < .001$
Enjoyment		$r = .70$, $df = 562$ $p < .001$

Table 2. Correlations between immersion, enjoyment and history coming alive.

Our results show that *Riot! 1831* was indeed a compelling and enjoyable experience with many people spending up to an hour walking round the square, interacting and listening to the content. We now describe some of the particular aspects that make mediascapes work well.

Magic Moments

Magic moments are those moments in an overall experience, which are deemed to be particularly moving and memorable and thus are those that people really value.[4] In the *Riot! 1831* experience the factors and situations, which led to 'magic moments', were:

The phenomenon of walking through a sea of voices. Factors such as having your hands free, sounds automatically starting up, a constant background bed of sounds and the rich inter-weaving of the various sound files all helped to make this a rich new experience.

Unexpected connections between the physical and virtual worlds. Sounds or narrative that have a strong relationship with a spontaneous event in the real world are very powerful. For example if you are hearing about looters raiding the

---

[4] Hull, R., Clayton, B., and Melamed, T. Rapid Authoring of Mediascapes in UbiComp '04 Workshop 'Ubicomp in the Urban Frontier'

homes of the merchants and at that point you hear the sounds of building work going on.

The synaesthetic confusion caused when you are not sure if a sound is real or virtual. For example if you hear a dog barking or footsteps you are not sure if the dog is in the real world or the virtual world.

Seeing the world with new eyes. Many people reported a sense of history coming alive and seeing the square in new light *"I work in the building over there, so next time I look out the building I am not going to look out in the same way"*

Sharing a private moment in a public space. A magic moment can occur when the shared context evokes a gesture or feeling of social bonding giving a shared private moment within a public space.

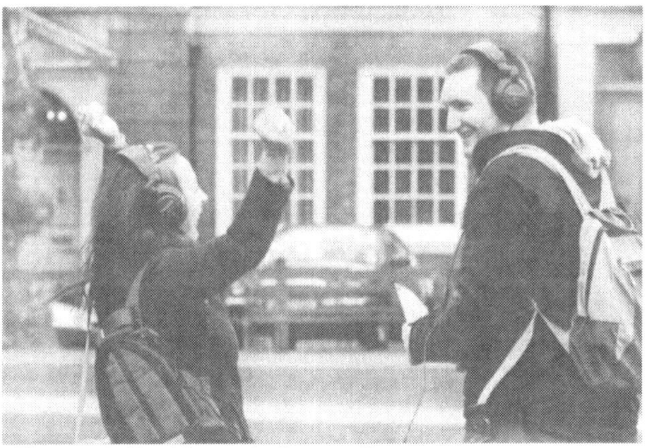

Figure 16: Participants in Queens Square

Social Bonding

To experience *Riot! 1831* it was necessary to wear headphones. The headphones that we lent were quite large and covered the whole ear, although they did not completely block out peripheral sounds. Obviously the use of headphones cuts down the ease with which people could communicate with one another. However it did not mean that the experience was completely unsocial. The headphones and the backpack provided a form of identification by which participants could notice each other. Some described this as being in a form of club because it was a loose bond between people you might encounter in the square who otherwise would be complete strangers:

"Yeh we were bumping into other people and they were smiling and saying oh try over there and that kind of thing. You know you are speaking to people you wouldn't necessarily speak to and that kind of thing. Or necessarily interact with."

Groups of people who tried the experience would glance and smile at each other as they went round and would sometimes swap tales of what they had each heard and talk about it afterwards.

Awareness and enjoyment

Mean enjoyment rating scores indicate a universally favourable response to the experience. However we did find that the way in which people became aware of *Riot* had an effect on how much they enjoyed the experience (Table 3).

Communication Mechanism	Mean Enjoyment Rating	Number of Respondents	Percentage
Newspaper	78	60	12%
TV	77	125	25.1%
Word of mouth	73	179	35.9%
Work/University	68	59	11.8%
Internet / email	70	38	7.6%
Saw it / poster	72	29	5.8%
Radio	85	9	1.8%

Table 3. Mean enjoyment ratings for how people heard of Riot!

People who found out about the event through work, university, word of mouth or internet, rated their enjoyment significantly lower than the people who had found out about *Riot!* via the newspaper, TV or radio. This finding may reflect the importance of awareness in building expectations of the experience. The newspaper, TV and radio coverage of the event provided a fairly accurate portrayal of the experience. For example in the TV news footage a reporter is shown wearing the backpack and walking round the square whilst the audio from the event is played. This provides an accurate portrayal of the user experience and expectations set can easily be matched. However because this is a new medium the language to describe the phenomenon has not yet emerged and so participants had difficulty conveying an accurate sense of the experience. This meant that people who heard about the experience through word of mouth might have had a confused idea of what to expect or even an "over-hyped" set of expectations about the experience. As the new medium becomes established we can expect to see new terminology and ways of describing mediascape experiences emerging out of the community of practice of mediascape design.[5]

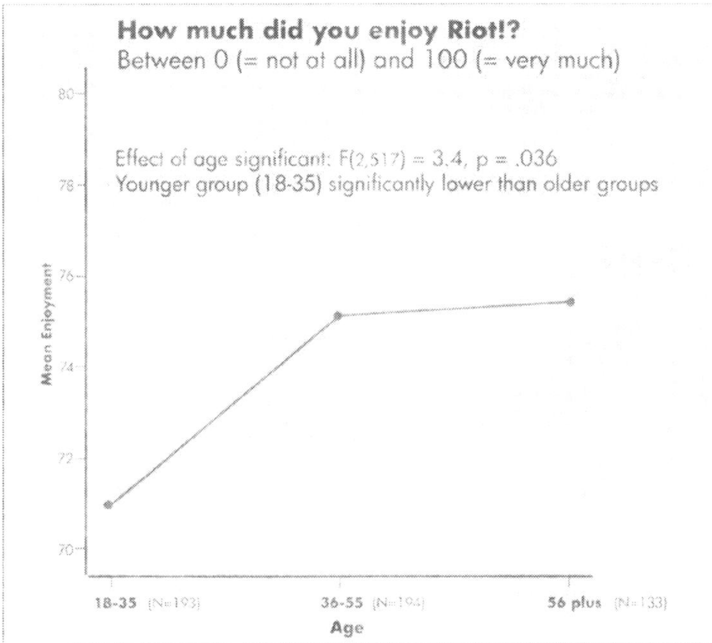

Figure 17: Mean enjoyment ratings by age group

We also found a significant difference in enjoyment ratings for different age groups with 18-35 year olds rating the experience lower than older age groups. There could be several factors that contribute to this:

People in that age group may be more familiar with technology mediated experiences and be more critical of "technical glitches" The content may not have been as appealing to that age group. They may have had expectations of more visual media.

---

[5] Reid, J., Geelhoed, E., Hull, R., Cater, K., Clayton, B. Parallel Worlds: Immersion in location-based experiences in *Proceedings of CHI* 2005

Movement round the square

Faced with a large square and little or no instruction as to where to walk many people felt very uncomfortable. As one man reflects "Well you are used to a set route there are always arrows, even if you go to IKEA they route you round." It took around ten minutes for people to get used to the phenomena of walking wherever they liked and triggering sound regions as they went. During those ten minutes they reported being very confused and unsure what this was about or where they were supposed to go:

> "I was a little bit confused at first. I don't think I really found my feet until I went into the centre. And then you really got the feel of it, looking at the square and hearing what went on."

From studying the trace files we identified four main categories of traversal shown in *Table 3*. A random walk where no clear pattern or structure was evident and then three structured categories: paths, circles and police march. Paths are structured from the physical layout whereas circles and police march were user-invented structures. It was noted that approximately 50% of the visitors traversed the square in a random manner whilst the other 50% showed some kind structure to their traversal, mainly that of walking on the paths.

In addition to being able to track the routes that people took; we could also analyse how much of the content they heard. Each region had up to three different sound files associated with it and so to hear all of the content a participant would need to revisit a region three times and listen to all of the content each time.

We inferred from the strategies that people took to walk round the square that people wanted to feel that they had heard a bit of everything. Many people spent over an hour in the experience with the average visit being three-quarters of an hour. *Figure 18* shows an analysis of one person's visit. We can see that the central regions, shaded light-blue, were visited more than once, where the majority of the content was heard.

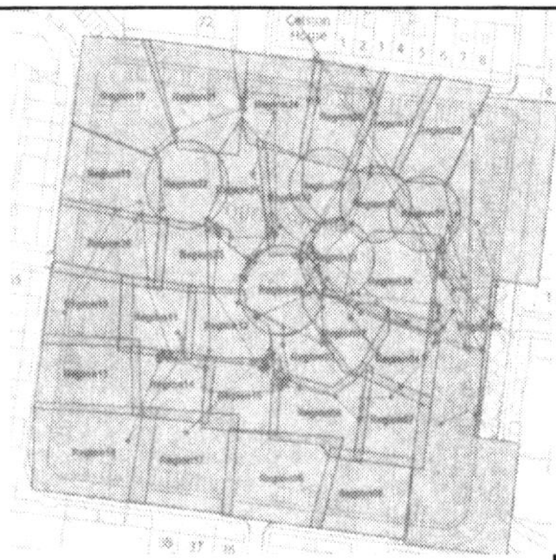	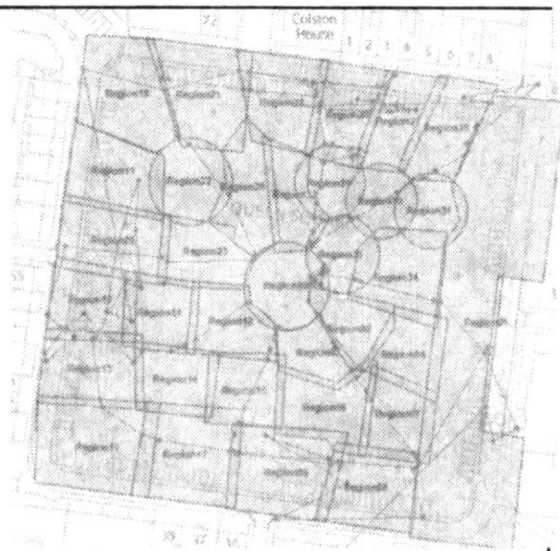
Random No clear pattern or structure to the route chosen.	Paths Follows the physical paths in the square, without moving on to the grass.
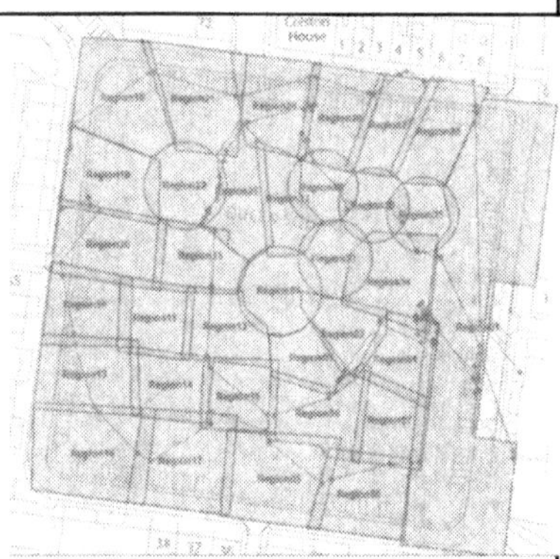	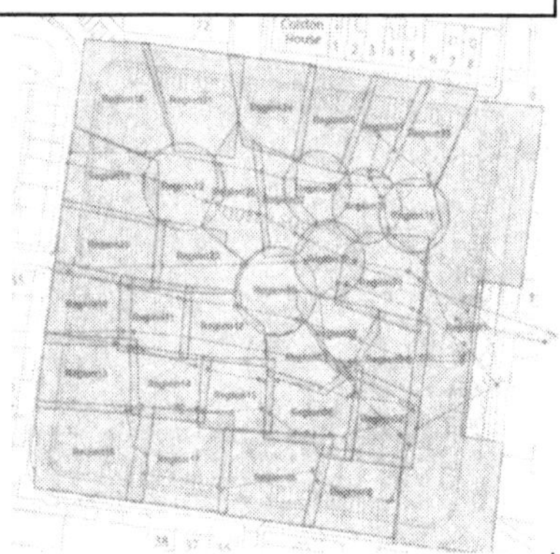
Circles In a circular traversal usually the outer perimeter is circled and then the person spirals inside walking around in increasingly smaller circles	Police March The police march traversal uses a criss-cross pattern that can be either horizontal or vertical from one side of the square to the other

Table 4. Four main categories of traversal of the square

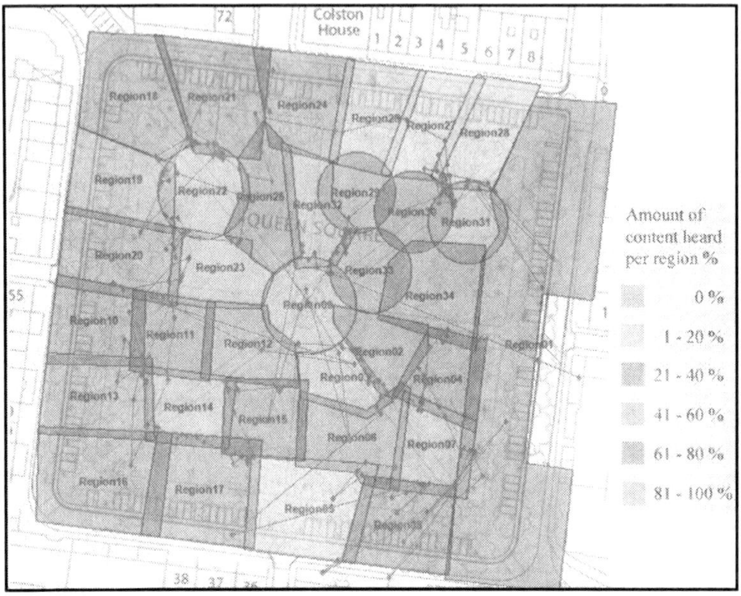

Figure 18. Regions visited and amount of content heard.

Immersion

Immersion is a powerful phenomenon that is frequently cited as a measure of success for games and virtual reality experiences. In locative mediascapes participants need to be aware of the environment so that they can move around the physical space and interact with others. Rather than thinking of immersion as a constant state we have observed that immersion is a transient state that can be fleeting or can last for several minutes. The circumstances that move people between immersive and non-immersive states are therefore an important consideration for future designs. In *Riot* the events that immersed people were when a powerful new sound started up, a familiar place was mentioned, a regional accent resonated, a familiar character was recognised or when a physical object in the present day could be related to the current event being described. Design for immersion in mediascapes should therefore concentrate on the mechanisms and flows that move the participant in and out of immersive states.

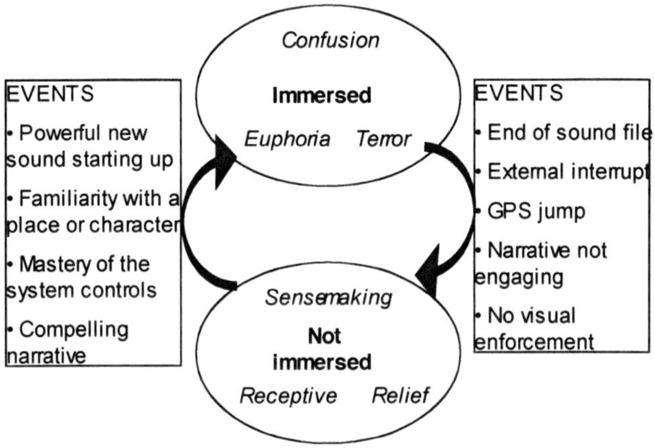

Figure 19. Model of flows in and out of immersion.

Conclusion

We have discussed the role of the mobile audience in the new medium of locative mediascapes. Participants in a mediascape are both audience to the designed content and actors in the experience of others. How people act and what external artefacts they need to carry or wear have an effect on the social cues and behaviours in experiencing mediascapes. We have described some of the factors and situations that lead to magic moments in locative mediascapes. We have also described how people move, socialise and become immersed in aspects of the mediascape. We have also found that the way in which people hear about an experience can have an impact on how much they enjoy it, with a supposition that the more representative the description is the better expectations can be matched.

The current version of the development framework, which produced *Riot! 1831*, has been released for free public download for non-commercial use and is available from the Mobile Bristol web site for anyone to experiment with. The insights that we have on what works well in the new medium and recommendations for design are published in Experience Design Guidelines.[6] We hope that a community of practice will arise around the use of the platform so that we can develop the language, design practice and insights gained through using the new medium.

---

[6] Reid, J., Hull, R., Cater, K., Fleuriot, C. Magic Moments in Situated mediascapes *Proceedings of ACE* 2005

Acknowledgements

We acknowledge the writers Liz Crow and Ralph Hoyte, the sound producer Armin Elsaesser, the actors and the student helpers and the Mobile Bristol team notably Kirsten Cater, Ben Clayton, Constance Fleuriot, Erik Geelhoed, Tom Melamed, Paul Marsh, Stuart Martin and Phil Stenton.

Bibliography

Benford, S., et al. "Life on the edge: supporting collaboration in location-based experiences." CHI 2005.

Dovey, J. and Fleuriot, C. *Experiencing Mobile Media: Bristol Seminars* Presented at PLAN:ICA (open-plan.org) Feb 2005.

Hull, R., Clayton, B., and Melamed, T. "Rapid Authoring of Mediascapes in UbiComp '04." Workshop *Ubicomp in the Urban Frontier*. Nottingham 2004.

Reid, J., et al. "Parallel Worlds: Immersion in Location-based Experiences." *Proceedings of CHI* 2005.

– – –. "Magic Moments in Situated Mediascapes." *Proceedings of ACE* 2005

– – –. "Experience Design Guidelines for Creating Situated Mediascapes." *HP Laboratories Technical report*.

Vogiazou, Y., and B. Raijmakers. "Urban space as a large-scale group playground in UbiComp '04." Workshop *Ubicomp in the Urban Frontier*. Nottingham 2004.

Hopstory: A study in place-based, historically inspired narrative.

Valentina Nisi and Glorianna Davenport

Why should we embed narrative in public space? Whether we realize it or not, we are creatures of place. Place provides us with cultural context; it situates our understanding of who we are, where have we been, and where are we going. As we navigate place, we are invited on a voyage that spans time and generations of cultural evolution. Our knowledge and sense of place grows through familiarity; and yet we can only ever have a fragmentary knowledge of what has gone before, what constitutes the national, regional and local history of a place. As our mobility increases, our familiarity with any particular place declines. More often than not we hurry through a place, traversing space not as an experience but as a mean to some other end. Soon we cease to imagine the texture of the past, present and future of lives lived, and the past becomes cloaked in forgetfulness. How can we make our passage through space meaningful for future generations? Can we combine our skills as storytellers and modern digital technologies to add spirit, atmosphere and past lives to a location? Can we augment a place with a sense of historical narrative?

MediaLabEurope suggested itself as an ideal venue for experiments in place based narrative. MediaLabEurope (2000-2004) occupied a renovated hopstore building that has been historically part of the Guinness brewery complex on James's street, Dublin 4. Situated in the heart of the *Liberties* area, the hopstore serves as an historical and anecdotal referential point of the neighbourhoods. However, the everyday life of MediaLabEurope was significantly different from the traditionally brewery business of the Guinness complex. Open 24 hours a day, the building hosted researchers exploring the potential of advanced technologies and visitors who came to the Dublin lab to taste the future. Every few months the lab would host a day-long event. On those days the researchers would demonstrate their current projects to the visitors. These situations provided the opportunity to implement and evaluate some challenging ideas about location embedded stories.

In this chapter, we look at *Hopstory*.[1] This project takes *cinematic* story beyond the computational desktop and distributes it in a physical space. In this process the physical navigation of the space is transformed into a blind-editing process; the

---

1 See <http://www.valentinanisi.com/hopstory.html> and <http://www.mf.media.mit.edu/pubs/conference/Hopstory.pdf>

audience member assembles her own version of the narrative based on her progress through the space, but without choosing any particulars of the story itself. The project explores how story content can be linked to the setting in which the story is being experienced, allowing the space to express history and personality. In this chapter we briefly detail this experimental narrative, offer some evaluation and highlight principles that may prove useful for future place based narratives.

Hopstory

Hopstory is a distributed cinematic story designed for the Guinness Hopstore of Dublin. The story recounts a day in the life of four fictional characters that might have occupied the brewery hop store building in the 1930's. HopStory is novel in the way in which it marries character point of view and physical navigation of the audience through the installation; furthermore, the progress of the story creates a relationship between character time and the time the audience takes to navigate the space.

The story is built around the movement of the four characters through the building; at each stage in the story any particular character has a limited view of the building space. This view parallels what the audience sees at the time that they collect a particular scene from one of a number of brewery 'cats' distributed within the building. In this was the 'cats', represented by cat sculptures, are special characters in the story, indicating the locations where the story content is available. Each visitor traverses the space according to his or her desire to do or see some other demos in the space. During this time the story progresses and as they pass the various cat sculptures they can collect the scene tags that are currently with that cat at that time. In this way, without consciously thinking about the story, each visitor creates a different edit of the story according to the timing and navigation of their journey through space. When they desire, each visitor can go to the designated playback area and playback the scenes in the order in which they have been collected. *Hopstory* pushes the boundaries of digital storytelling outside of the computer screen into the real space, bounding our stories to the architectural structure of the building where the facts are taking place, distributing the story in time and space.

1. Methodology for Story Creation

The Story Structure

In the Hopstory narrative, the day-in-the-life stories of each of four characters converged into a day-in-the-life story of this brewery building. Built up through a succession of incidents that occur to the different characters in different parts of the building over the course of a day, the characters finally converge and each witness a final incident according to their own perspective.

While designing the Hopstory narrative we started from the classical elements of setting, character and events over time (plot). We adapted this basic approach to a physically distributed structure where a real space coincides with the story setting and story time is mapped to real time. During the storyboarding process the story structure was represented in a three dimensional grid, where time was mapped to the X axis and space to the Y as shown in Fig. 27.

Figure 20: Grid illustrating the spatio-temporal construct of the Hopstory

Characters are moved through this grid according to their role and actions over the course of the day. Within the constraint of each character's appropriate activity, we invented the sequence of events to progress the overarching plot to a climactic event, which is perceived by each character based on their physical location in the space. This event, while perceived by all characters, each from their own physical perspective in the space, is only a driving scene for a couple of the characters.

As these scenes were distributed through the space, it was inevitable that there would be times in which no action happened in a particular location within the installation. For these times, the brewery cats served the role plot independent character and related an historically inspired anecdote. These fragments do not progress the plot, but add flavour and atmosphere to the story.

1.2 The Content

From the beginning, the intention of the story authors was to provide the audience with a flavor of a day in the brewery building, when it was an active industrial building. We began by investigating the history of the building that would host the installation. This research informed the choice of characters and the development of perspective-based stories for each of the four characters.

The four fictional characters, a foreman, his daughter, a boy, and a planner who works for the Guinness Trust generate four different journeys through the hop store during this normal workday. The characters actions are informed by personal accounts of social conditions in the Liberties, the neighborhood surrounding the brewery, and the lifestyles of workers at the brewery in 1920. These accounts, reinforced by our access to the physical surround for the neighbourhood today, provided a concrete starting point for scene visualization and plot events. For instance, the girl whose father worked at the brewery would have lived close by in one of the small houses that were built by Guinness for the workers; the short distance from home to work place meant that the girl was able to bring her father his forgotten lunch. The planner on the other hand might have visited several locations of the estate in the course of the day. He would have used a bicycle to get around. The cynical brewery cat character was created after a word-of-mouth anecdote referring to a cat that used to live in the building. All the filmed material was then treated in post-production to create a vintage atmosphere, as shown in Figure 21 below.

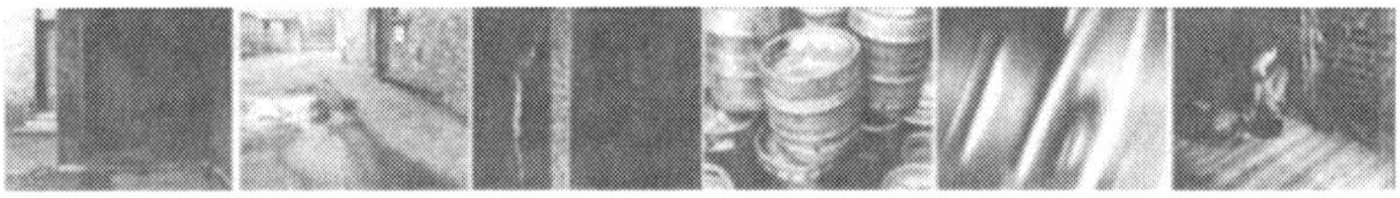

Figure 21: Selected frames from the from the Boy's character film footage.

In the story, plot events center around an accident caused by a combination of arbitrary actions from each character, providing many causal threads from which the audience builds connections. Each character enters the building at a different point in the day. Ambient scenes supplement the narrative, providing background historical information during the story times in which no character is present.

The narrative is composed of forty-eight story fragments that illustrate the lives of the four characters as they progress through their day. Ultimately, the story is experienced in the style of traditional cinema, through a linear succession of scenes played back on a screen. However, this final movie is assembled differently by each audience member, who interacts with the system, and each movie will possess a different number and combination of scenes, characters, or plot events.

The coherence of the audience experience therefore depends on how the scenes in their entirety are crafted. Story time always moves forward; this principle insures that no matter how many or how few scenes are collected story time will always progress. In addition to this principle, additional crafting helped the scenes flow from one to the other even as the ordering resulted in shifts in character perspective. A few of these techniques are discussed below.

Simple themes are referenced in multiple scenes to unify the character's stories. For example, a recurring theme is food. The foreman is very hungry throughout the whole story as he missed dinner the previous evening and forgot his lunchbox for the day. This hunger is mentioned repeatedly throughout his story, influencing his actions during the day. To connect to this theme, the other characters often refer to food and meals during the day, creating a common theme among the story fragments narrated by the different protagonists. For example, the foreman's daughter starts her day by bringing a lunchbox to her father.

The character's days are very different from each other. The foreman attends to the machinery while the foreman's daughter simply wanders through the building. Sometimes they can be seen at the same time in the same location or dealing with the same event from a different perspective, conveying the impression of a connected overall narrative. An example is the accident that happens in the Hop store in the late afternoon. Each character experiences it in some form, even if just auditory. This technique is used to emphasize the differences in perspectives among the characters regarding the same incidents. The accident involves the boy falling on the ground while climbing a stack of kegs and it can be experienced by the audience from each character's point of view: the boy first person perspective of falling from the stack; the foreman witnessing the accident and feeling guilty about having assigned the task to the boy; the foreman's daughter who is scared by the noise but curious about its source; and the surprised planner, who hears the noise and wonders what's happening on the other floors of the building.

The perception of the different story fragments as a whole is further facilitated by the presence of some characters in other character's scenes. For example, the boy arrives late into the hop store building. As he sneaks inside, he sees the foreman drinking a glass of beer with his men near a window. The viewer can see the men laughing from the boy's point of view. At the same time, the story experienced from the foreman's side does not mention the boy, because he cannot be seen. In another incident while hiding from the foreman behind a wall, the planner spots the boy. From the planner's point of view the encounter is of minimal importance, as he is completely absorbed in his work and barely notices the child. The boy would be just a shadow in the background for the planner, but from the boy's point of view, the incident is quite important. In the boy's scene, the planner walks by in front of him and stares at him for a second. The boy then puts a finger in front of his mouth in order to ask the planner not to tell the other workers of his whereabouts.

With these techniques, we attempt to ensure that an audience member receives a coherent and meaningful narrative regardless of the number and order of the scenes collected and independently of the presence or absence of any specific scene.

2 Technological implementation and The Hopstory Installation

2.1 The Technology

The initial development cycle for Hopstory was two months. At the time Bluetooth was not yet widely available. However *iButtons*, metal coins made by Dallas Semiconductor[1] to collect the story fragments provided a relatively inexpensive and easy to program technology that would allow us to realize the installation. Often used for security applications, *iButtons* have two parts—a small button or fob and small metal contact receptors. In the installation, receptors were embedded in each of the cardboard cat sculptures, and each audience member was given a fob.

This arrangement enabled the audience to collect and play back their story scenes through physical contact with the cat sculptures. By touching the fob to a receptor, contact is established and a small amount of digital information, in this case the two-digits scene ID, is transferred to the fob. This insured that the audience would be free to roam the space. Each receptor was connected to a local computer, which when the fob was touched, played a short audio segment—either from the current scene or, if no character was programmed to be at that spot, a random anecdote as perceived by the cat.

[1] Button Overview: <http://www.ibutton.com/ibuttons/index.html>

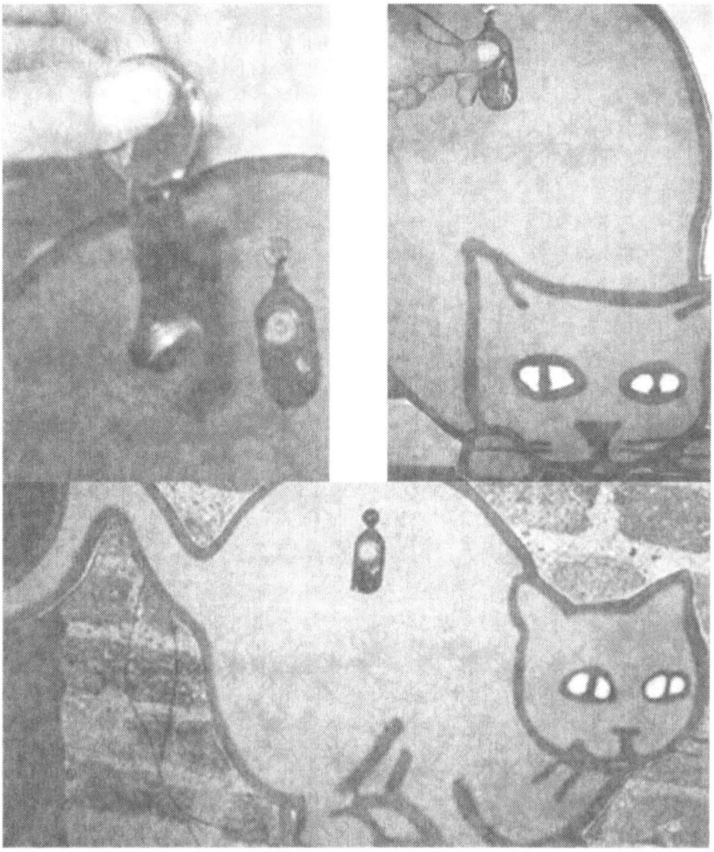

Figure 22: iButton fob and Hopstore cat sculpture with receptor

When the audience finished their journey, they could touch their fob to a final receptor and the list that they had assembled would be downloaded to the play back computer. The scenes were played back seamlessly by the playback application that was written in Isis.[2]

[2] Isis web site: <http://web.media.mit.edu/~stefan/isis>

2.2 The installation

As David Rokeby[3] points out in his essay, *Transforming Mirrors*, a story space can be conceived as a navigable structure or world:

"The navigable structure can be thought of as an articulation of a space, real, virtual or conceptual. The author structures the space with a sort of architecture, and provides a method of navigation. Each position within the conceptual space provides a point-of view … Exploring this structure presents the spectator with a series of views of the space and its contents. The sequence in which the spectator experiences these vistas forms a unique reading of that space."

In the case of the *Hopstory* the architectural metaphor can be taken literally, as audience members ordered their own scene list based on their own decisions about how to journey through the space. While using the fob was simple enough, some introduction to the installation was still required. We settled on the following scenario. As the visitor entered the space, she picks up a key ring with an attached i-Button and card, which contained the following instructions:

> "When you see a cat, approach it. Press your iButton against its silver disc for a few seconds. As the story progresses over time, you can visit any cat more than once. The story starts again every quarter hour. When ready to see your story, bring your iButton to the playback projection area. Press it onto the cat you find there to view your personal version of the Hopstory."

One important issue regarding location-based stories is whether fragments of the story should be played out when the visitor is at the space in question or whether all the scenes should be collected for later replay. In the case of *Hopstory*, we chose the latter solution, in the belief that if the story was offered as a fragmentary experience, the sense of the whole would have to compete with other information given out in demos and random conversations. Therefore, rather than viewing each scene in full at the moment it is encountered, the audience saves up their collections for later viewing, at the playback area. Only a short audio cue is played back on location, manifesting the presence of the character just encountered and reassuring the audience members that a story clip has been uploaded to their collection.

This solution turned out to be revealing, sociable and fun as in the playback area audience members are able to compare their assembled movie with those gathered

---

[3] Rokeby, D. Transforming Mirrors: Navigable Structures. <http://www.interlog.com/~drokeby/mirrorsnavig.html>

by others in the form of a public projection. As audience members compared story paths, they learned from other collections more about the character's experiences during a day in the hopstore. This style of viewing emphasizes the cinematic experience of narrative, and stimulates reflections upon the differences between public installations as a framework for collective cinematic experience.

3 Reactions to the first draft

Hopstory was demonstrated at an exhibition held at MediaLabEurope in January 2002. Throughout the day of the event as visitors entered, we handed them an instruction card and an iButton. Visitors freely roamed the floor, collected scenes, and returned to play them back at the designated playback station. As they returned, they shared with us their impressions and learned more about the installation.

In general, the audience was interested and curious about the distributed structure of the narrative; they found the concept of a location-based narrative fascinating. A variety of visual aids were used to explain how the story was distributed in space and time, as shown in *Figure 23* below.

Very few audience members had problems with the technology. The interaction was simple and most users were able to interact with the system with no difficulties. The audience enjoyed sculptures of the brewery cats; they were easy to find, appealing and even amusing.

The separation of story navigation and story viewing provoked mixed response among the users. While some had difficulty relating the collected fragments to the experience, others enjoyed the fragmented experience. They pointed out that not having to stop and view the story at every node created a less disruptive experience that contributed to a more coherent plot when retrieved at the end of their journey.

Different visitors collected story fragments in different ways. Some visitors did not realize that the story only lasted around 20 minutes and collected fragments only when they happened upon a cat sculpture throughout the day. When these visitors came to the playback area to cash in on their story, we usually had to refresh their understanding of how the experience worked to help them relate to their collected fragments. Other visitors focused fully on the Hopstory installation, actively searching for all the cats; these visitors collected story segments rapidly and then immediately came to the playback area to review the story they had collected while navigating the floor. They had a better understanding of the story experience. Few wanted to go back and experience more stories and encounter different characters as they had noticed from the public projection of others, that different story clips could be encountered. Finally, the visitors passing by the playback area stopped and watched movies that others had collected and wanted to talk about the demo.

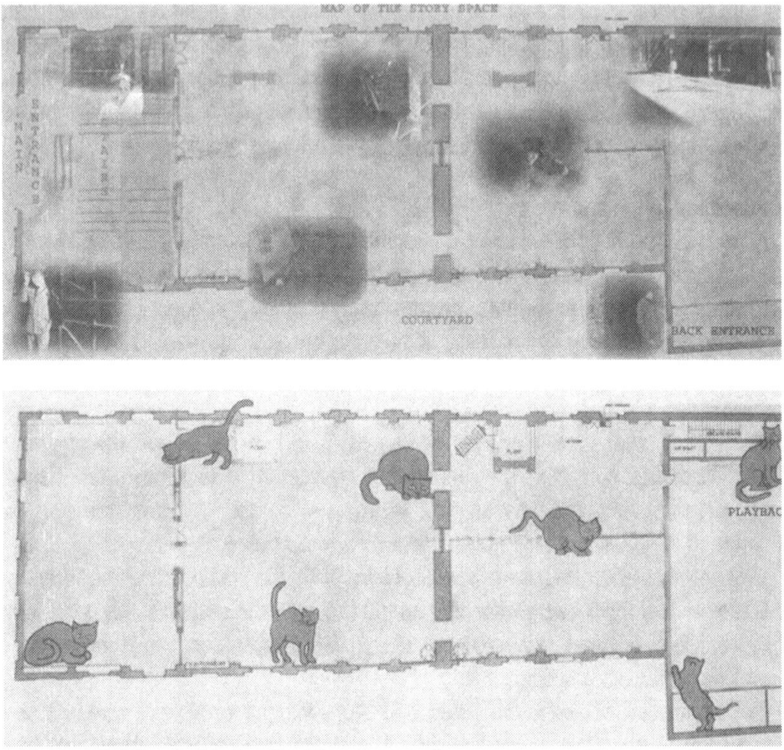

Figure 23: Visual aids for the Hopstory audience, representing the cat locations on the hopstore floor.

One advantage of using iButtons for interaction with Hopstory was the non-intrusive quality of this technology. iButtons are portable and undemanding, leaving the audience free to make choices about their level of engagement with the story at any given time. They could remain in a bag or pocket, easily accessed if a participant chose to engage with Hopstory. They stored story bits so that a conversation could be held during collection, and viewing could occur later when convenient.

With the Hopstory, we place the selective interactive act of the audience in the navigation and collection of story fragments before the viewing. Consequently each audience member is empowered with the agency and personal responsibility to explore the physical environment and by doing so collect story fragment in order to assemble a complete personalized narrative. The story experience comes in at a second stage. When the interaction with the location is concluded then and only then the final result can be viewed. Each participant's story will differ from those constructed by other audience members and each will be made public in at the projection station. This way, Hopstory while maintaining the characteristics of story sharing, is linearly presented and preserves a loosely arched structure. It also allows the audience to experience agency in navigating the story world and collecting personalized versions of the narrative before viewing takes place.

4 Modification for the next generation

A second version of the Hopstory was designed and implemented following shortcomings observed during the first installation event. A seamless interaction between the story and the audience was implemented using Bluetooth enabled iPaqs instead of the iButtons. Due to the radio range of Bluetooth, the content needed to be spread out through a broader space. The second version of Hopstory, therefore, was implemented across the three floors of the hopstore building. A map of the section of hopstore on the iPaq screen helped the audience in orienting themselves while exploring the building in search for the cats, shown in *Figure 24*.

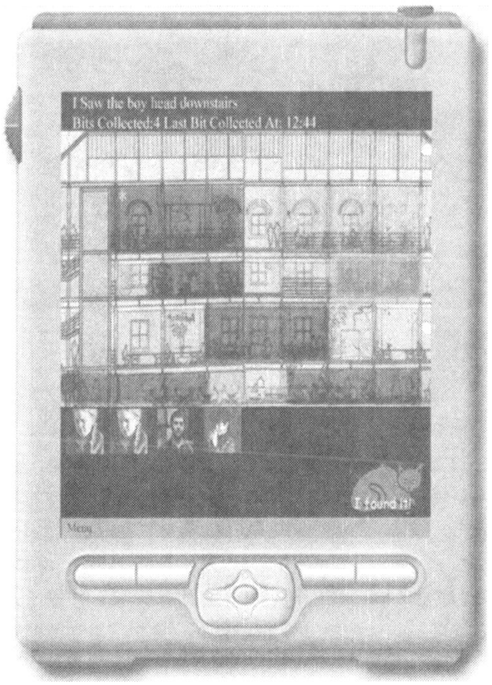

Figure 24: Graphical interface for Hopstory II

An audio signal is being used to draw attention to the location (The visual contact with the cat sculpture is not always happening as in Hopstory v.1). The cat meowing guides the audience in the direction of the cat sculpture. Once in range, an ad-hoc network is created via Bluetooth transmissions. This network allows the user to receive a different scene of the Hopstory at each device-cat interaction. While the story clip is downloaded on the device and an icon representing the character just encountered appears on the screen. The audience member is also given a text-based message on where this particular character is headed next, enabling participants to follow one character at the time. This method can possibly be used strengthen the perception of the story plot and maintain coherence of character perspective throughout the collected fragments.

In this implementation, the viewing of the story was still kept for the end of the journey. At the playback area the story clips were assembled in software as in version one and presented to the audience in a public screening projection, combining the potential of mobile wireless technologies with the public screening paradigm of traditional video-based narratives.

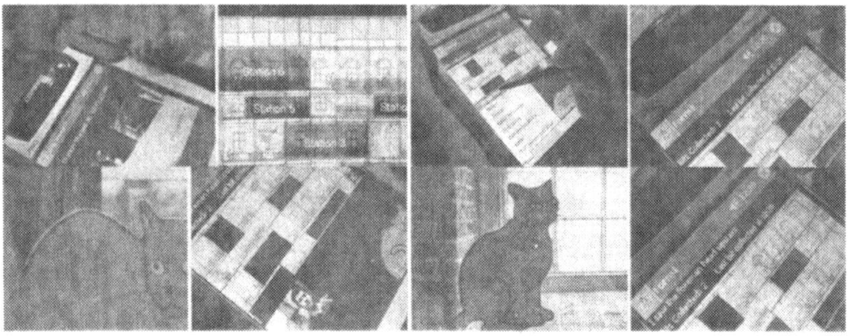

Figure 25: The Hopstory II installation

Acknowledgments

We'd like to thank our actors, actresses, and voices for Hopstory: Giovanni Dini, Susi Harris, Steven Davy, Colin Harris, Mark Bent, Felix the cat, Eamonn Doyle, Ronan Swift, Jenny Langley, William Hederman. Also, for help with the audio production and postproduction: Eamon Doyle and Herve Gomez. For software counsel: Stefan Agamanolis, Cian Cullinan, Jamie Rasmussen, Matt Karau and Ian Oackly. Linda Doyle for being part of the initial brainstorming processes of Hopstory 1. Sean Flanaghan who implemented Hopstory and Mads Haahr, for constant support and advice.

The Media Portrait of Liberties: A Non-linear Community Portrait

Valentina Nisi, Mads Haahr, Glorianna Davenport

1 Introduction

Why do people enjoy listening to and empathizing with stories? Stories are a common thread across cultures, a universal mechanism for sharing information about others and us, imparting lessons and having fun. Stories connect us to place and time, providing a significant portion of the cultural legacy that we pass on to future generations. Stories also serve as an important social function as a stimulus to discussion and debate.

Because of the importance of stories to culture, new technologies are rapidly adopted for the purpose of storytelling. Cinema is probably the most prominent example but there are others, such as computer games, many of which contain relatively sophisticated narratives. In the Western world, the popularity of such screen-based media appears to almost universally have surpassed that of live performance (e.g., theatre or oral storytelling). As a result, a significant amount of the stories told now occur in the two-dimensional world of the screen rather than in a 'real' environment. One may lament or celebrate this development, but it is hard to dispute. Today, another technological shift is taking place, which is having an impact on storytelling: technological advances in the areas of mobile and ubiquitous computing, location-tracking systems and wireless network technology are pushing stories beyond the screen and back into the real space.

This chapter describes the *Media Portrait of the Liberties* (MPL) project,[1] a hands-on investigation of a new, digitally mediated form of narrative that makes extensive use of mobile computing technology. MPL is an evolving collection of historically inspired stories drawn from written accounts of the rich inner city area in Dublin, Ireland known as 'the Liberties.' The objective of MPL is to provide viewers with a nuanced and evocative sense of place as they walk the streets of this striking neighborhood. The project also functions as catalyst for members of the Liberties community to contribute new stories, potentially enriching and evolving the portrait. Project development encompassed a number of diverse activities, including the collection and production of reenacted stories in form of video clips, each of which related to a particular location within the Liberties. In tandem with this effort, a handheld display system was developed. This system is location-aware and capable

---

[1] Media Portrait of the Liberties website: <http://www.tcd.ie/nisiv/research.html> (accessed on July 1st 2005.)

of delivering the video content mediated by this awareness. Essentially, the system only allows a video clip to be viewed when an audience member is co-located with the events depicted in that clip. Co-location could mean to stand in the same street as in which a family drama had once taken place or to look at the ruins of a once-grand building. In addition, an information-architecture was developed, such that the clips are arranged in a highly connected network. This structure allows viewers to experience them in a variety of orders, depending not only on the physical path they took through the city, but also by way of topics or plot elements that attracted their interest along the way. Plots were developed around multiple themes and characters by creating a series of narrative fragments that eventually intersected with one another. We believe this system provides an innovative and creative artistic artefact in the form of a unique community-focused story web, and that it is of interest to both local community members and to casual visitors to the area.

Methodology for Story Creation

Researching and Reconstructing the Stories

The MPL project is a collection of stories based on the life and history of a community of people living in the same neighborhood. Factually based stories appeal to us as because they are directly connected with what we perceive to be real events and history, such stories can provide us with inspiration for our own lives, prompting us to recollect anecdotes and memories. Like a spider web, real stories lead into each other; connect characters and themes providing a natural hyperlinked structure that can be used as the basis for an interactive modular narrative. While researching the content for the location-based narratives, we entered in contact with the local writer Maireen Johnston. Her book "Around the banks of Pimlico" [1985] provided us with the material that forms the basis for the scripted media fragments. In her book, local anecdotes and characters are described through a social and historically detailed narration; they render a warm and engaging picture of the Liberties area from the early eighteenth century until circa nineteen fifty. Leo Monhogan, a former local tourist guide, generously made available his collection of old photographs of the Liberties. Charlie Hammond, a member of the local Maryland Residents Association, offered papers, booklets and anecdotes on the history of the area. He told us stories about the river Poddle, a river that runs underground through the Liberties and was the main water supply for the city after the Liffey. It is why breweries and distilleries, laundries and weavers of silk and poplin flourished in the area producing a labour intensive and thriving industry. The enthusiasm and participation of these local people was a genuine sign of the desire of the community itself to share their sense of history through this project. "Social history is rich in this area. I believe is important for this community to understand their own neighborhood" commented Charlie.

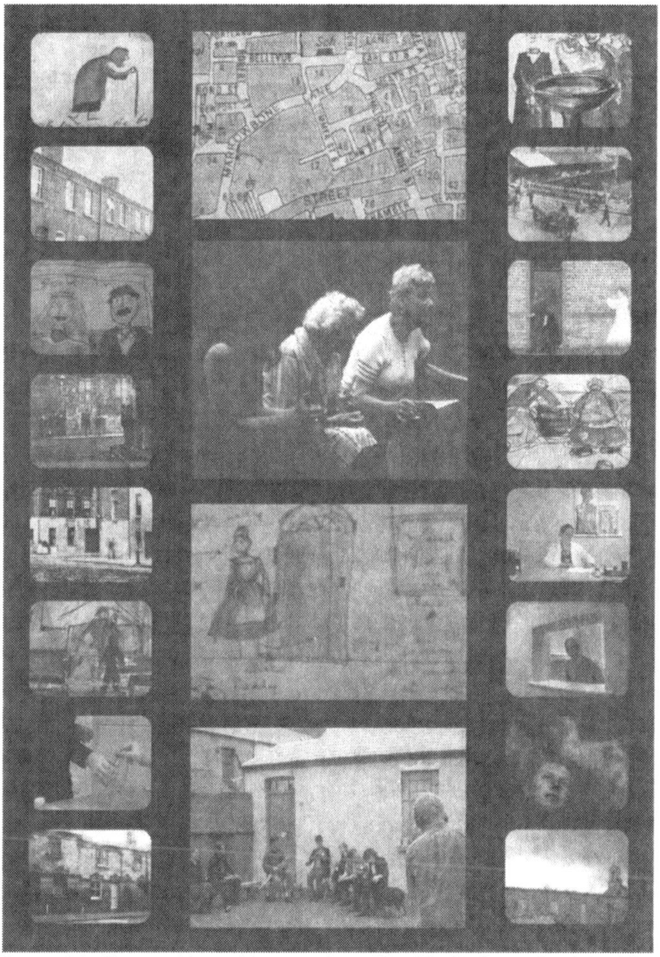

Figure 26: Composition of Images from The Media Portrait of the Liberties

An initial set of twenty short stories was selected from Johnston's book for production. These ranged from Johnston's family stories about moving around the neighborhood to socio-historical anecdotes relating, for instance, the origins of the area and its ancient Abbey dedicated to St. Thomas.

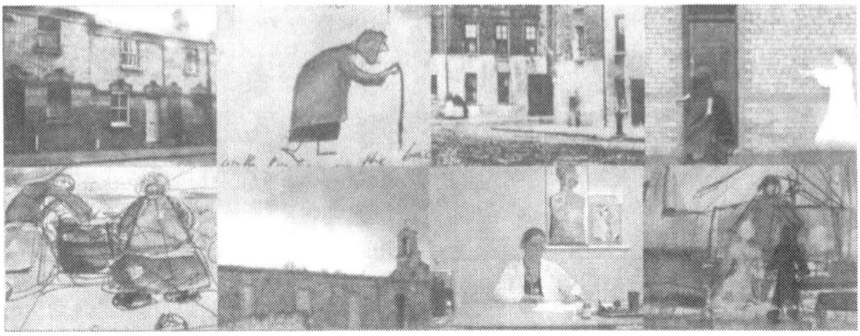

Figure 27: Frames from the stories videoclips

Designing the Story Structure

The stories have been crafted as units of a modular structure where story fragments are interchangeable. The classical arched plot-based story structure has been exchanged for a more fragmented one where each scene is a self-contained anecdote that references a particular location in the Liberties neighborhood. This approach places fewer requirements on adjacency of story material and insures that the scenes can be experienced in any order. The resultant hyperlinked structure consists of the collection of story fragments and the different types of links that connect them. The reader can end the experience at any time, without needing the closure that is typically expected in stories that have a strong story arc. This kind of structure is designed to remain open and evolve, allowing new stories to be added at any time.

The Liberties story fragments are organized in a structure following a content model that takes into account the social network of the community and the geographical area. Variables that determine a different story experience of each audience member include themes, such as Guinness stories (related to the famous brewery that resides in the Liberties area), historical anecdotes, the user profile and the user's position in the real space. MPL is grounded in a real place. The influence of this fact on a user's experience is twofold. The audience can merge the story settings portrayed in the video clips with the surrounding real space, adding it to their mental image of the story and augmenting the real space with the story's content and aesthetic. This may add atmosphere and warmth to the neighborhood, bringing it to life. In addition to that effect, relationships between places can be used to trigger memories and recollections of an audience member who is intimately familiar with the area.

The spatial distribution of the stories is used as one of the navigation criteria for the narrative, providing the readers with a map of the story space that illustrates the relation of the content to specific locations, as well as strengthening the spatial-temporal immersion of the audience transporting them into the scene. Spatial-temporal immersion takes place when the distance between the position of the narrator and his/her audience, and the time and place of the narrated events, are reduced to near zero [Ryan 2001]. We believe that by placing the story fragments in the location where they happened and allowing the audience to experience them there, we reduce that distance and increase spatial-temporal immersion.

The Production Phase

A variety of methods were employed in the production of the stories, according to their different characteristics. The stories ranged from ghost stories over descriptions of architectural changes in the area to individual portraits of local characters. In creating the audiovisual media segments we used a mix of video, animation and photographic media and a narrating voice that related the anecdotes in the first person (which coincided with that found in Johnston's book). Harnessing the friendly and open atmosphere of the neighborhood, film sets were staged in the Guinness Hopstore (at the time transformed into Media Lab Europe) and around the Liberties area. Locals and colleagues from the Media Lab would reenact the stories using costumes and in settings inspired by the descriptions in Johnston's book. One of the most interesting sets was the reinterpretation of the traditional music session that took place every Sunday in "Mickey Murphies Yard" during the nineteen-forties.

The traditional musicians Cormac Cannon, on the *uilleann*, or Irish, bagpipes, and Finbarr Naughton on the fiddle, played in the courtyard of the Pimlico cottages for an afternoon. Local people came along and engaged in the action, participating as local characters in the filming. The session itself was also an opportunity to informally chat to locals about the project and verify their enthusiasm for providing new content to the story collection, reviving memories, anecdotes and a sense of pride of being from the Liberties area.

A different kind of story, referencing buildings that do not exist anymore, were produced using photographical material collected during the research and some of the watercolour sketches produced for the storyboarding process. The varied methodology added expressivity and freshness to the stories, and highlighted some ideas for how to develop a framework under which to guide the community members in the feedback collection phase, when they will be able to provide and produce their own stories to add to the collection.

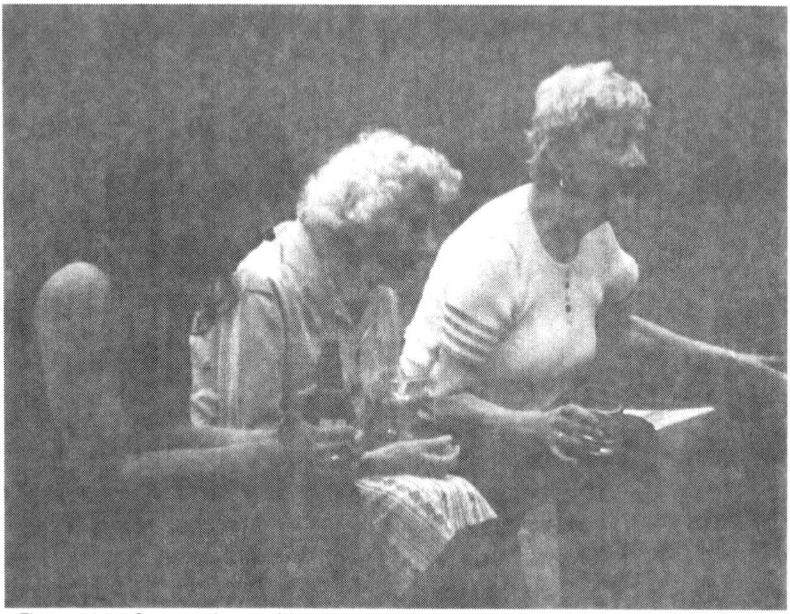

Figure 28: Community residents participating in the in the "Mickey Murphy's yard" story production. Courtesy of Rob Bourke

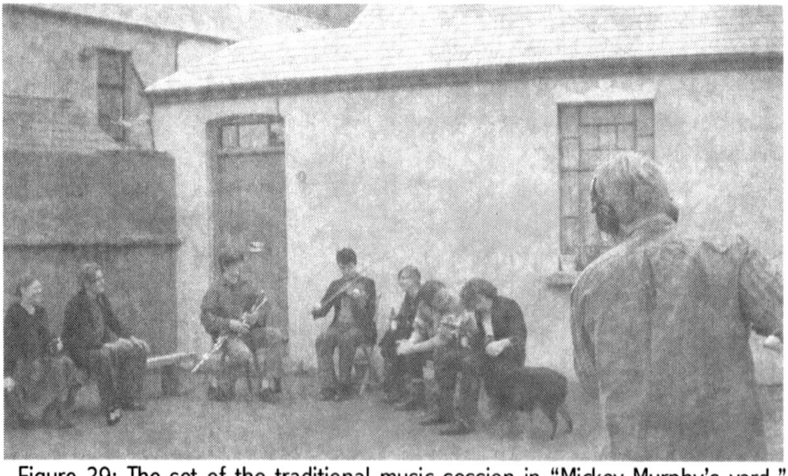

Figure 29: The set of the traditional music session in "Mickey Murphy's yard." Courtesy of Rob Bourke

2 Technological Implementation

MPL is running on the Pimlico software, a bespoke platform developed by Ian Oakley under the Carmen project-funding framework[2] supported by the Irish Higher Education Authority (HEA). Two versions of Pimlico exist, one for wireless mobile devices (iPAQs) and one for desktop PCs. Portable devices are the primary target platform. An iPAQ equipped with a Global Positioning System (GPS) card enables the audience to view the stories depending on the their current position on the streets. Access to the story content occurs through the appearance and disappearance of selectable icons. Each icon represents a particular story and is positioned on the digital map on the iPAQ's screen, in the location where the story is set. Once an icon is selected, the corresponding audiovisual story fragment is played full screen. At the end of the video clip the digital map of the area is returned to the screen.

Each story clip is linked to other clips through a set of relationship between related topics. This set of relationships has been modeled in the XML markup language. Four types of topics are supported: characters, locations, themes and times. These topic titles relate to the content of the stories video clips. After a video clip has been played by the audience, all topic related clips are displayed on the map, indicating to the audience member where to position themselves to access related content. Relationships in each of these four topics can be activated or deactivated separately. It is possible to view clips related by theme alone, or just by character and location, or by all relationships. For instance, all clips containing a character called Honora are related to one another by the character relationship set. Some of these clips also feature other characters, for instance Honora's son Michael, and are therefore also related to all clips that he appears in. When this practice is applied to all four topics, it forms a rich web of interconnections between the video clips. Furthermore, it is easy to place new content in the web as the connections between content pieces happen through meta tags describing the themes, characters and locations for each story. Each story is described by a set of keywords that link to the content to other pieces tagged with the same keyword [Davenport & Murtaugh]. Selecting a few topics that relate to the story is sufficient to richly connect it to the web.

---

[2] CARMEN project website: <http://www.dsg.cs.tcd.ie/?category_id=-35> (accessed on July 1st 2005.)

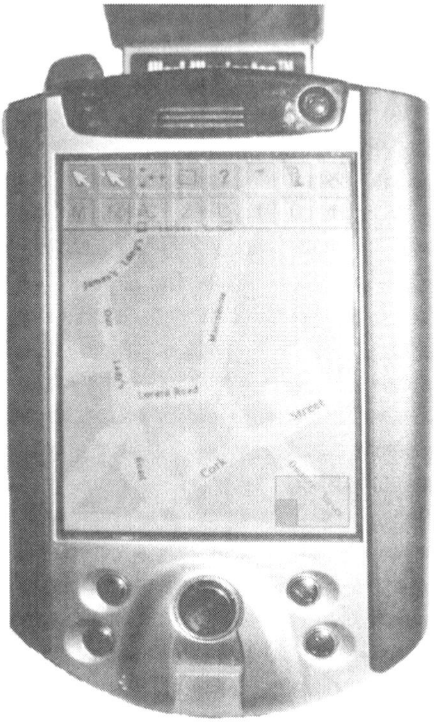

Figure 30: The iPaq interface.

3 User Study

The process of storytelling always involves a story, a narrator, a medium and an audience. To complete the project, a multifaceted user study was designed to capture the audience response to the experience. Three different kinds of audiences have been targeted: media experts (people with experience in digital media technologies), community members (people that live in the area and consider themselves members of the Liberties community) and non-residents (people from outside the Liberties neighborhood, Irish-native and non-Irish). The methodology for evaluation is qualitative, involving observation and semi-structured interviews [Silverman 2001] Seven people for each category were sampled and their comments recorded through a wearable Dictaphone as they were touring the Liberties neighborhood with the iPAQ in search of stories. Four users from each group experienced the Liberties project in pairs while the other three toured on their own. This method was used to investigate what kind of exchanges the experience would prompt among audience members, in particular for community members and tourists. As the people were walking around the streets experiencing the stories, we observed them at a distance (so as not to influence their choices), recording their trail through the Liberties and wrote notes about their behavior. At the end of the tour, the audience members were assembled for a focused conversation about their general feeling about the experience, about the location-based aspect of the stories and about the technology used.

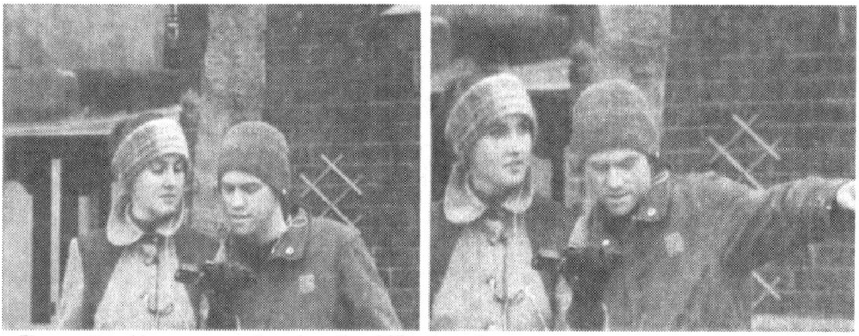

Figure 31: Users experiencing the MPL in pair.

Preliminary results have already shown general interest and engagement with the project by all participants. Audience members from different user groups demonstrated different focii and interests in the project. The next section describes the issues brought up by each audience group and the three main categories that emerged from the collected data. At the end of each section we summarize the findings within each category.

Community Reaction

Community members were selected from the "Living Heritage" class organized by the local community centre of Saint Nicolas of Myra on Carman Street, in the heart of the Liberties area. The contact was made through John Gallagher, head of the community centre and who was of fundamental help in organizing the sample audience. The community members were enthusiastic about experiencing the project. Valentina Nisi followed them around the Liberties area, together with John Gallagher, who often told the group stories on the spot and prompted conversations about the neighborhood. Going around the Liberties with local people was a great insight into the community atmosphere and a very different experience from when observing audience members from the other categories. People on the streets had a very friendly attitude towards other community members. Often conversations with passers by would start spontaneously in the street with topics ranging from questions about what the group was doing, to local news and events happening in the neighborhood. The Liberties is a poor inner city area, and it was clear that both the authors and the audience members who are not native to the area were not able to relax and enjoy the experience as much as the community members.

The first four participants of the study (Adele, Aileen and two sisters: Liz and Mariel) were born in the area. Liz and Mariel's family worked for the Jacobs biscuit factory (a well known firm that started its business in the Liberties in the forties) for three generations. Adele and Aileen are also descended from families rooted in the Liberties for two or more generations. Memories were triggered by some of stories for each of them. An example is an anecdote about a rude dispensary officer in South Earl Street Health Centre. From residents memories triggered on the spot, we got to know that, until a few years ago, he was still performing his job with the same rude manners portrayed in the story.

Another location that triggered memories and stories in community members was Braithwaite Street. Liz and Mariel's family used to live there and they knew the family of the man in one of the anecdotes. At one time, he was a familiar face in the Liberties, walking the streets selling coal from a horse drawn cart. Liz and Mariel told that when they were children they used to play in the shed where he stored the coal, even though it was dirty and full of rats. "His hut and stable for the horse were only recently demolished" Liz and Mariel remembered. Adele and Aileen on the other hand, recall him shouting: "Coal" as his horse and cart criss-crossed the Liberties.

When returning to the San Nicolas of Myra community centre for the focused discussions, members of the "Living Heritage" class would spontaneously join the conversation and contribute with their comments. The comment included stories about the poor conditions of the area in the past, and how men used to run to the pubs and bars as soon as they received their salary; how women were not allowed to enter the pubs until recently, and used to have to ask John Gallagher to go to retrieve some of their husbands' salary before it was all spent on drinks; how people from the area still know today which houses originally belonged to the Guinness workers and which to the Jacob's factory workers, despite the fact that the Jacobs biscuit factory is not located in the area anymore and Guinness now employs few Liberties residents. These sessions clearly indicated the interest of the community in participating and contributing to MPL with their own stories and expand the database with old and new anecdotes about their neighborhood.

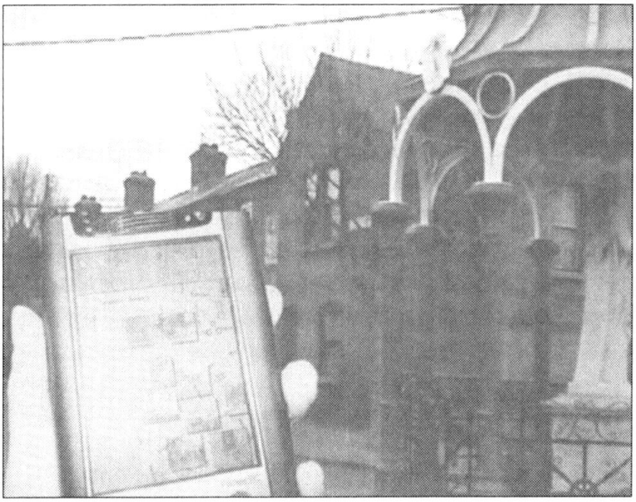

Figure 32: iPaq interface displaying story icons in location

Non-Residents Reaction

Non-residents ranging from native Dubliners, to foreigners living in Dublin, showed a variety of reactions to the MPL. People from outside Ireland generally were drawn towards more historical information. They enjoyed listening to stories that would tell them about the origins of the area and socio-historical anecdotes about the Liberties's population and the area's architectural structure, such as the stories about the Courthouse of the Liberties, or the Brabazon family, descended from the Lords of the area. Oscar from Cuba expressed a desire to know specific dates and facts about monuments such as the statue of the Jesus on Gray Street. John Gallagher was there when Oscar made the comment and was able to answer the question immediately by telling about the origins of the statue and how it used to be a fountain. This type of interaction between the people experiencing the tour is interesting as it shows how the system can prompt memories and anecdotes about the neighborhood and stimulates a dialogue between locals and visitors.

Some non-residents could not connect to characters like Honora and her family moving around the Liberties from house to house and job to job. They found those kinds of stories too fragmented to be able to synthesize them as a whole. Others reported the feeling of being dropped in the middle of a story, missing out on the beginning. We attribute such comments in part to the fragmented nature of the narrative, but have identified a number of other possible reasons for this experience. For example, not being from the area and therefore not being able to connect the story fragments to similar stories that might have occurred in relatives' lives, was also an important factor. The lack of a cultural reference frame may have diminished their interest in and capability to relate to these kind of story fragments.

This conclusion was supported by an interesting comment reported a person originally from the North side of Dublin. He recognized that similar stories would exist even in his own neighborhood. This cultural reference sparked his interest in the adventures of the Johnston family, struggling through economic highs and lows, moving premises around the neighborhood every few years and constantly dealing with the everyday problems of health and accommodation. On the other hand, Bart, a native from Belgium, visiting Dublin for his second time just for a few days, enjoyed the Johnston's family anecdotes most. He found them warm and entertaining and was not disturbed by their fragmented nature.

Media Experts' Reaction

Media experts had a variety of different responses to the experience. The technology was sometimes reported as distracting from the story experience. Reasons ranged from the fact that they were engrossed in understanding the interface issues, or because they were unfamiliar with the Liberties neighbourhood and therefore the navigation of the real space was a demanding task. One of the users, a German writer with multimedia experience, felt the stories were too fragmented to form a coherent narrative structure, especially if considered from a classical narrative point of view. He found the underlying narrative structure too weak to form a coherent story and reported that the characters lacked motivations and psychological depth. Users more familiar with the fragmented nature of many interactive narratives, such as two former colleagues from Media Lab Europe, found the fragments pleasing and reported that the relationships between the fragments were engaging and motivated them to seek related story fragments.

The media experts often observed that the link between a story and a specific location was not always immediately obvious. (In fact, this comment was reported by audience members in all three user-categories.) Due to the GPS resolution (approximately 15-10 meters), the audience was not always situated in exactly the right physical location when a story became available. Furthermore, the audiovisual media fragments did not always provide precise pointers as to their setting. Suggestions to resolve this problem included starting each video clip with an image of the location in which the story is set instead of having the location visualized somewhere in the middle of the clip, according to the demands of the story being told. Another suggestion was to use the narration to direct the audience's gaze in the appropriate direction. For instance stories could commence in the following style: "Can you see the red brick building opposite South Earl Street no 8? That used to be the local Health centre." By explicitly naming and describing the important locations, the audience would familiarise themselves with the setting before the beginning of the story, without missing on the visuals nor feeling confused about what they should be looking at.

4 Lesson Learned

From the data collected during the interviews with the audience members from all three different users groups, three main categories of comments emerged: observations regarding the stories, the interface and the technology used.

Story Design

The modular structure of the story collection evoked different responses. Some people enjoyed the fragmented nature of the narrative and found it engaging and challenging, motivating them to find more story parts to complete the picture, and confronting their notion of traditional narrative structure. Typically, they enjoyed the anecdotal self-contained style of the story fragments. Other users found the fragmented nature of the narrative confusing and frustrating. These users expressed a desire to be able to experience the fragments in a more linear, traditional way. There were also comments about the simple treatment of the characters. The main suggestion for strengthening the characters was to have them introduce themselves and contextualize their position in the Johnston family, informing the audience in which relationship they stand to the author narrating voice. Unfortunately, this was the approach that was actually implemented in the project. The narrator, who adopts the point of view of Mareen Johnston, the last surviving member of the Johnston family, talks in the first person and specifies the relationship of the characters to herself. For example: "Honora, my great, great, grandmother used to live in Pimlico...." From the comments we received, it is clear that this technique proved to be insufficient to describe the complex web of Johnston ancestors who feature in the story fragments. Furthermore, the relationships among stories were not obvious to everybody. Suggestions to strengthen these links ranged from displaying trails on the map to highlight continuity between stories themes, to using the narrator to explicitly make connections with other stories and to pass on suggestions as to where the audience should go next to collect related stories. A timeline was also proposed, to facilitate to experience the stories in chronological order.

Comments on the Interface Design

The interface was reported to be very easy to understand and to use (see Figure 33). Users appreciated the difference between the hand drawn map we adopted and the more traditional survey style of map. The radar view and cursor indicating the user position on map were also useful and easy to understand. What led to a less favorable response, in particular from the media experts, were the markers that indicated the presence of the stories. A frequent comment was that the dots that show story locations on the map could have been more informative. For example, dots of different colors could be used to illustrate whether stories had been previously viewed or not. Many people were unable to remember which stories they had already seen, even when they returned to a previously visited location. Another frequent comment was that the icons that appear on the map to indicate story availability were too big. They "cluttered" the screen rather than offering useful information, limiting the visibility of the map and were of little descriptive value. This comment was expressed by a high number of users. The majority commented that they would have like to be able to switch off the icons in order to be able to see the map more clearly. However, one user offered a contrary opinion, He reported that he really enjoyed the icons richness, and found them colorful and pleasant.

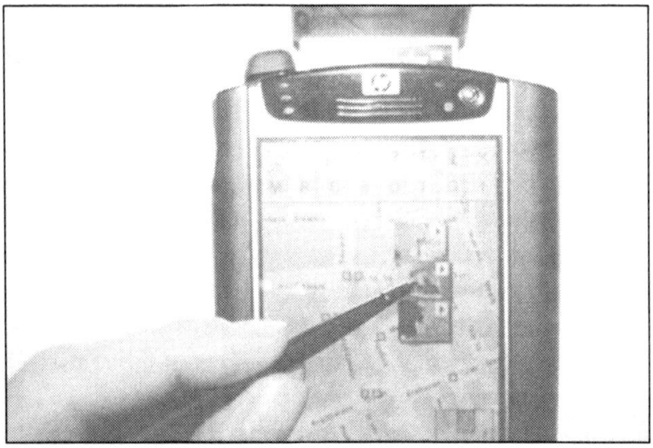

Figure 33: iPaq interface: Story Icons and stories placeholders (dots)

Comments on the Technology

From the experience of observing the audience, and from their comments, we feel confident that the technology used in the project (a GPS-enabled iPAQ) has been shown to be both simple to interact with and reliable. This is an important conclusion, as some of our target audience has limited experience with computer technology, and many mobile devices can be challenging to use. The simple interface we developed plays a considerable part in this success.

The choice of an iPAQ for the experience was also successful. As a miniature portable cinema screen, the iPAQ delivers good quality audiovisual media that allowed the audience to engage with both the audio narration and the visual interpretation of the story. Adopting iPAQs as delivery platform—a device that had to be explicitly rented or requested in order to try the experience—was a design choice made in order to stress the uniqueness of the experience. By ensuring the project was only available under certain conditions and for an already interested audience, we were able to make the experience more akin to attending the theatre or visiting a gallery than more casual activities like watching TV or surfing the web. Through the purchase of a ticket and allotment of time to dedicate to the experience, an audience member chooses to engage with the project and pays a deeper, more focused attention to the work. If we had made the project available for mobile phones and other more popular and widely available mobile devices, we expect it would have become more accessible but not necessarily more appreciated. The fact that locations were not always immediately recognizable led to a mixed response. Some audience members were curious, others frustrated. The problem can only be partly attributed to the resolution of the GPS technology, which spans around 20 meters. Specific filming techniques could be combined with technologies, which offer more accurate localization, such as Bluetooth beacons or RFID tags to ensure that stories can only be retrieved in the exact locations to which they related.

The simplicity of the interface made it easy for most audience members to focus on the stories. Understandably, people unfamiliar with the area had more difficulties. Rainer and Sara, a German and an American who had been living in Dublin for about six months found it hard to navigate around the new area, relate to the stories and the real locations simultaneously and coordinating where they had already been with where they wanted to go next. One of the media experts, Alison, found using the technology a distraction from the story experience and would have liked to have more time to become familiar with the device and its functions so as to be able to focus on the story experience more.

It is interesting to note that community members paid less attention to the technology and focused on the content much more than the media experts. The non-residents could become engrossed in or distracted by the system, but the Liberties community members immediately focused on the content.

5 Reflections and Conclusions

Through this initial broad user study, we have established that the project succeeded in enhancing a neighborhood by making stories about the area available to the public in the places in which they are set, through a mobile location-based narrative system. All people who participated to the user study indicated that they would like to have similar experiences available in other cities or in their own neighborhood. People agreed that the place acquired atmosphere and warmth after the tour and that the experience added and strengthened the sense of the place. For community members the experience was twofold. On one hand they were able to discover new stories and anecdotes about the history of their neighborhood, a process they found extremely rewarding. On the other hand they were also prompted and stimulated to recollect and tell their own stories about the area. From these preliminary responses we can see the project successfully functioning as a catalyst for community awareness and the recollection of individual memories that could comprise a rich social history. The study also pointed out some important aspects of the story design. Comments from media experts in particular highlighted the need to develop more sophisticated strategies for linking story fragments together and to strengthen the characters that appear in the stories. For some people the fragmented nature of the narrative ensemble was a positive aspect of the project. It engaged some audience members in a sort of treasure hunt for story pieces and a bigger image portraying the area character in general did emerge. For people oriented towards more classical story structures, the fragments felt incomplete or too loosely related to each other.

The portable and intimate characteristics of the medium lend themselves to a closer relationship with the audience, very different from the aesthetics of cinematic public screening. Audience members appeared to engage well with the stories, but there is still much to explore in the area of a specific cinematic language for small portable screens.

The interface design was generally felt to be intuitive and easy to use. The navigation aids were all useful except the icons indicating story availability. They were perceived as cluttering the screen and not adding much information to the interface. The dots indicating story location could also have been more informative, for example if they had been colored differently depending on whether the story had been already been viewed, which would have allowed people to avoid involuntarily returning to already visited locations.

The study of the reactions of different types of audience group to the project has been extremely useful in identifying how people use and feel about location-based story experiences. Major strengths and weaknesses of the work have been uncovered and explained. It is clear that the project has potential for enhancing spaces, transforming them into something richer for both community members and more casual visitors. A strong positive feedback from the community itself showed potential in the direction of community storytelling and the recollection of memories by local residents.

Story sharing and empathizing though has been a reaction expressed from all sampled audience members, demonstrating the importance of stories in our culture and the need to continue to engage in combining narrative tradition with emerging technologies.

Acknowledgements

The authors wish to thank Ian Oakley for his invaluable help in many aspects of the project. Thanks to all the actors, narrators, musicians and everyone who helped in the production of the media story videoclips. Thanks to MediaLabEurope and Trinity College Distributed System Group for supporting and funding the project.

In addition, we are very grateful to the Irish Higher Education Authority (HEA) for funding the Context-AwaRe Media Environment for Narrative (CARMEN) project, a collaborative research project between Media Lab Europe and the Distributed Systems Group (DSG) at Trinity College Dublin's Department of Computer Science.

Bibliography

Davenport, G., and M. Murtaugh. "Automatist Storyteller systems and the shifting sands of story." *IBM Systems Journal* 36.3 (November 1999), 446-56.

Johnston, M. *Around the Banks of Pimlico*. Dublin: The Attic Press, 1985.

Ryan, M. L. *Narrative and Virtual Reality*. Baltimore and London: The John Hopkins University Press 2001.

Silverman, D. *Interpreting Qualitative Data*. London, Thousand Oaks, New Delhi: Sage Publications, 2001.

Loca: 'Location Oriented Critical Arts'[1]

Drew Hemment, John Evans, Mika Raento, Theo Humphries

Loca is an interdisciplinary project on mobile media and surveillance. *Loca* explores the shifting boundaries between art practice, the event, and data systems.

> "In the inverted logic of the post-Orwellian city, *Loca* agents, software and human, decrypt the hertzian passages of its own inhabitants. For better or for worse?"
> [Steve Dietz, Artistic Director, ZeroOne]

Loca: Set To Discoverable is an arts-based group project on grass-roots, pervasive surveillance which seeks to expose the disconnect between people and the trails of digital identities they leave behind. The premier full presentation at ISEA2006 and ZeroOne in August 2006 combined art installation, software engineering, activism, pervasive design, hardware hacking, SMS poetry, sticker art and ambient performance.

Loca is an exercise in everyday surveillance, tracking digital bodies in physical space. It examines what happens when it is easy for everyone to track everyone, when surveillance can be affected by consumer level technology within peer-to-peer networks without being routed through a central point. The *Loca* project walks the knife-edge of locative media, itself involving surveillance. But it does not create a new surveillance potential, it only reveals what was already there.

Pervasive surveillance is potentially both sinister and beneficial. New ways of organising media and of communicating with each other become possible when the context of the media and the user is known. But also as a consequence, ever more can be surveilled and ultimately controlled. *Loca* sought to lay bare this truth of mobile social software, to highlight how it can function as surveillance, without collapsing this ambiguity. It is an experiment that neither blindly celebrates the technology, nor claims that the technology is inherently bad. It aims to raise awareness of the networks we inhabit, to provoke people into questioning them, and help people equip themselves to deal with the inherent ambiguity of pervasive media environments.

---

[1] See <http://www.loca-lab.org>

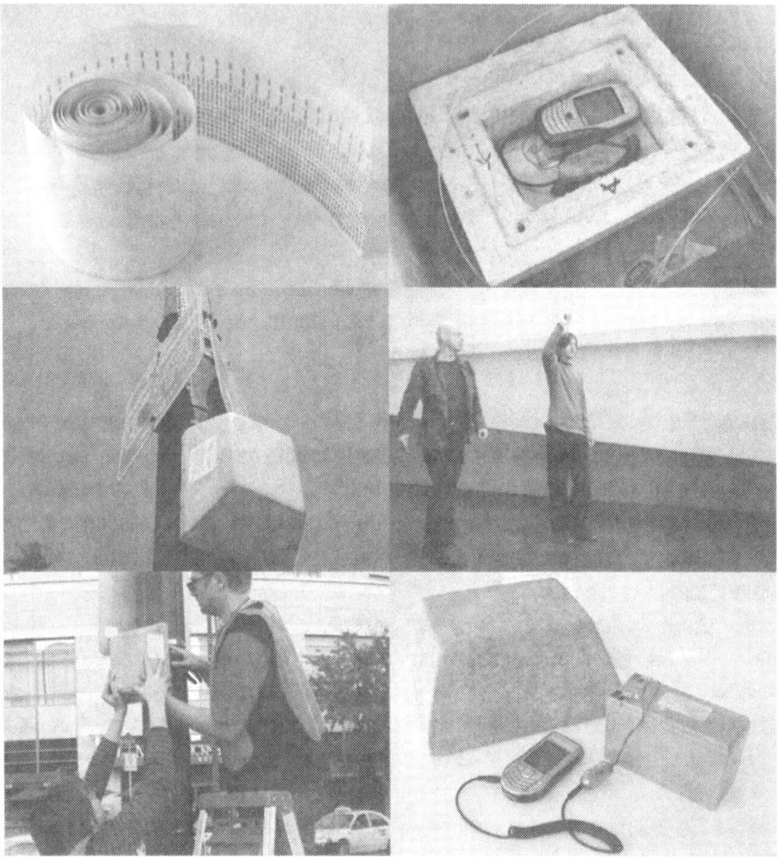

Figure 34: Images of Loca at ISEA2006 and ZeroOne in August 2006

The Loca Node Network

One element of *Loca* is a node network, through which *Loca* observes people's movements by tracking the position of the Bluetooth-enabled devices that they carry. One principle of the project is that people should be able to participate through their own mobile phone without their device needing to be modified in any way, either through installing *Loca* software or by altering settings. *Loca* deploys a cluster of interconnected, self-sufficient Bluetooth nodes within inner city urban environments. In San José the nodes were encased in concrete and fixed to street lights in the park, at pedestrian crossings on road intersections, and buried in the earth by a popular bar terrace. Others were deployed in hotels, cafes, and popular destinations such as cinemas. They were hidden in flowerpots, underneath a chaise longue, and in the foot of the podium used by a cinema ticket collector. The *Loca* team was constantly visible about the city centre in bright orange workman overalls, climbing ladders, moving in and out of hotels, deploying and maintaining the network. A production line was created to recharge spare car batteries, and constant maintenance was needed to replace batteries and reset the software.

The Loca art group were then able to track and communicate with the residents of San Jose via their cellphone without their permission or knowledge, so long as they had a Bluetooth device set to discoverable. Over seven days more than two thousand five hundred people were detected more than half a million (500,000) times, enabling the team to build up a detailed picture of their movements.

The *Loca* node network enables people to explore pervasive surveillance environments in a performative way. A person walking through the city centre hears a beep on their phone and glances at the screen. Instead of an SMS alert they see a message reading:

> "We are currently experiencing difficulties monitoring your position: please wave your network device in the air."

Loca engaged people by responding to urban semantics, the social meanings of particular places:

> "You walked past a flower shop and spent 30 minutes in the park, are you in love?"

People were sent messages from a stranger with intimate knowledge of their movements, written in such a way as to leave them unsure whether or not they had unwittingly joined a social network called *Loca*. Over the course of the week the tone of the messages changed, the all-knowing friend turning out to be one friend too many: "coffee later?" changing to "r u ignoring me?".

The purpose of the messages is to make the presence of the Loca network known, and to illustrate the types of data that can be gathered and the inferences that can be drawn from it. These messages could highlight people's daily routines e.g. "You have been here for an hour," reveal the 'others' within the network e.g. "You can be seen by 4 other devices," or even be used to control people's behaviour e.g. "Please wave your phone in the air." The messages also made the Loca project accountable for what it had done, and provided channels for feedback and further enquiry.

Loca aims to lightly touch large numbers of people. The aim is not complex interaction, but subtle affect, and only a minority of people will receive it, let alone give it any more than passing thought. *Loca* is like a picture glanced at sideways, a message caught in the corner of the eye, or a mosquito swatted on the arm. People who accepted the Bluetooth connection received a longer message, which explained the background to the project, and gave directions to the *Loca* stand at the exhibition venue where people could scan their device and receive a personalised printout of their movements. At *ZeroOne* some individual logs were over 100 metres long. As the first project of its kind, the system was fragile, and some higher levels of interactivity could not be achieved. None-the-less, the system was operational for the full length of the exhibition, during which more than half a million data points were recorded.

Loca does not ask people in advance; it does not want their permission. Not to obtain people's permission was a difficult decision, but it was felt necessary for the validity of the project—which sets out to make visible the limits of what is technically, legally and ethically possible, and to put people's responses to pervasive surveillance environments at the centre of the debate.

Stickers: An Ad Hoc Sticky Network

A second element of *Loca* involves stickers. The experience of the node network is intangible: it is to do with what is not seen. The stickers make visible the traces of digital identities that people leave behind. The stickers are visible to all, not just to those with a Bluetooth device, making the pervasive surveillance transparent. At ISEA2006 trails of stickers appeared across down town San José.

The *Loca* stickers enable people to become *Loca* "agents," playing on the double meaning of secret agent and software agent. They are able to act out one of the functions of the *Loca* network—to become a small piece of surveillance code—introducing an element of participatory, urban play. People scan for Bluetooth devices using their own mobile phone or laptop, by using Bluetooth in the usual way. On the *Loca* stickers they record the name of any uniquely named device they detect, and the time and date of detection. The stickers can then be stuck at the point of detection to leave tangible evidence of the presence of the digital identity, the time at which it was detected and the detection event itself. The process is simple, and can be undertaken by anyone with a Bluetooth device.

An important question about 'responsibility' arises when individuals use their own name for the name of their device. The *Loca* node network does not publish any personal information. But a person's name written on the sticker is personal data. Is the 'publisher' of that information the user of the device, the person writing on the sticker, or the *Loca* team who produced the stickers for them to write on?

SJPD: "Booked in Evidence"

On the final day of ZeroOne a node left in the downtown Sainte Claire Hotel was taken away by San José Police Department as a suspicious object and "booked in evidence." This took place on the same weekend in August 2006 that three Palestinian-Americans were arrested for possession of 1000 cellphones, which the authorities suspected were to be used for surveillance or as bomb detonators. Technology is often imprinted with our hopes and fears, and it says a lot about contemporary America that the SJPD chanced upon Loca's genuine DIY surveillance, but it was the people with Arabic names who were detained. The police revised the status of the node from "Evidence" to "Found Property" and the artists were able to retrieve it, complete with a log of data on people's movements at the police station, whether they were officers, criminals or simply people passing by.

The Loca Pack: Surveillance Code

All *Loca* code and tools are made publicly available in the form of the *Loca* Surveillance Pack. The *Loca* Pack provides for both surveillance and counter-surveillance. It makes *Loca* transparent, providing blueprints and code. It also informs people about the scale of the threat. Fear is used by politicians to create docile subjects; we want to show people that they can take back control.

"Don't do harm to anyone. We will find you."

Responses to Loca

Loca was interested in people's responses. These varied across the different project presentations, which included San Jose plus earlier proof-of-concept presentations in Helsinki and London. Some learned for the first time through the project how Bluetooth functioned, and were compelled to switch their Bluetooth device to undiscoverable. Responses ranged from being intrigued, to shocked, to dismissive.

Previous projects on surveillance, such as ZKM's CTRL:SPACE exhibition, have highlighted how we today live between two polarities of scopophobia and scopophilia—fear of being observed mixed with a seemingly endless desire to reveal ourselves, as epitomised by both reality TV and social software. *Loca* found that these responses are often intertwined, and that locative media can inspire each in equal measure.

A discussion on *Loca* in Seoul, South Korea highlighted how debates on "privacy" can seem alien outside of Europe, and yet the issues of "disclosure" that Loca explores were found to be relevant to the participants in Asia, just framed in a different way. Most critical of *Loca* was Shami Chakrabati, Director of *Liberty* and one of the UK's leading civil liberties advocates, who discussing the project on BBC World

Service's Digital Planet, was concerned about the distress it could have caused the public.

Background

Loca is an anticipated accident. The project was initiated in 2003, out of an interest in how surveillance and social control emerge as a residue or unforeseen effect of otherwise virtuous information systems and network technologies. Then it sat in waiting for the accident to happen.

The accident was when the "Aware ContextPhone"[1] collaboration started generating surveillance data that was unforeseen by its designers. When users published media to Aware directly from their phone, using software called 'ContextPhone', it automatically annotated this media with contextual information derived from the phones actual surroundings, e.g. time, GSM cell-ID (an approximate geographical Locator), and the bluetooth environment, i.e. a list of the Bluetooth devices around at the time. The premise of Aware and ContextPhone (themselves developed by Loca participants) is that the social context of the media can be used both to situate the media and to help organise it. The Bluetooth information would allow queries like 'show me all the pictures captured when I was in the vicinity of that person', which would be useful in a wide variety of contexts—if, for example, someone wished to gain an overview on an event at which they were present. This contextual information leads to unforeseen consequences, such as the 'accidental' tracking of people present during the media arts festival ISEA2004 in Helsinki. If someone wanted to reproduce what Aware/ContextPhone had been found to do, but for commercial gain or unethical ends, how hard would it be to implement technically and legally?

Loca examines the surveillance potential of different consumer platforms. In San José it focused on Bluetooth for a number of reasons: because Bluetooth has been designed in a way that is problematic for privacy management; working with Bluetooth rather than GSM makes possible some independence from the mobile phone companies; and because Bluetooth is the first 'everyday' network technology that enables people to be tracked, and to track each other, within the physical environment. The privacy trade-off found in many contemporary forms of surveillance (that you must incrementally give up ever more of your privacy in order to access new services) is common to all network technologies; but here it is not just data, but also bodies in space that are being tracked. (WLAN is similar, but is not always on and is less mobile; GSM tracking remains largely the preserve of the

---

[1] See <http://aware.uiah.fi and http://www.cs.helsinki.fi/group/context/>

mobile phone companies; RFID—at this time—is still not established in the consumer domain.)

Loca works independently from the mobile phone companies and other service providers so that it is clear to participants that Loca is not swayed by commercial interests in technology and also to show that the project can be done in a low-cost way. Each node is built using readily available, cheap parts, and is encased in concrete in order to be deployed in the urban environment. Loca does not need any special privileges nor to break the law—nothing stops one from Bluetooth scanning; in fact it is part of the protocol, whereas GSM is prohibited. All you need to participate—to watch or be watched—is a Bluetooth device.

An aim for *Loca* is to make people aware that they have agency, that they can avoid being tracked by turning off their device or switching their Bluetooth device to 'invisible'. But *Loca* also sets out to reveal the limit of this agency.[2] With all technologies that are susceptible to pervasive surveillance techniques, the only way to opt-out of the surveillance is to switch off altogether, which is often impractical, and means losing the benefits of that technology. This was not inevitable, and we need to ask why these technologies are not privacy preserving: why, for example, do all network technologies use permanent unique IDs; who made those decisions, on what agenda, who has it benefited? Equally, computers that are invisible are bad for privacy: do you want the things that are tracking you to be hidden? *Loca* advocates the development of countermeasures and of better privacy management provisions in policies and protocols. An issue with Bluetooth is that Bluetooth scanning is anonymous. Should not the person or device doing the scanning have to provide their identity before they obtain the identity of the devices that they are scanning? Many such measures will involve a cost, so unless an argument is made and demand

---

[2] "LOCA ... is a surveillance project that talks back. A crew plans to plant some 30 Bluetooth scanners encased in concrete blocks at bus stations, in shops and at other busy spots in San Jose. Carrying a Bluetooth-discoverable phone within 25 feet of the scanners can trigger the receipt of a surprisingly intimate message like: 'You walked past the flower shop and spent 30 minutes in the park. Are you in love?' Such notes are not sent via text messaging but through a subversive technique called Bluejacking, in which a Bluetooth device's name is replaced with a short message meant to be picked up by neighbouring devices Part of the point is to catch people by surprise, jolting them out of their daily rituals with a Dada-worthy prank. Another goal, said Mr. Hemment ... is to show people how much data they may be revealing every time they turn on their phones. 'In an office you can shut the door for privacy,' Mr. Hemment said. "In conversation you can hide a facial expression. But with the new digital technologies, you may have no idea how much you're giving away." If this sounds more like an educational project than a work of art, Mr. Hemment does not seem to mind. He said he doesn't make hard and fast distinctions between the two: he considers LOCA a policy-minded research effort with the art serving as its public face." [Jori Finkel, 'An Exhibition Where Paintings Are So Last Century', New York Times, August 6, 2006]

exists, then it will not happen. Loca highlights the asymmetry involved in this process, the lack of reciprocity between the person scanning and the person scanned, and enables people to experience the unsettling distance between disclosure and connection.

Loca explores peer-to-peer surveillance, and yet, like many such projects, it is peer-to-peer only to a point. Surveillance data is generated independently on each node, but then that data is relayed between the nodes and a server via the GSM network. This does not compromise the principle, however. The surveillance is independent, a server is only used for convenience within this project as it simplifies implementation, and the data could be relayed between nodes in alternative ways, but with less mobility, or higher cost. This would lead to a new set of parameters, alternative questions, and a change in the nature of the project.

Locative Media is principally concerned with the context of location. Projects, which can be labelled as 'Locative', use or create technologies that enable users to log and/or publish this contextual media. Users of such systems reveal personal information that is pertinent to the project, but, importantly, third parties can repurpose this information. In failing to address this issue many locative projects leave themselves open to criticism over the potential re-use of such personal information. The critical point is that such projects ask and/or require users to give up this information for a perceived benefit, but do not address the (often unforeseen) consequences of these actions—and the principle unforeseen consequence of Locative Media's demand for logging location and time is that it creates systems susceptible to various forms of surveillance.

Locative Media has yet to fully address its own critical context. Loca seeks to make a contribution to these debates, while at the same time critically assessing its own methodology and the risks of its approach. How can creative work with surveillance technologies add to or distract from traditional campaigning strategies? Does it risk getting the public used to a new control technology prior to its deployment in a coercive way? How may both positive and troubling sides of a new technology be simultaneously explored?

Loca asks: how do people respond to being tracked and observed? How ready are people to observe others? Who is the user, and how? What does it mean to participate in this project? Do we get fear of surveillance, disinterest, scopophobia or scopophilia? What kind of behaviour is this technique suited to map, and to what behaviour is it not suited? What kinds of behaviour can evade this form of surveillance? How does the contextual information we can detect (such as Location, time spent in one place, etc) relate to people's everyday experiences of the environment? What happens in-between physical, embodied space and the digital space of abstract data? What is the relationship between the embodiment of the mobile user and the abstraction of the data we capture?

Pervasive surveillance has the potential to be both sinister and beneficial, simultaneously. The intent of Loca is to equip people to deal with the ambiguity and to make informed decisions about the networks that they populate.

Invisible Topographies

Usman Haque

(Update, with permission, of an article originally published in *Receiver*, December 2003)[1]

Like the banks of the River Nile, the boundaries of art and architecture are constantly changing. They expand and recede, cover new territory and expose old presumptions. As their boundaries shift, the distinctions between art and architecture become less clear. Architecture, the design of spatial experience, and art, the production of cultural experience, for several centuries have not shared as much common ground as they do now.

The overlapping territories of art and architecture have come about in large part because of technological developments that upset conventional understandings of spatiality. Traditional dichotomies between audiences and performers, designers and users, architects and occupants are less evident than they used to be. Wireless technologies in particular have challenged our relationship to designed space because they encourage us to think not of static silent structures that surround us but rather of fluid dynamic fields beyond the edge of our natural perceptions, fields within which we are all consumers and all contributors.

Mobile communication has relied equally on scientific and cultural evolution and has prompted a distinct shift in the way we relate to space and the way we relate to each other. Apart from issues arising out of being in contact virtually anywhere, anytime, the mobile technologies through which we conduct our daily lives and businesses have made us far more aware of the electromagnetic environment that envelops us.

Electromagnetic waves exist just about everywhere in our atmosphere. Visible light is a form of electromagnetic radiation, as are infrared and x-rays. Even radio and television broadcasts and the microwave wavelengths of mobile phones are part of the electromagnetic spectrum. Electromagnetic waves can be detected and measured using a gaussmeter; the intensity of an electromagnetic field varies greatly with the distance from the source and the material qualities of the substances (walls, air, windows) in between.

---

[1] See <http://www.receiver.vodafone.com>

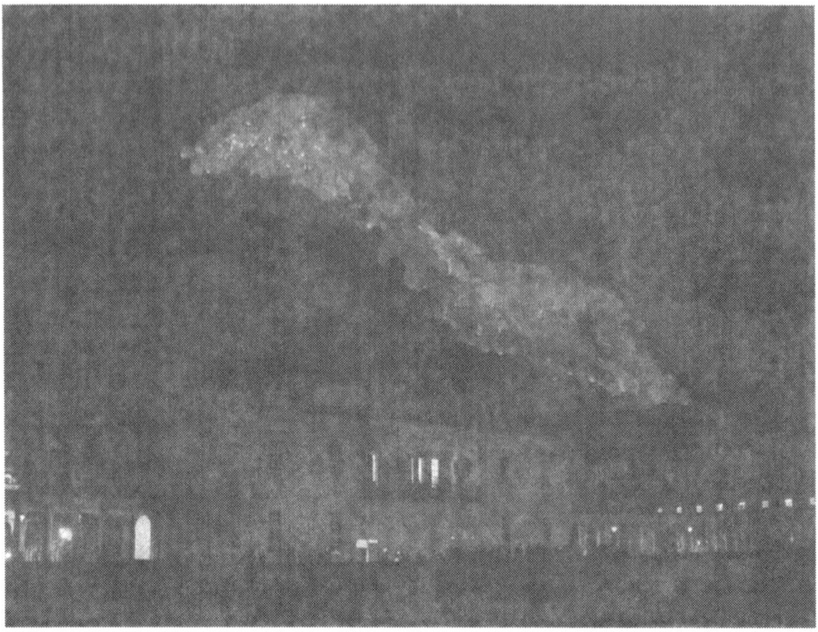

Figure 35: Sky Ear takes off

While we have been concerned about the health effects of electromagnetic radiation (from power lines or mobile phone handsets), these waves often exist as natural phenomena in the form of radio waves emanating from distant stars, gamma rays coming from elements here on earth or even electrical waves from inside our own skulls. Humans have only recently begun contributing to the cacophony with their pagers, medical devices, televisions broadcasts and mobile phones. Industrial design theorist Anthony Dunne has called this abundant, invisible territory, a topography that is altered in shape and intensity by both natural and human-constructed landscapes, "hertzian space." He has observed that hertzian space is often ignored by designers saying, in Hertzian Tales, that the "material responses to immaterial electromagnetic fields can lead to new aesthetic possibilities for architecture" [Dunne 1999].

Most of us are familiar with buildings where signal strength is variable or rooms where mobile calls are possible in one corner though it is impossible in another even to make a connection. There are areas in some buildings where one mobile company's network penetrates while others are unable to. These interactions with built fabric are even clearer with the implementation of wireless computer networks, for example the 802.11b standard that is increasingly being used in homes and offices to create Local area networks. In such networks the positions of furniture, the thickness and material qualities of walls, doors and windows or the distance from a base station contribute to the varying quality of a network connection. These variations in field strength suggest a richly textured ethereal cartography of hertzian space, a real (i.e. non-virtual) space that affects us but which we only know about through use of our instruments.

Our mobile phones have, in the form of signal strength indicator, sensors that measure this invisible electromagnetic radiation as we drift from cell to cell of a mobile phone network or from room to room among the varying densities of a building's construction materials. These simple devices, acting almost as gaussmeters, indicate the network durability of our phones and prompt us to imagine mapping out the contours of mobile phone microwave intensity (or hertzian microclimates) around furniture in a room or buildings in a city.

Through such use of our phones we have become more aware of the interactions between what might be called 'hardspace' and 'softspace'. Hardspace, like the hardware of a computer, refers to the solid, static structure of our built environment: the walls, the floors, and the roof. Softspace, again paralleling computer software, refers to the ephemeral qualities that make up our experience of space, including smells, sounds, temperatures and of course electromagnetic waves.

We are aware that hardspace affects our ability to make mobile phone calls because it has a direct effect on the electromagnetic intensity of the signal; what we often forget however is that the very act of making a mobile phone call also changes hertzian space because the phone itself becomes a transmitter of microwaves. We notice a materialisation of this process while talking on a landline: a few seconds before a nearby mobile phone makes or receives a call a distinct galloping beat is heard on the landline handset as the impinging microwaves penetrate the wired connection and ripple Local hertzian space. Flashing stickers and accessories can be added to mobile phones to light up when a call is made or received. Such inexpensive devices are passive inductors of particular microwave radiation and function as consumer-friendly indicators of increased electromagnetic intensity.

These phenomena have helped blur boundaries between art and architecture because operating creatively in hertzian space requires the exploration of concepts and correlations that have never before been encountered, particularly with regard to their effect on cultural activity. We begin to read cities as more than just a collection of destinations—they are overlapping fields of differing experiences and logics. We are able to explore a ghostly poetic ecology that exists just beyond our familiar perceptual limits (see, for example, *Lost Cat* and *Stampede*[1] by Ben Hooker

[1] Flirt: <http://www.dunneandraby.co.uk/designing/FLIRT/FLIRT.html>

and Fiona Raby, in which people track a cat or a herd of reindeer moving through hertzian space; *Urban Tapestries* [2] by Proboscis, a framework for Location-based mobile content; *Spatial Memory Architecture*[3] by Hoshi Takuya (ISEA 2002) NAGOYA where people store memories in three dimensional hertzian space throughout a city; *Remain in Light* [4] by Haruki Nishijima, where an electronic insect "net" is used to capture electromagnetic waves discovered while walking round a city).

The fact that such explorations often have no immediate commercial application tends to make us place them in the domain of art. However, since there is an explicit effect on the nature of our relationship to space, and since architecture is in some sense the choreography of spatial relations, it would suggest that operating within hertzian space entails architectural processes and conceptualisations.

To investigate some of these themes, I have developed several projects involving mobile phones, the first of which in 2000 was a performance/installation entitled *Japanese Whispers*.[5] In this project, 20 mobile phones were laid nose-to-toe in a circle. Calls between the phones were initiated in a variety of patterns (neighbour to neighbour or across the circle) and the ambient sounds and voices of participants were input into the mouthpieces to be propagated through the phones and mobile phone network. The resulting feedback loop delayed and distorted the sounds through the iterative process of being digitised, transmitted, output and re-digitised. While not exploring the features of hertzian space itself, the aim of the project was to look at how the communication process is affected by translations from the physical medium to the hertzian medium and back again.

More recently, in 2004, following on from previous investigations and experiments into the qualities of softspace (including smell, sound and temperature) I developed project *Sky Ear* to explore the hertzian habitat above Greenwich Park in London. London has a diverse and vibrant hertzian culture, with mobile phone calls overlapping text messages, combining television broadcasts with garage door openers that interfere with radio transmissions which transmit from wireless laptops, etc.—a tumultuous electromagnetic space. With this project, I wanted to give form to this space, to make visible the invisible. I also wanted to pursue the idea of a temporary performance/installation that only comes to life through interactions with the audience, corresponding to my architectural concern with placing creativity in the hands of the occupant.

---

[2] See <http://urbantapestries.net/>

[3] See <http://vis.vision.ss.is.nagoya-u.ac.jp/isea/text/program/E/exhibitions.html>

[4] See <http://www.fundacion.telefonica.com/at/vida/paginas/v4/eharuki.html>

[5] Japanese Whispers <http://www.haque.co.uk/japanesewhispers.php>

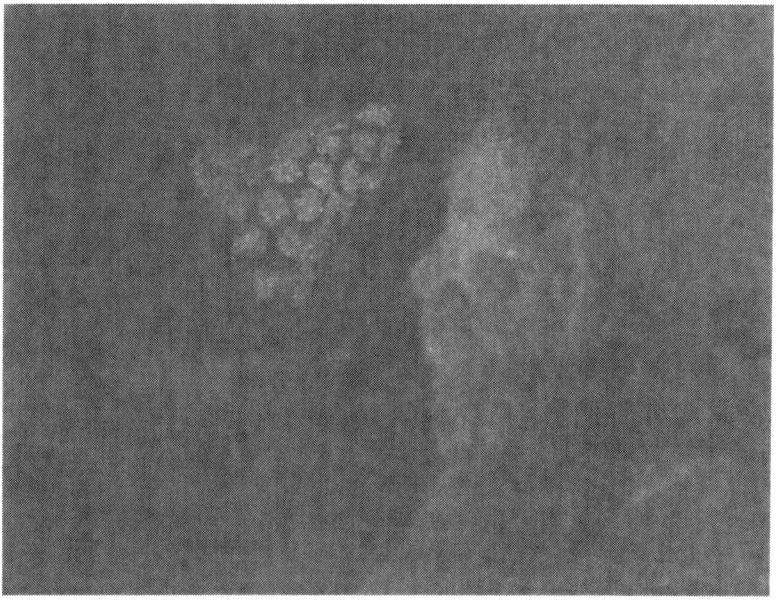

Figure 36: Sky Ear communication

In order to chart this unexplored territory, shortly before dusk on September 15, 2004, the *Sky Ear* cloud was released from its ground moorings and slowly floated up into the sky sampling the electromagnetic field fluctuations as it rose, rather like a vertical radar sweep. The 30m-wide cloud, made up of 1000 glowing helium balloons enclosed in a carbon fibre frame and monofilament sandwich net, was embedded with 37 mobile phones. The balloons each contained miniature sensor circuits that responded to fluctuations in electromagnetic activity at a broad range of frequencies and caused brightly coloured LEDs to illuminate the balloon surfaces. The cloud was tethered to the ground by 3 cables and held aloft at a height of about 60-100m where it remained for several hours, fluctuating in response to the wind environment much like a glowing jellyfish.

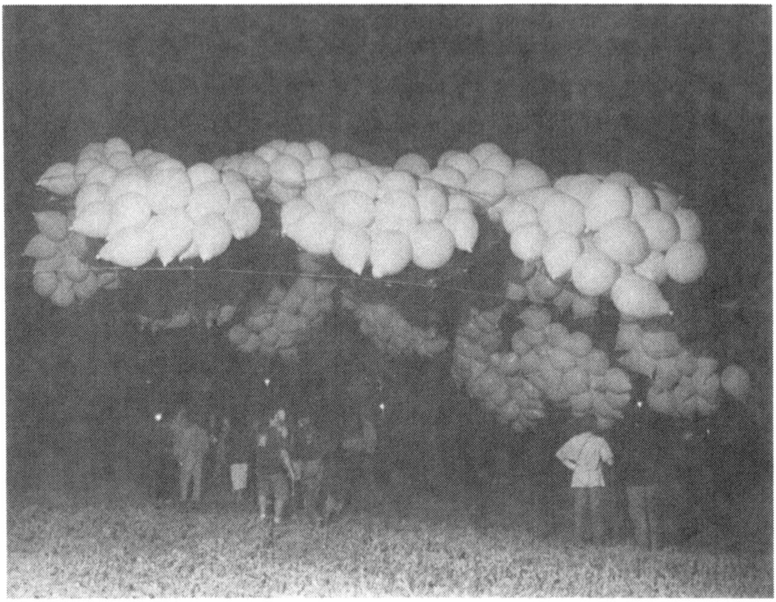

Figure 37: Sky Ear Launch

As visitors to the event called into the cloud to listen to the natural electromagnetic sounds of the sky (including whistlers and spherics), their mobile phone calls changed the Local hertzian topography; these disturbances in the electromagnetic fields inside the cloud altered the glow intensity and colour of that part of the balloon cloud. Feedback within the sensor network created ripples of light reminiscent of rumbling thunder and flashes of lightning as the balloons formed a "B2B (balloon-to-balloon) network," resulting in cellular automata-type patterns. As an "ear" to the "sky," the project is as much an auditory experience as it is a visual spectacle. *Sky Ear* gives the impression of extending visitors' hearing into the atmosphere and into the electromagnetic spectrum. The combination of people's calls to listen to the spherics and whistlers together affect the visual patterns in the balloons.

The cloud helped show both how natural invisible electromagnetism pervades our surroundings and also how our mobile phone calls and text messages delicately affect the new and existing electromagnetic fields. As an art project, *Sky Ear* encouraged people to become creative participants in a hertzian performance; as an architecture project, *Sky Ear* made visible our daily interactions with the invisible topographies of hertzian

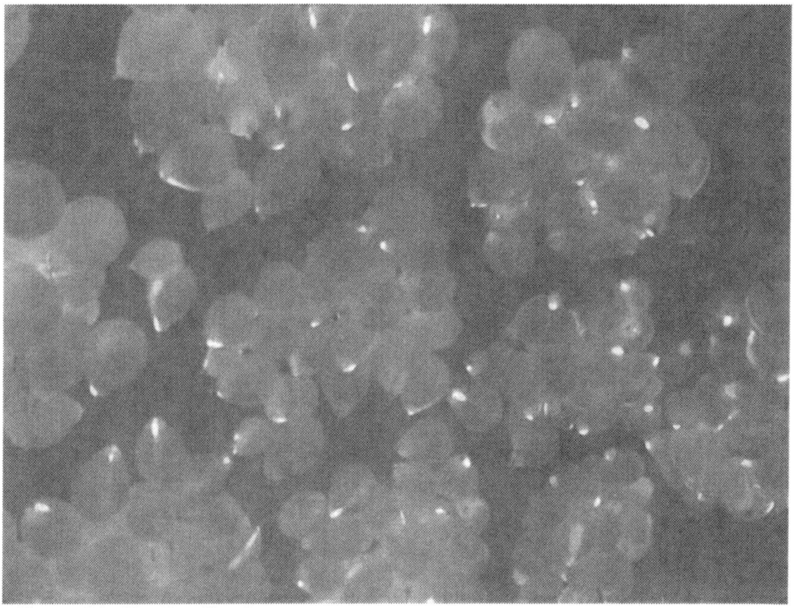

Figure 38: Sky Ear Cloud

Acknowledgements

Sky Ear is a project by Usman Haque, Haque Design + Research, with financial assistance from the Daniel Langlois Foundation. However, many dozens of people have worked to make the project possible and full credits may be found at <http://www.haque.co.uk/skyear/credits.html>. It has had two major launches, in Switzerland at the Belluard Bollwerk Festival (July 2004) and another in Greenwich, London, hosted by the National Maritime Museum (September 2004). Published with the kind permission of receiver magazine at *www.receiver.vodafone.com*.

Bibliography

Dunne, Anthony, and Giles Lane. *Hertzian Tales: Electronic Products, Aesthetic Experience and Critical Design.* Computer Related Design Critical Design Unit, Royal College of Art, 1999.

Wifi-Hog: The Battle for Ownership in Public Wireless Space

Jonah Brucker-Cohen

When technologies are first introduced, hype usually follows. The hype naturally disperses over time, but when news begins to spread about how people are using the technology, the hype machine begins to resurface. When I first heard about wireless internet (or 802.11b) back in early 1999, I ignored it. This was a technology that seemed very far off, as no computers were yet equipped with wireless receivers (except a few Personal Digital Assistants (PDA)). I heard about people using the technology in classrooms and hospitals, with very particular applications that seemed too particular for any mainstream adoption. A few years later, wireless internet (now affectionately called "Wi-Fi") began to resurface as reports of projects and public community networks began to sprout up. Wireless was becoming cheap, pervasive, and simple to implement.

After hearing about the hype and projects, I began to notice something else that was happening in and around wireless nodes and their deployment. In August 2002, Slashdot ran an article about "Starbucks vs. Personal Telco Project (PTP),"[1] a battle that was quietly taking place in Portland, Oregon's Pioneer Square. This was a challenge over public obstruction of wireless space, where corporate signal was out-blasting the pre-existing community signal. Peer-to-peer had two 2 T1 connections with off the shelf routers setup providing free wireless access to anyone in the square. A few months later, Starbucks who partnered with T-mobile, set up in-store satellite Internet access that broadcast on Channel 1 within the store and around the square. Channel 1 is the default connection found by most consumer wireless cards. As a result, since Starbuck's signal was stronger and its connection speed was faster than Peer-to-peer, the once free network that pervaded the park had to close down. The struggle over claiming ownership of public spaces with wireless nodes was in full swing. On a trip to the NYC Wireless[2] headquarters, I heard a story about how Verizon (a major telecommunications company in NYC) had started to put high-power wireless access points (APs) on the tops of all of their pay phone booths in the city. These blanketed every block of the city and were only available to customers of Verizon's DSL service. NYC Wireless had set up a free node from their office which was meant to reach the street below, but Verizon's corner payphone node was interfering with it. The problem was further reaching than I thought.

---

[1] Slashdot, August 2002: <http://yro.slashdot.org/yro/02/08/20/0431202.shtml?tid=98>

[2] NYC Wireless: <http://www.nycwireless.org>

In 2003, I began working on a project called "Wifi-Hog" that was a direct reaction to the claim of ownership that corporations and individuals were placing on public wireless space. The project consisted of a laptop connected to a Portable Video Jammer (PVJ), and some custom circuitry that communicates to software on computer. The software was comprised of a packet sniffer (such as Carnivore)[3] and wireless stumbler (such as NetStumbler[4] which allows the software to find open networks) which monitors incoming packets from an open node. The idea was to only allow traffic originating from the Wifi-Hogger's IP address to access the network, otherwise the PVJ is switched on, blocking others from connecting to the open node. Since most Wi-Fi networks operate on the un-licensed 2.4 GHZ band, jamming this spectrum is not illegal. There are over 100 websites that advertise and sell the PVJ, so finding one was relatively easy.

Wifi-Hog is a tool that enables control over a specified network by someone who is not the network's administrator and looks specifically at what happens when these seemingly open networks are made exclusive and competitive. Since these networks exist as private, public, and corporate monitored services, there is also confusion about rights ownership over networks in public spaces, thus *Wifi-Hog* is specifically reacting to the lack of an "Acceptable Usage Policy" of wireless networks. As mobile technology has entered public space and brought private conversations and interactions along with it, an interesting rift was forming between what is deemed acceptable usage. In a sense, *Wifi-Hog* exists as a tactical media tool for controlling and subverting this claim of ownership and regulation over free spectrum, by allowing a means of control to come from a third-party.

As mobile and wireless devices become more ubiquitous, free and public wireless nodes have gained high penetration. Free nodes are popping up in public parks, airport terminals, libraries, schools, and other venues worldwide. In addition to sanctioned spaces for the nodes, private nodes without encryption are leaking from offices and houses onto city and rural streets. Activities that exploited and actively seeked out these networks began to materialize. Some examples include the WARchalking[5] and WARdriving phenomenon (where you search for open nodes on city streets and mark their location with chalk) and artist interventions like "Noderunner"[6] and Blast Theory's "Can you see me now?"[7] which integrate urban street players with wireless connectivity. As the networks grew, especially in dense

---

[3] Carnivore: <http://www.rhizome.org/carnivore>
[4] NetStumbler: <http://www.netstumbler.com>
[5] WARchalking (Wireless Access Router): <http://www.warchalking.org>
[6] Noderunner: <http://www.noderunner.com>
[7] Blast Theory, "Can You See Me Now?": <http://www.canyouseemenow.co.uk>

urban spaces, signals from private, public, and commercial (or paid) nodes began to interfere with each other. This spectrum overload brings up even more questions about how jurisdiction of signal is defined and who has precedence over others.

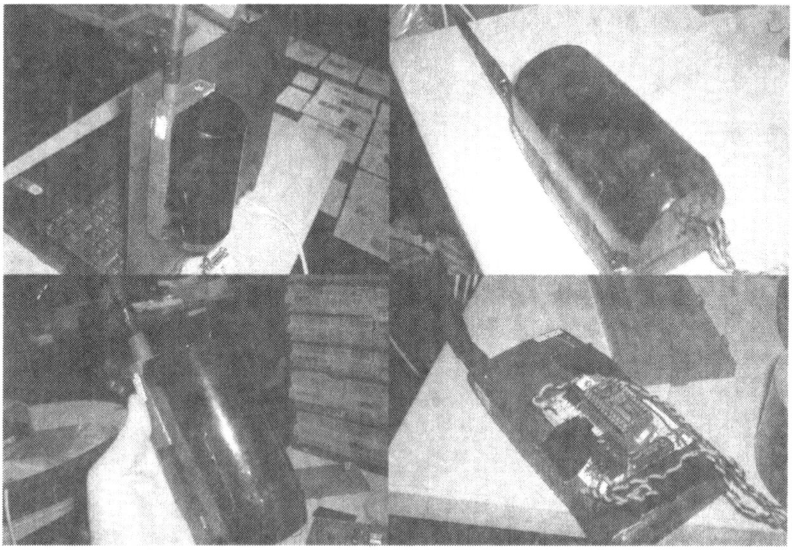

Figure 39: Wifi-Hog under construction

Looking specifically at free wireless access points, *Wifi-Hog* is also a reaction to the public spaces they inhabit. *Wifi-Hog* is a personal tool to enable both private interaction in public space as well as social obstruction and deconstruction of shared resources. The idea has some historical precedent in the area of property acquisition prior to the introduction of state-controlled zoning laws, when at the time land was a public resource that had to be regulated due to misuse and territorial disputes. An example of this type of territorial dispute occurred the United States in the late 19th century. The Homestead Act of 1862 provided that unoccupied public land be transferred to a homesteader after five years of residence. This was an act sanctioned by the US government to create a system of land grants to encourage settlers to develop the then uninhabited West. In effect, the Homestead Act was a pay off for settling in the region.

My aim with *Wifi-Hog* was to investigate how wireless networks could fall into this predicament since they can leak or pervade from private to public spaces. Imagine if you had lived on the land for 3 years, it was still in the public domain, but you had invested your life into it, and someone came along and fenced off the land with a barrier you could not penetrate. In this case you do not have any legal right to the land, but you still feel as if it is yours since it has been in your custody for 3 years. This is a scenario closely linked to *Wifi-Hog 's* premise that a public wireless network maybe be partially owned or controlled by someone, but it can nevertheless be taken away and controlled. This containment issue might also allow for third parties to disrupt or interfere with them. The project is intended, thus, to send a clear message to groups attempting to claim ownership over a public space by demonstrating that their network can be easily jammed and controlled by others. An example of its use might be for an individual to use *Wifi-Hog* to disrupt a corporate signal and let a weaker, but free node exist in the same space. This signifies a loss of control by providers and sparks a challenge to their "land-grabbing" attitudes.

Since the project was introduced, most of the reaction from the wireless and media arts communities has been negative. This mostly stems from misunderstandings of why the project exists and how it was presented. Most people were upset that I was "advertising" the PVJ as something that could disrupt all of the progress and work that had been done to create open networks. My focus at first was to disprove the fact that wireless was leading us into a "utopian" world where networks would be everywhere and people would work harmoniously beside each other. I see this as a simplistic view that fails to see the conflicts of ownership and the complex integration and use of wireless in public spaces. Some thought that my project created rifts in the "community nodes" that existed such as London's Consume.net[8] or NYC Wireless's wireless parks, since I was promoting a disruptive tool. From a discussion on the NYC Wireless list, some comments about the project were made evident by an anonymous poster:

> "If I remember the way NYC Wireless, etc started out, the very act of putting up public wireless nodes was to exert territoriality—we were claiming the public parks as free Wi-Fi zones, and betting that these would deter pay providers from locating there. To a large degree this has turned out to be an accurate prediction. We were also trying to re-contextualize networks within local places, grounding them in real urban communities rather than having them exist in some kind of an abstract non-geographic cyberspace. I have to agree that this project doesn't seem to be terribly sophisticated, and is very reactionary. It is a yes/no proposition, without any selectivity. You might just as well just be climbing atop the maintenance shed in Bryant Park and plugging / unplugging the antenna lead." [August 2003]

---

[8] Consume.net: <http://www.consume.net>

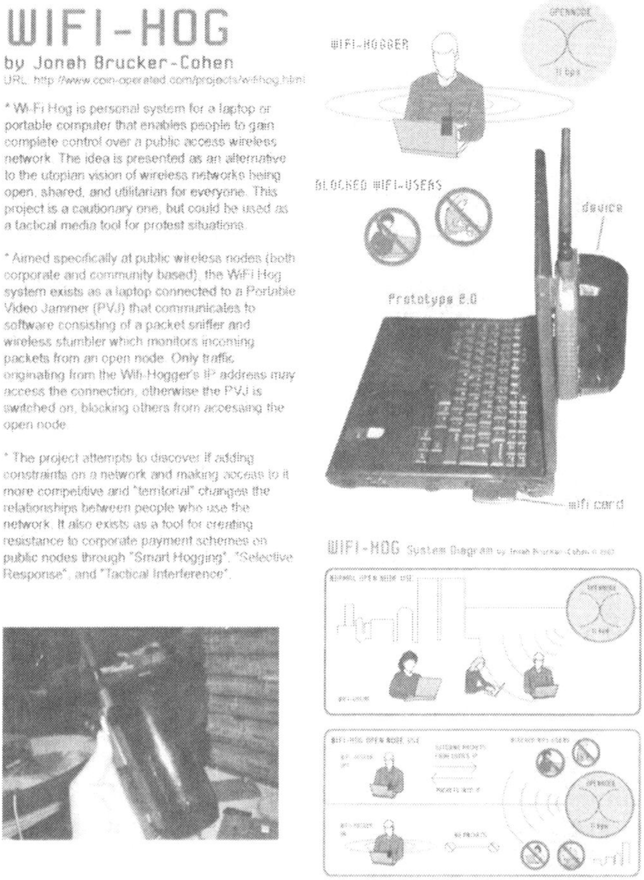

Figure 40: Wifi-Hog schematic

Despite the mixed reactions and confusion surrounding the point of the project and its execution, the problem it addresses remains important. As spectrum overcrowding becomes more common in cities, the conflict between for-pay and free nodes will reach a critical point. Companies will have to enforce strict delineation of their signal strength so that free networks cannot impede on their business models and vise versa. Projects like *Wifi-Hog* are clear and critical reminders that wireless networking is still a young technology that displaces architectural and social boundaries. This distinction is important for the future of wireless and the communities that support its development.

3.2 The Creative User: The User as Co-creator

Puppeteers, Performers or Avatars: A Perceptual Difference in Telematic Space

Paul Sermon

The telematic performer

My work in the field of telematic arts explores the emergence of user-determined narrative between remote participants, who are brought together within a shared telepresent environment. Through the use of live chromakeying and video-conferencing technologies these divided audience participants enter a video installation and initially suppose they entering a passive space: sitting, standing or sometimes lying within it. Their presence within the space is recorded live on video camera and mapped in real-time, via a chromakey video-mixer, with an identical camera view of another participant in an identical installation space: combining two shots of live action by replacing a blue or green back drop in one image with the image of the other. The two spaces—which can be any geographical distance apart—are linked via an internet videoconference connection, making it possible to combine these telematic installations and their performing audiences, between almost any two locations in the world.

This is essentially how all my installation projects function, but what is most surprising for the intended viewer is that they form an integral part within these telematic experiments, which simply wouldn't function without their presence and forced participation. By observing their body within a telepresent space (represented on self-view video monitors in front of them), the audience-participant rapidly becomes a performer, or preferably an actor within these spaces. The user/actor ascends a rapid learning curve and begins to control and choreograph the human/avatar representation of themselves in a new telematic space, in combination with another physically remote, role-playing user. The narrative that unfolds here would appear to be self-determined by the user, but what is essential in such experiments is the architecture of the installation. As an artist I am both designer of the environment and director of the narrative, which I determine through the social and political context in which I choose to play out these telepresent encounters.

Locating the performance

Emotional exchange in the telematic space is highly dependent on location and interface. Whilst I have chosen to use the bed or sofa as a meeting place, other artists in the field of telematic arts have relied purely on the dynamics of site specificity. In the seminal telematic installation "Hole-in-Space," produced by Kit Galloway and Sherrie Rabinowitz, in 1980, geographically remote public audiences were instantly transformed into performers in the first networked narrative performance in a social context of its kind. Initially it appears to be a random choice of locations for this public intervention from the point of view of the user/actor; but it increasingly becomes clear that these cities and locations were chosen by the artists for very specific social and political reasons; creating a networked narrative within an extremely dynamic context. Kit Galloway and Sherrie Rabinowitz describe the work as follows:

> "Suddenly head-to-toe, life-sized, television images of the people on the opposite coast appeared. They could now see, hear, and speak with each other as if encountering each other on the same sidewalk. No signs, sponsor logos, or credits were posted—no explanation at all was offered. No self-view video monitors to distract from the phenomena of this life-size encounter. Self-view video monitors would have degraded the situation into a self-conscience videoconference."[1]

If you have ever had the opportunity to see what the award winning video documentation captured then you would have laughed and cried at the amazing human drama and events that were played out over the evolution of the three evenings. Hole-In-Space suddenly severed the distance between both cities and created an outrageous pedestrian intersection. There was the evening of discovery, followed by the evening of intentional word-of-mouth rendezvous, followed by a mass migration of families and trans-continental loved ones, some of which had not seen each other for over twenty years."[2]

"Hole-In-Space" relies on the US cultural cliché of East Coast—West Coast indifference; confronting the pedestrian passes-by in New York and Los Angeles and bringing them up on a telematic stage to tell jokes: *" Question: how many New Yorkers does it take to change a light-bulb? Answer: None of your fucking*

---

[1] See <http://www.mediaartnet.org/works/hole-in-space/> (20.07.03)
[2] See <http://www.ecafe.com/getty/HIS/> (20.07.03)

business"; sing songs: *"New York, New York"* and play games, even charades on one occasion. Viewers were instantly transformed into performers in an East Coast meets West Coast soap opera.

Continuing to focus on location or stage, we can also look at the work of Sergei Eisenstein. However, not in reference to his films, but to his lesser known theatre productions with the Proletcult Theatre in the 1920's as a theatre designer and later as director at the Director's Studio of Vsevolod Meyerhold in 1921. This is where most of the issues relevant to us now were developed, described in his "Mon-tage of Attractions"; which established a new principle of dramaturgy, producing extremely strong effects on the audience by means of combining posters, slogans, circus, variety show, gymnastics, scenery and theatrical effects. In 1923 Eisenstein staged "Gas Masks," a play about the employees of a gasworks, where he actually moved the play entirely out of the theatre and staged it in the Moscow Gas Factory. The play, which depicted life in the gas factory, ended each performance as the new shift came to work—thereby breaking the bounds of theatre in very similar ways to our current experiments with network-performer locations. Although Eisenstein considered it to be a failure in terms of film/theatre realism, his intention to embody the viewer within the performance is entirely that of my own work. One in which I consider the site-specific nature of the stage set to be of paramount importance in the telematic installation. Eisenstein described the performance in his notes:

> "In Gas Masks we see all the elements of film tendencies meeting. The turbines, the factory background, negated the last remnants of make-up and theatrical costumes, and all elements appeared as independently fused. Theatre accessories in the midst of real factory plastics appeared ridiculous. The element of 'play' was incompatible with the acrid smell of gas. The pitiful platform kept getting lost among the real platforms of labour activity." [Eisenstein 16]

Eisenstein's use of the Moscow Gas Factory as a determining social narrative relates directly to an installation I produced in collaboration with Andrea Zapp in 1999, entitled "A Body of Water" [Zapp 83]. Similarly to Eisenstein's "Gas Works" this installation was entirely situated in a disused coalminers changing and shower room, and was also reliant on the sound and smell of coal dust and running water in order to determine the context of telepresent showering and the users/actors performance within it. Commissioned and produced for the "Connected Cities" exhibition "A Body of Water" connected three sites in one single telematic space. Live video images of the gallery visitors at the "Wilhelm Lehmbruck Museum" in Duisburg Germany were chroma-key mixed with images of visitors standing in the changing rooms at the nearby disused "Ewald/Schlaegel und Eisen" colliery in Herten. The video image of the combined audiences was simultaneously back projected onto a fine wall of water in the shower room, providing a third interactive space—literally by walking through and touching the projected participants in the water—who were connected by a video camera that distributed real-time images of the water screen projection to all three sites at once. In this collaborative installation with Andrea Zapp it was our intension utilise the site-specific sounds and smells of this context in order to develop and reinstate the ceremonial like camaraderie taking place amongst the showering miners. "A Body of Water" allowed our live performing spectators to reenact and experience this simple, yet complex act in the form an open-ended narrative.

The visual interlocutor

Essential to all the interfaces in my installations is the use of non-verbal communication. By not using sound the user/performer is forced to communicate in a melodramatic style akin to a silent movie—bringing about an enforced use of gesture and body movement in order to communicate with the fellow participant. By restricting verbal communication the participant is further distanced from their telepresent-reflected performer role, which allows a far less self-conscious experience in the space. Whilst the silent melodrama was introduced for precisely this reason, it also refers directly to Charlie Chaplin's comments on the end of cinema when sound was first introduced—highlighting that the inadequacy of a technology is its single most creative potential. The following extract from Margaret Morse discussing *Telematic Dreaming*, describes the process of induced mime and mimicry extremely accurately and has been influential on me in further opening up this discussion around user-defined narratives:

> By not transmitting sound, Sermon has chosen to explore the visual and kinetic codes of proxemic relations, that is, the relative distance of human bodies in private/social exchange, rather than verbal exchanges. A cyberspace couple on the bed can interact in any way gesture allows. The dematerialization of gestures and objects tendered, far from undermining their meaning, makes images and actions naked of anything but symbolic meaning and all the more powerful therefore. Thus, the stage has been set for an exploration of the effect of symbolic acts on the psyche. …[3]

[3] <www.methodsnetwork.ac.uk/redist/pdf/act18_sermon.pdf>

Specifically with the bed interface, the body has been turned into a visual interlocutor that is not only without the ability to have speech but also dominance, strength, smell and on occasions gender—depending on the clothing and concealed physique of the participant. Without these characteristics the performer is immediately persuaded to accept and interpret communication through touch with an otherwise complete stranger in a highly emotional and sensitive relationship. Morse further explains it:

> "...Paul Sermon's experiments with "telematics" or "telepresence" continue research that began in the late 1960's using satellites to link live interaction in sound and image between two or more sites. This strand of experimentation also has predecessors in the closed-circuit video and installation art of the early 1970's. Artists of the time experimented not only with "narcissism, but with temporal and spatial displacements of body and its image that reveal the gap between a body and its imaginary self or "identity." Sermon's work is the site of a collective imaginary, a public "family" reunion, albeit as a surreal composition of bodies without a counterpart in physical reality, akin to the condensations Freud identified in dreams. What the "live" mixture of bodies in Sermon's work exposes is the far from explored field of human relations as they have become inflected with and transformed by technology."[4]

The narrative environment

The installation *Peace Talks* was commissioned by FACT Liverpool for the BitParts Exhibition at Worcester City Art Gallery in February 2003. In this installation I took the theme of the United Nations negotiations room as its determining narrative environment. Needless-to-say, in the political climate of the time the participants fully engaged and debated issues concerning the looming invasion of Iraq. Again by chromakeying two remote sites the performers shared the same telepresent space, but in this particular installation I chose to construct two identical environments based on Ames's distorted room experiment from the 1960's—onto which an 'unreal' computer game model of a United Nations negotiations room was mapped. The two users were placed within a shared telepresent UN environment—four walls and table. However, one participant appeared to be a great deal smaller than the other and when walking around the room their scale altered in accordance with the monocular perspective of the Ames's room optical illusion.

---

[4] <www.methodsnetwork.ac.uk/redist/pdf/act18_sermon.pdf>

In addition to this disorientating perception of space the immersive nature of this piece was further enhanced by the use of head-mounted displays that physically denied the user/actor any visible connection to the outside installation space. The two performers no longer had the ability to control their movements in distant monitors as in previous installations; instead they viewed their movements via a fixed camera position alone. In effect , removing their eyes from their head and investing them solely within the camera, on which they have to rely to navigate within the space; resulting in complete telepresent immersion. As one participant exclaimed after 20 minutes of dialogue "I've just realised I'm talking to myself here, people outside must be wondering what an earth is going on." For such an installation theme and context to function as an open narrative structure, I was forced to introduce the capability of speech, which was exploited to the point that participants began to lose the sense and importance for time, place and other appointments. Complete strangers were desperate to talk and share there feelings about the looming situation, for up to half an hour on occasions, making comments such as:

> "I feel I could make some decisions here that would be viewed later as quite uncharacteristic"

A topology of body, space and location

In drawing towards a conclusion I would like to use a quote from Machiko Kusahara, referring to the topology of body and space in my telematic performance, specifically the bed interface which was first realised as "Telematic Dreaming" in 1992 in Finland.

> "Within this series of works that employ the teleconference mechanism Telematic Dreaming surely has the most powerful impact because of the dissimilating effect of the bed, a sign shared by everyone. By putting audience participants in that familiar situation from TV drama of getting into bed with someone one has just met, this work drives one, or the member of the audience before one's eyes (the performer), into a state of bewilderment. Members of the audience are placed in the position of the actor, who plays out a bed scene on stage or before a camera, or the voyeur who peeps in on the acts of others. This is a secret act taking place in a public space, and that public space is a virtual space that does not exist in reality. Furthermore, despite the fact that the body is the only means of communication therein, the body of the other party is ghost-like, without substance. This contradictory situation not only confounds the audience, but also, after first releasing them from the logic and restrictions of daily life and dismantling the various elements of signatory identity and the biological

environment of the body, it enables experimentation with and enjoyment of the role the body plays in communication. The virtuality of the space enables it to maintain both theatrically and the context of daily life at the same time. In the interstice between material physicality and an informational space in which electronic signals collide, Sermon reverses the meanings and sensibilities tied to daily life and provides us with an opportunity to think about the essence of communication." [Kusahara 127]

Machiko Kusahara points out that the telepresent body is both simultaneously present and yet at the same time virtual. In the domestic context of daily life we slip back and fourth between the roles of performer and viewer and here we return to Lacan and his mirror stage—an observation my own daughter made a few weeks ago when she told me about her new friend who has the same colour hair as her, the same clothes as her and the same name as her. When I asked her where her friend is, it was "Lacan's mirror" that she showed me:

> "... he experiences in play the relation between the movements assumed in the image and the reflected environment, and between this virtual complex and the reality it reduplicates—the child's own body, and the persons and things around him." [Lacan 2]

At home with Jacques Lacan

This lead to the development of a further piece of work developed for The Biennale of Electronic Arts Perth (BEAP) in 2004 that included an arrangement of rooms, including Lacan's bedroom, living room and dining room. The final room and interface the user confronted before exiting out the back door, was the bathroom—and its mirror. What initially appeared to be a normal mirror lacked one essential truth—the viewer's own image. A momentary illusion that was broken only when the user realised that the mirror was in fact a window into an identical room. Whilst the actor become accustomed to accept their existence in telepresent forms throughout the installation, they were finally denied the most simple telepresent truth they expect from a mirror. A process Lacan identifies by putting the notion of the real and the virtual into question.

> "We have only to understand the mirror stage 'as an identification', in the full sense that analysis gives to the term: namely, the transformation that takes place in the subject when he assumes an image." [Lacan 2]

By representing the domestic reality inside the installation as a fabrication of the technological apparatus outside. "At home with Jacques Lacan" attempted to present all realities as a construct of language.

The double consciousness experience

What actually made *Telematic Dreaming* most successful for me, was not the latest ISDN or Internet videoconferencing technology of its time, but the interface—the bed. The bed was the most psychological charged complex piece of "technology" in that whole piece. Perhaps today I am still looking for other such interfaces that function, just like the bed—as a portal between the human avatars we control as performers inside the matrix and ourselves. This dilemma aside—my ultimate aspiration is as always to go beyond the keyboard and screen-based interface and experience telematic communication within an immersed telepresent location.

In January 2001 I exhibited *Telematic Dreaming* at the Wroclaw Media Festival in Poland, for the twenty-first time since its realisation in 1992. Due to certain technical complications I was forced to exhibit the two beds in the same exhibition space, they where placed 25 meters apart across an open gallery space making it possible to have sight of the other bed. This was the first time ever that the two participants/performers where able to see each other in both a present and telepresent state simultaneously. The acute feeling of having two bodies in this installation set up was extremely alarming; by simply placing my hand on a pillow on one bed it was simultaneously placed on another bed 25 meters in front of me. The experience brought pure clarity to the notion of a double consciousness that Roy Ascott defines as:

> "The state of being which gives access to two distinctly different fields of experience, or two distinctly different views of the same event, or two separate or remote locations, at one and the same time." [Ascott 87]

Bibliography

Ascott, Roy. "Gesamtdatenwerk. Connectivity, Transformation and Transcendence." *Ars Electronica: Facing the Future.* Linz: MIT Press, 1998.

Eisenstein, Sergei. *Film Form—Essays in Film Theory.* Trans. and ed. Jay Leyda. New York: Harvest Books, 1972.

Kusahara, Machiko. "A Topology of Body and Space." *The Museum Inside The Network,* Tokyo: ICC/NTT Publishing, 1995.

Lacan, Jacques. "The Mirror Stage." *Écrits: A Selection.* London: Routledge, 1989.

Zapp, Andrea. "net.drama://myth/mimesis/mind_mapping/." *New Screen Media—Cinema/Art/Narrative.* Ed. Martin Rieser and Andrea Zapp. London: BFI, 2002.

Mobile Feelings: Wireless Communication of Heartbeat and Breath for Mobile Art

Christa Sommerer and Laurent Mignonneau
(IAMAS Institute of Advanced Media Arts and Sciences, Gifu Japan in collaboration with France Telecom Studio Créatif, Paris)

Abstract

"Mobile Feelings" is a mobile art project where users can send and receive body data over a wireless communication network. Specially designed "Mobile Feelings" devices allow remote users to feel each other's heartbeat signals and breath over distance.

The system explores novel forms of intuitive and non-verbal communications that go beyond the conventional transmission of voice, sounds and images used in standard mobile communication. "Mobile Feelings" enables intuitive bodily communication between remote users by exploring the emotional quality of touch and breath as some of the less explored communication senses.

Background

Research in the field of "affective computing" concerns itself with analyzing human emotions by measuring physiological signals and extracting corresponding affective patterns. This is done for example through analyzing facial expression, gestures, voice modulations or a change in autonomic nervous system activity such as accelerated heart rate or increasing skin conductivity [Picard and Klein 2002,14,2]. "Wearable Computing" [Mann] is one of the possible applications of effective computing and "Affective Jewelry" [Picard and Healey 1997] can for example include low-cost biosensors into earrings, rings and even shoes to measure the wearer's emotional state. In 2001 France Telecom Studio Creatif developed the "Multimedia Scarf"[1] which contains a tactile screen, a telephone and a camera through which users can connect to the Internet to send video mail, videophone communication, listen to music or watch a film. All electronic parts were hidden in the lining of the tissue. Other versions of smart cloths, called also CreateWear™ were recently introduced at the 3-GSM World Congress in Cannes.[2]

Media artists have long explored the sense of touch for tele-virtual communication. In 1992 Norwegian media artist Stahl Stenslie and Kirk Woolford devised a

---

[1] France Telecom Studio Creatif: <http://www.studio-creatif.com/Gb/Vet/ Vet02Prototypes03Fr.htm#>

[2] 3GSM World Congress, <http://www.3gsmworldcongress.com/fashion/default.asp>

"cyberSM" body suit that connected users in Paris and Cologne [Stenslie]. Users on these Locations wore special "cyberSM" suits and by touching images of virtual bodies on a screen they could send the touch coordinates and its intensity to the remote user. As a result both users could perceive strong physical feedback including vibrations and shock onto their bodies, which created a strong sense of presence and connection between the remote users. In 1995 Japanese media artist Naoko Tosa et al. demonstrated a system called "Networked Neuro-Baby" where users between Tokyo and Los Angeles could remotely shake hands by squeezing a specially designed "Handshaking Device" that measured the handshake's pressure and relayed the position and pressure data to the remote user through a force-feedback interface.[3] In 1995 Australian media artists Stelarc devised a system called "Ping Body/Proto-Parasite," where on-line users could remotely control Stelarc's body, which was connected to the Internet with various sensors, motors and actuators attached to his muscles and extremities.[4]

Conceptual Consideration

With the rapid advances in mobile communication technologies, ubiquitous computing, ad-hoc networks and wearable devices these fields have become full-fledged research areas. Most research applications in mobile communications and ubiquitous computing however concern themselves primarily with the transmission of image, voice and sound information. Considering that human communication is not only based on conscious communication of information but often also includes unspoken, intuitive and sensual information exchanges, we set out to construct wireless communication devices that let users communicate in a very intuitive, emotional and private fashion. The sense of touch still remains one of our most private sensations for which we still lack a concise descriptive language [Heller and Schiff].

Another motivation for this project comes from the consideration that mobile phones have transformed our social and individual lives in such a radical fashion (as described by Sadie Plant[5]) that we all have become to accept a decreasing sense of privacy in exchange for connectivity and mobility. To explore this relationship

---

[3] Tosa, N., Hashimoto, H., Sezaki, K., Kunii, Y., Yamaguch, T., Sabe, K., Nishino, R., Harashima, H., Harashima, F. 1995. Networked Neuro-Baby with robotics hand (An automatic facial expression synthesizer that responds to expressions of feeling in the human voice and handshake), demonstration presented at Siggraph'95, Los Angeles.

[4] Stelarc. 1995. "Parasite Visions: Alternate, Intimate and Involuntary Experiences," at: <http://www.stelarc.va.com.au/articles/index.html>

[5] "On the Mobile: the Effects of Mobile Telephones on Social and Individual Life," Study report for Motorola Inc

between privacy and ubiquity and connectivity, we created the "Mobile Feelings" project, which was first showed and published at the Ars Electronica 2003 in Linz, Austria [Sommerer and Mignonneau 2003, 258-261].

Mobile Feelings—Description

Research on "Mobile Feelings" started in 2001 as a collaborative project between the authors, the IAMAS Institute of Advanced Media Arts and Sciences and the France Telecom Studio Creatif. In 2001 a joined patent was applied.

Mobile Feelings Devices

We constructed 6 egg-shaped "Mobile Feelings" devices (shown in Fig. 48) which each contain various sensors, actuators, micro-controllers, batteries and a wireless communication module.

Micro sensors, Actuators and Microcontrollers

Each of the "Mobile Feelings" devices hosts various miniature sensors, such as a pulse sensor, a touch sensor and a breath sensor as well as a micro-motor, a micro-ventilator, 2 white LEDs and a rechargeable battery. In addition, a 16 Mhz microcontroller, 3 pre-amplifiers and 1 amplifier are contained in each device as well.

Wireless Communication Module

Each "Mobile Feelings" device also contains a Bluetooth module, which can either establish direct connections between the 6 devices in a range of 10 meters or communicate to a nearby Bluetooth-enabled device (such as a PC or PDA), which connects to the Internet or to a mobile phone network using a Bluetooth-capable phone. This allows all 6 devices to wirelessly communicate to each other and send the information over the Internet or telephone network to remotely Located users. An image of the complete "Mobile Feelings" device including sensor, actuators and communication module is shown in *Figure 41*.

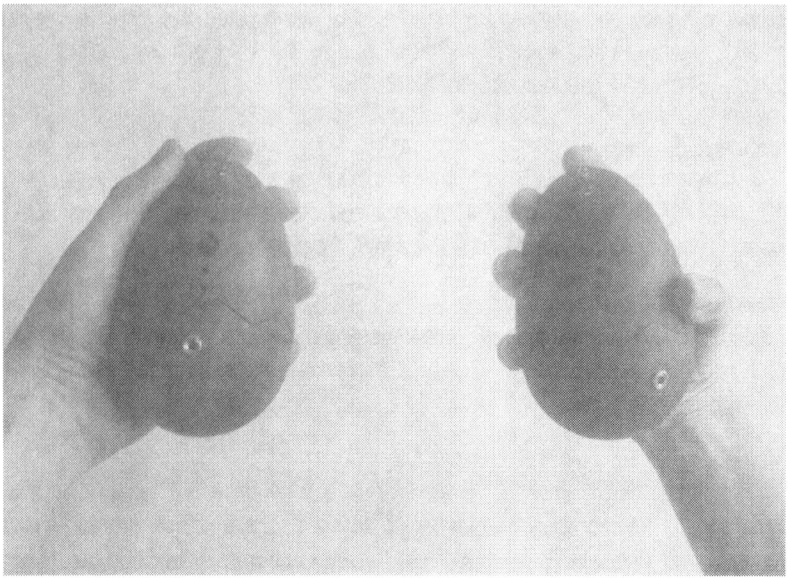

Figure 41: Two "Mobile Feelings" interface devices which enable users to wirelessly transmit and receive each others' heartbeat and breath.

Mobile Feelings Communication

As the user picks up one of the "Mobile Feelings" devices and places his/her finger on the pulse sensor on top of the egg-shaped interface, a LED light starts blinking: it shows the strength and frequency of the user's own heartbeat. When the second remote user also holds a "Mobile Feelings" device, his/her pulse data are captured and visualized by the LED on his/her device as well.

Communicating via heart beat

Each device holds a second LED, which also shows the heartbeat frequency and strength of the remote user. This provides both users with visually feedback about their own pulse and that of the remote user. In addition to seeing their heartbeats via the blinking LEDs, both users feel a strong rhythmic pulsing in their palm, which corresponds to the actual heartbeat of the remote user. This haptic sensation is created through an actuator consisting of a micro motor that moves a small piece of metal, creating the sensation of pulsing. No sound is generated from this motor, as the pulse rate frequency is fairly low.

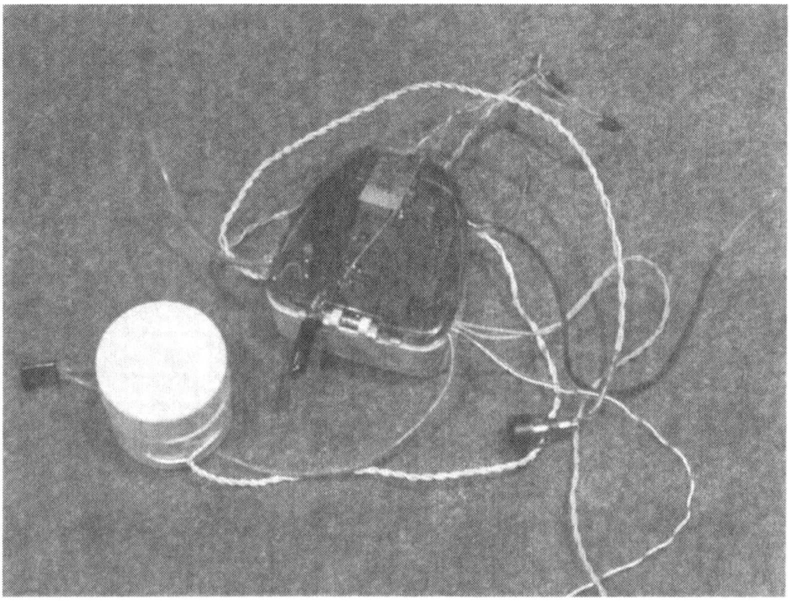

Figure 42: "Mobile Feelings" interface device containing a pulse and a touch sensor, breath sensor, a micro-ventilator, a micro-motor (inside a yellow box), 2 LEDs, microcontrollers and a Bluetooth module (both inside the blue box). Size around 5 cm.

The exact frequency and strength of the remote user's heartbeat is received via the wireless Bluetooth module at the devices' CPU, which relays this information to the actuator's movement. This results in an immediate haptic feedback of the remote user's heart beat and its frequency in the form of a rhythmic pulsing which varies from user to user depending on his/her heart beat and even emotional and physical state. The communication speed between sending the pulse from one user and receiving the pulse, as haptic feedback for the remote user is almost instantaneous, as the communication speed between the devices is around 38 KHz. Both users can thus feel a strong sensation of bodily connection, which feels like "holding each other's heart in their hands" and feeling the other users heartbeat and its strength through these devices. Figure 43 shows a snapshot of two remote users at the European Media Art Festival in Osnabrück [Sommerer and Mignonneau 2004, 68-71] where this system has been presented in 2004.

Communicating via breathing

In addition to sending and receiving heartbeat data, each "Mobile Feelings" device also hosts a breath sensor as well as a micro-ventilator. When a user breathes onto his/her device, the heat of his/her breath is captured and analysed and sent to the remote user's "Mobile Feelings" device. There, the breath data are transformed into a small wind that comes out of the device through a small micro-ventilator. It instantaneously blows as small wind into the face of the remote user, creating an additional effect of bodily connection.

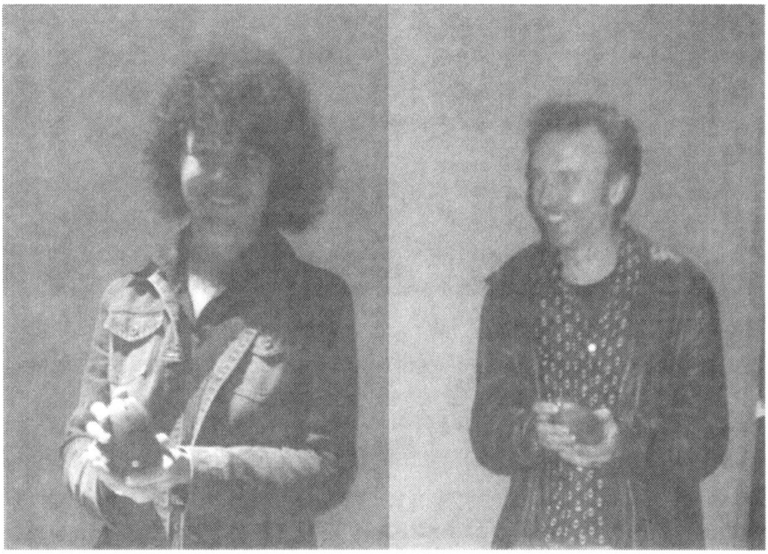

Fig. 43: Two remote users as they communicate with each other through their heartbeats.

User Evaluations and Observations

Users who communicated with each other through the "Mobile Feelings" devices described their experience as very unusual and slightly unsettling, as in real life one usually does not touch others much, let alone breathe at them. Many users found the experience also to be very comforting and sensual, reminding them to touching a lover, a child, their mother or other persons with whom we usually share private feelings through touch. Young users of opposite sex also reported that the found the devices to be good "flirting tools" as it lets them feel and touch each other without having to talk to each other.

Another observation we made during the exhibitions of the system [Sommerer and Mignonneau 2003, 2004] was that when users concentrate on the sense of touch, they often reduce other sensory input channels such as vision and sound. It almost appears as if the sense of vision and sound is so predominant in our daily live interaction and communication, that we have to specifically focus on the sense of touch when we fully want to feel something as strong as another person's heartbeat. Gault describes touch as a very strong "break-in" sense: coetaneous sensations, especially if aroused in unusual patterns, are highly attention-demanding [Gault 155-159].

Poupyrev *et al.*, who explore the implications of using tactile feedback for mobile interfaces describes the sense of touch as having "a strong emotional impact. Running a finger into a splinter, touching a cat's fur, or immersing into some unknown sticky substance all bring intense, though very different, emotional responses. Touch is fast, needs little conscious control, allows for information encoding, and produces strong emotional responses" [Poupyrev, Maruyama and Rekimoto, 51-60].

Future Work and Applications

"Mobile Feelings" is an art project that explores novel forms of communication, which include sharing touch and breath over a mobile communication network. The resulting experience has proven to be very unique as it lets complete strangers share private body sensations over a mobile network. While the system clearly breaks the conventional boundaries between private and public space, the potential of the system lies also in its strong emotional impact when used in special setting such as flirting or creating a bodily connection between remote users in a public setting. A future application of this system could include personalized "Mobile Feelings" devices that can carry the heartbeat of loved ones whom one can "feel" when being alone or in stress. Similar research for stressed children has recently been presented [Parness, Guttman, and Brumback, 93-95], showing that the sense of touch will become a more valuable sense for future mobile communication. Further applications of our technology and idea could also include mobile network games as well as the expanding field of mobile[6] and wireless[7] art.

Finally, out current research investigates how to capture and transmit further haptic sensations, such as materials and textures and also how to capture and relay olfactory information via mobile communication networks to create an even stronger connection and sense of presence between remote users.

---

[6] Mobile Connections <http://www.mobileconnections.org/>

[7] Medosch, A., 2004. "Not Just Another Wireless Utopia - developing the social protocols of free networking" in: Future Sonic conference, Manchester, <http://www.futuresonic.com>

Bibliography

Gault, R. "Progress in experiments on tactual interpretation of oral speech." *Journal of Abnormal and Social Psychology* 14 (1925).

Heller, M. A., and W. Schiff, eds. *The Psychology of Touch.* Hillsdale, NJ.: Lawrence Erlbaum Associates, 1991.

Mann, S. "Wearable Computing: A First Step Toward Personal Imaging." *Computer* 30.2 (February 1997).

Picard, R., and J. Healey. "Affective Wearables." *Proceedings of the First International Symposium on Wearable Computers*, October 1997.

– – –, and J. Klein. "Computers that Recognise and Respond to User Emotion: Theoretical and Practical Implications." *Interacting with Computers* 14.2 (2002).

Sommerer, C., and L. Mignonneau. "Mobile Feelings." *CODE—The Language of Our Time.* Ars Electronica, Hantje Cantz Verlag, 2003.

– – –. "Mobile Feelings" *EMAF2004 European Media Art Fesival catalogue.*

Stenslie, S. "Wiring the Flesh: Towards the Limits and Possibilities of the Virtual Body." *Ars Electronica'96. Memesis. The Future of Evolution.* Vienna/New York: Springer Verlag, 1996.

Poupyrev, I., S. Maruyama, and J. Rekimoto. "Ambient Touch: Designing tactile interfaces for handheld devices." *UIST'2002. ACM.*

Parness, A., E. Guttman, and C. Brumback. "Squeeze Me: A Portable Biofeedback Device for Children." *UbiComp 2003*, the Fifth International Conference on Ubiquitous Computing Proceedings.

The Living Room
Victoria Fang

"The Living Room" is a narrative puzzle installation. Players take part by moving panels with LCD monitors in an effort to solve a 'whodunit' murder mystery. When all three panels are positioned correctly, video 'memories' are triggered with which you are able to unravel the mystery. Through the position change of the panels, one's experience with the space changes in every scene. Ultimately, the installation functions as a giant physical puzzle that houses a narrative puzzle.

Concept

This project arises from interests in narrative gaming and theatre. It is a spatial entertainment experience designed to deliver an entertaining and cathartic experience that challenges, frustrates, and rewards. The project also aims to encourage a compelling spatial interactive experience through its narrative and game-play, using the advantages and opportunities provided by digital video and wireless technology. "The Living Room" tells a story through location-triggered video clips within the structure of a puzzle-based game, turning the space into a storyteller.

Narratives superimposed or intertwined with our real world are becoming increasingly prevalent, from wireless urban games to alternate reality games to SMS graffiti. These narratives arise from the changing technologies and audiences, which allow them to exist. As technologies evolve, new forms of entertainment will emerge. This project aims to participate in this progress by combining the elements of traditional entertainment with emerging technologies in a unique way. It explores the possibilities that wireless and RFID technologies provide and applies them to entertainment and play. The concept of spatial narrative is not new, but we are finding new ways to introduce these narratives with the technologies that are emerging.

"The Living Room" is a spatial movie experience—the user is in front of the movie, between the movie, around the movie, and physically shifting the movie. While the movie itself is primarily linear, players' participation to progress the fragmented narrative results in changing their surroundings, both physically and narratively. Each segment of the film forces players to interact with the space in a way that compliments the content of the scene they are watching.

Experience-Designed

In game structure, "The Living Room" is similar to a text-based adventure game, with segments of narrative intercepted by puzzles. Players begin by entering the 12'x12' space and receiving a set of general rules. The game imposes a loose role upon the player, as the child witness of a murder. The design choice to make the player assume the role of a child allows the user to be a silent yet important presence in every scene and demands very little active role-playing, aiming to appeal to a wider audience. The space contains 21 childlike drawings scattered across the floor. These are the clues that players must connect and group together in order to solve the mystery. The screen panels move feely about the space, in any compass direction and rotation. The interface is a vehicle for storytelling but does not attempt to disguise the players' physical reality—in other words, the experience is not a virtual reality experience. To the contrary, the transformation of the space due to the positioning of the units is part of the designed experience.

Through experimentation and teamwork, players discover the game system and work their way through the narrative by identifying and remembering clues to trigger video sequences. When the panels are positioned over the three correct clues, the corresponding scene begins to play. Each of the scenes is viewed through the three moving panels as if through windows, taking the players through the events of the day of the murder. Players discover new clues within the scenes, which they can use to unlock future scenes. As the puzzles get more difficult, the spatial layout of the screens becomes increasingly non-traditional and encourages more interaction with the physical space. This spatial layout creates unique tensions with each scene. For example, players might feel trapped between the screens during an argument, or frantically running around the screens during the murder, or calm while watching a panoramic dinner scene. All scenes take place in the same room (the living room), giving the players greater knowledge of that room as the game progresses. Similarly, the objects in the floor drawings correspond to where they are viewed in the space of the room, allowing a virtual room to arise from the empty space. Both narrative and temporal information is location-based, delivering a fragmented narrative experience through space and time that the player must piece together.

"The Living Room" uses familiar mechanisms to guide the players through the experience. An investigator acts as the game guide, providing hints as players try to work through the puzzles. The story is that of a 'whodunit' murder mystery, with straightforward motives to enable quick recognition of plot points and the instinctive desire to solve the mystery. The use of puzzles to progress the narrative is meant to rouse this instinctive desire as well, creating a tension that seeks resolution and satisfaction. What arise from this project are both expected and unusual interactions and problem-solving techniques, as described in the real-life observation of the installation.

Experience-Observed

The installation ran in both New York, at the Parsons MFADT Thesis Show in 2004, and in Berlin at the Transmediale festival in 2005. In its first iteration, the installation was run on an appointment basis, with 2-5 players and my presence as moderator. In its second iteration, the installation was run in a more freeform manner, with numbers varying from 1-20+ and my sporadic presence as a moderator or as an anonymous observer. What follows are my observations regarding user interaction in both instances.

As with any interactive piece, the initial engagement with the piece relied on a sense of curiosity and a certain amount of dedication. This was even more important with "The Living Room" due to the project's puzzle structure, and this was probably the most difficult interaction to control. Not surprisingly, most users had no interest in instructions or rules, and preferred to learn through experimentation or in-game guidance, despite the unfamiliar interface. With a moderator, players took their time in experimentation, in the safety of knowing there was help available. These players also had more dedication to the experience, knowing that they would be able to complete all the puzzles, even if they were to get stumped at some point. Without a moderator, there was a responsibility left on the players to discover and engage with the system. This happened either through experimentation and/or observation. The most successful interactions arose when players were intrigued but not overwhelmed by their frustration. If players felt that the puzzles were too difficult (and likewise, too simplistic), or if they did not know what response to seek, they quickly lost interest and did not engage. When a satisfying balance of intrigue and reward was met, players took a dedicated interest in the piece, and usually saw the entire piece through to completion.

Players interacting with the piece were constantly working together, with a dialogue and agreement necessary to allow puzzles to be solved in a logical and systematic way. Since all three units had to be placed correctly to trigger a scene, players had to work as unit to agree on what solution to try, and to keep track of which solutions had been eliminated. They wandered the floor in thought, examining the available clues, conversing with one another about potential solutions. Sometimes, one player might take the initiative to move a unit and try a solution. More often, a solution would be suggested and a few players would move units simultaneously to cooperatively expedite the process. Successful solutions were met with satisfied looks or remarks, while unsuccessful solutions led players immediately to scanning the floor for different clue combinations. When a newcomer disrupted the system, players were quick to object, immediately forcing the newcomer to assume a role within the community of players. To describe the types of interaction that I observed, I am able to identify three main personalities: the protagonist, the reluctant protagonist, and the observer.

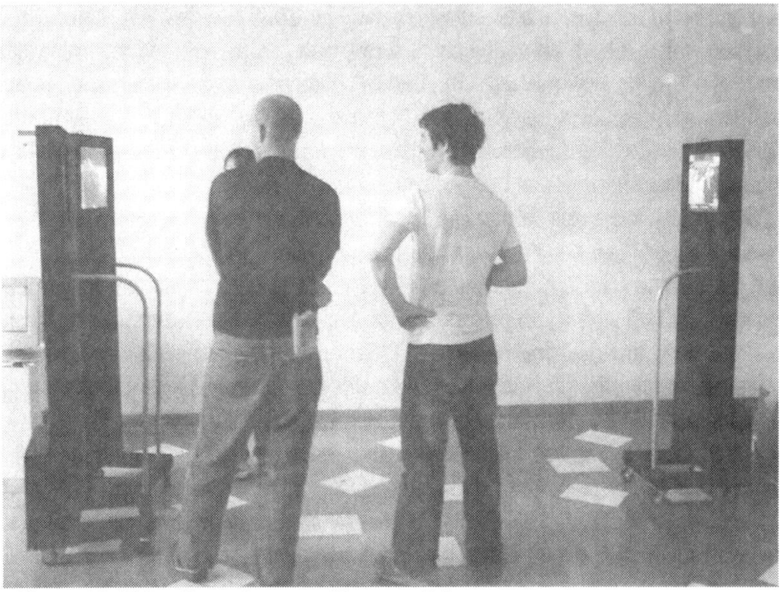

Figure 44: Living Room in use at Transmediale

The protagonists often began the experience through experimentation, unintimidated by the unknown system. The boldness of these players was crucial to getting the experience under way. These players took the active lead in moving the units, in an immediate effort of trial and error. Eventually, these players would end up seeking ideas or willingly incorporating suggestions from onlookers. Suggestions were offered up amidst friends and strangers, with the most accurate suggestions often arising from the reluctant protagonists. These players were hesitant with their ideas, only offering them up when all other options had been exhausted, and many times, when they could not stand it any more. They would remain quiet observers while bolder players threw out suggestions, waiting for someone to offer up what they knew to be the answer. I watched several instances when a player of this type would repeatedly offer up an idea almost inaudibly, to themselves or to their companion, until they had gradually increased the volume to an insistent suggestion. These players were also more hesitant to actually move the units, preferring to tell one of the protagonists where to move them. Ultimately, however, they would do so themselves when no other option arose. Observing these players offered the most demonstrative evidence of the puzzle instinct, wherein the tension created by their desire to solve the puzzle urged them to step outside of their normal comfort zone. In the community of players, the attention would then shift, with players expecting the newfound protagonist to provide their insight on the next puzzle. The reluctant protagonists became active members of the collaborative effort, in a role that they would not have naturally assumed. Sometimes these players would step back away from this role, but more often, they shared the active role with the protagonist players, playing out the rest of the game. The third player type: the observer, watched without getting involved, out of politeness, intimidation, or laziness. Some of these players found the experience too involved to actively participate. Some of these players were waiting for the round to end so that they could become the new protagonists.

Players moved organically amidst one another to adapt to the ever-changing space. Once a video was triggered, audience members naturally shifted to view the video content. Players would sit on the floor so as not to block the view, lean around one another or lean around the units to catch a glimpse. Heads would turn back and forth as the action shifted, with audience members pointing when they saw something of interest. The space changed not only with the physical layout of the units or the digital video content playing, but also with the arrangement of bodies and attentions shifting amidst the screens.

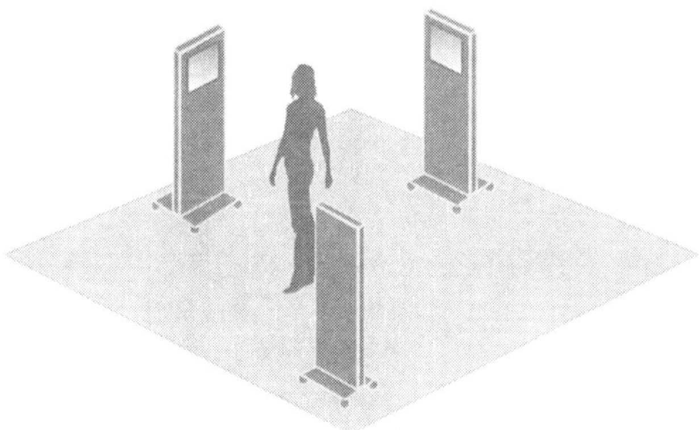

Figure 45: Schematic of arrangement showing wheeled screens.

The project successfully pulled strangers together in a collaborative effort, with ideas being sought and offered. While some people initially watched from a disengaged perimeter, the spatial nature of the piece forced movement, even amidst the most passive of spectators. When one group completed the narrative, they were often more than willing to start off the next group with some hints and instructions. People would return with friends and became improvised moderators. Onlookers that joined halfway through a session would be more inclined to take initiative as protagonists in the next round, wanting to experience the story from the beginning with insight into the system that might otherwise have intimidated them.

Conclusion

In its designs as a spatial narrative, "The Living Room" managed to involve its audience and create interesting and rewarding spatial interaction. Some of the interactions observed were intended and designed. Others arose out of the real world implementation of the project. Despite the unusual interaction and presentation of content, players naturally adapted to the environment and to each other. They tackled the learning curve of an interface they had never seen before and successfully shifted gears, both mentally and physically, to adapt to a spatial narrative environment. The experience resulted in roles being formed within a collaborative community of friends and strangers. Players occupied the space of "The Living Room," organically moving with the physical, narrative and social shifts, as they worked together to progress the experience.

tunA and the Power of Proximity

Arianna Bassoli

Introduction

Once upon a time there was Globalization. Internet and mobile phones liberated individuals from the constraints of physical proximity and allowed them to communicate with others all over the World. A perpetual feeling of connectedness [Katz and Aakhus 2002] convinced millions of people that they could never be too far from each other and that the freedom of a virtual space could help strangers socialize with unexpected ease. Powerful Information and Communication Technologies introduced new means of interaction and added a multitude of facets to the traditional face-to-face form of interpersonal communication.

Nowadays, while the new media revolution seems far from an ending, part of the technological progress is taking a centripetal direction. New proximity-based ICTs are in fact introducing an overlap between the virtual and the physical space where people currently interact, with the result of generating the oxymoron of *mediated* face-to-face communication. Short-range wireless technologies are partly responsible of this process of remediation, where socio-technical networks can be created among co-located strangers and acquaintances, sharing digital resources and communicating through mobile devices. Bluetooth and WiFi do not only increase the ubiquitousness of ICTs and give access to the World Wide Web, but they also provide a link between people who are at a relatively short distance from each other, opening up a whole range of unexplored potentialities in terms of human-human interactions.

Proximity-based Social Technologies

Sceptics could easily suggest that it is redundant to support through ICTs a level of communication that is already available and codified. In other words: why would people in physical proximity want to communicate with each other in a virtual way? The answer to this question is still unknown, but practitioners and research centers are currently investing time and money in exploring this new opportunity. Commercially, the gaming industry is pioneering, by having recently introduced portable consoles that encourage user to play and chat with their co-present friends,[1] while stakeholders of the mobile phone industry are slowly trying out a way

[1] See for instance the new portable Sony Playstation and Nintendo consoles

to connect strangers over Bluetooth through mechanisms of personal profiling.[2] From a research perspective, a wide range of interesting projects can be grouped under the umbrella term of 'proximity-based social technologies'; these projects[3] try to explore potentials and limits of these ICTs before they are being commercialized, and attempt to answer the question of their social desirability. Mostly implemented on existing mobile devices, proximity-based social technologies (PBST) introduce individuals to their surroundings as virtual entities in possess of digital resources.

To partly understand the importance of conducting research in this field and to approach proximity-based social technologies from a closer angle I will introduce a project that has been developed at Media Lab Europe, called tunA.

tunA: Sharing Music on the Move

tunA[4] is an application that runs on Wi-Fi-enabled and Pocket PC PDAs; it works as a standard MP3 player, but it also displays a list of other people in range, providing some information about them (i.e. icon, nickname, song currently playing). tunA allows browsing the playlist of others and to listen to the same songs in synchronicity. An instant messaging feature is also provided, in order for the users to be able to exchange messages about music or any other topic. Finally tunA gives the possibility to bookmark the songs other users are playing, to keep a record of interesting new artists being discovered while moving around.

The tunA prototype has been evaluated through a user study where qualitative and quantitative methods have been applied to understand the usability of the technology and a potential real world usage of it within a specific community. The user study has been conducted within a university college in Dublin, where, following a survey and an interface evaluation, a field study has been performed with six students using the application throughout a day of campus life. While results from this study cover a wide range of topics, I will here focus on particular aspects that can relate the use of tunA to PBSTs.

Are you talking to me?

The concept of proximity between users interacting through PBSTs is mainly defined by technical constraints. Theoretically, the range of WiFi used in peer-to-peer mode to connect two mobile devices is of 100/ 200 meters; nevertheless, within the tunA user study students were not able to spot each others on their PDAs when at more than 30 meters of distance. While this could change in the future

---

[2] Consider the Nokia Sensor application

[3] Two among others: Serendipity developed at Medialab and Jabberwocky developed at Intel Labs

[4] See:<http://web.media.mit.edu/~stefan/hc/projects/tuna/>

because of further technical progress, the short range implies now that mediated interactions can happen only among people who can also see each other, and this will provoke interesting new social dynamics.

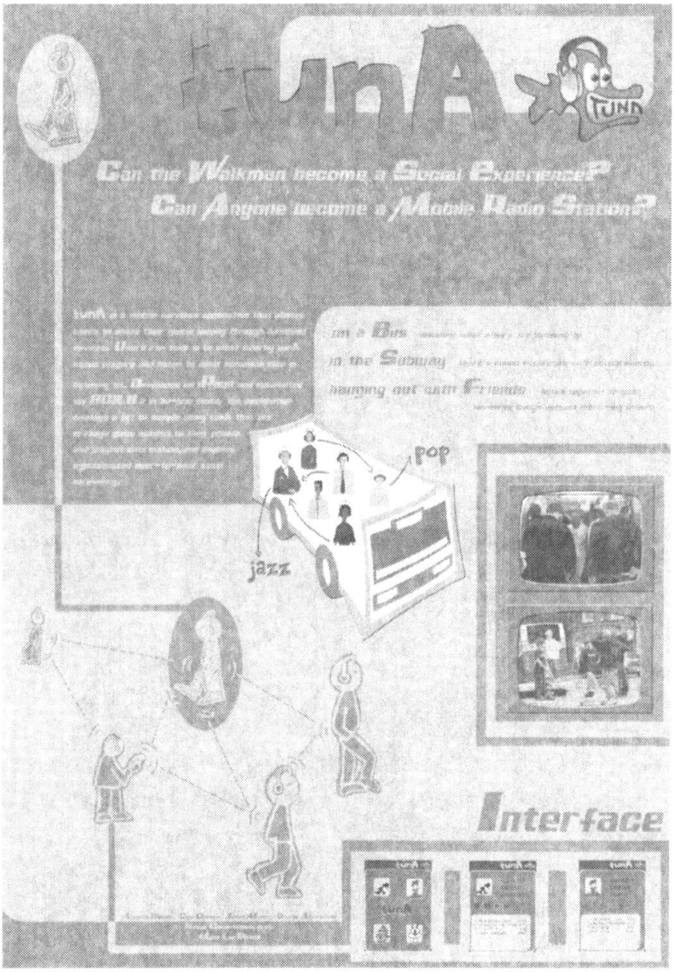

Figure 46: Poster advertising tunA

During the trial students were represented by a nickname and a generic icon that would not make explicit their identity. Despite the proximity, they did not always know whose songs they were listening to when they were connected to another device, and, instead of asking each other face-to-face, they would have preferred to use tunA to send messages, which at the time was not possible because of technical problems.

"There was one time or two times maybe that I managed to tune in into someone and I didn't know who they were, and I would have sent a message to them ... I just think it would have added another dimension to it."

Unexpectedly, even between friends the opportunity to send messages to each other seemed very appealing. The recent success of SMS and Instant Messaging suggests that text-based communication is becoming increasingly important and that it might partly replace voice-based communication over distance. Results from the tunA trial show a further possible tendency, which implies an integration between face-to-face interactions (based on voice, gesture etc.) and a virtual level of text-based communication between co-located users.

'It could have been like trying to have a bit of a laugh with someone, just make fun or whatever. Like if Alan (friend) had a good song, like 'well done' or whatever, or have a laugh about crap tastes in music, like that sort of things. I think if you are out (from the college) it would be more: 'hello', this kind of stuff, 'I like that band as well' or bla bla bla."

Reasons for connecting

Communication happening through PBSTs could nevertheless become popular mainly between people who already know each other, or within a trustworthy environment such as the workplace or a university college. While the popularity of social network websites such as Friendster, Orkut and the Facebook shows that there is a general desire to make new friends and that mediated interaction can succeed in supporting this process, the physical proximity of users might constitute a barrier when we consider PBSTs. The tunA user study supported the concern that people might have problems with giving out personal information to strangers nearby.

"Well I think I'm a bit paranoid about stuff like that. It would depend also on the context, like here in college I wouldn't have problems to put my information, but outside I would probably try to be more anonymous."

"I think less (information to people nearby), cause if like you run across a psychopath in the Internet, they cannot do anything, unless you're stupid and you give out your address, while if somebody is right beside you, it could be more dangerous I suppose."

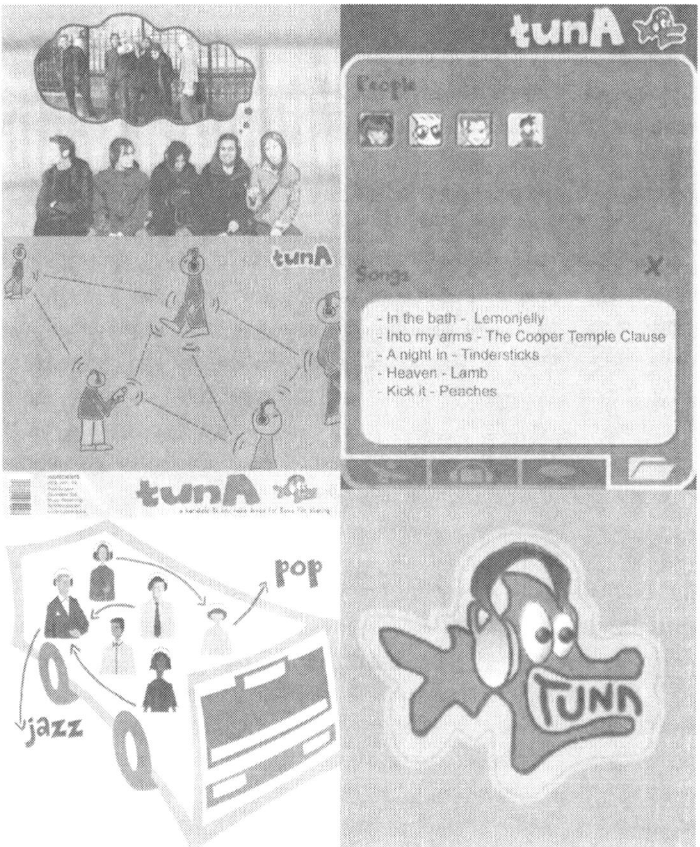

Figure 47: TunA details

Despite the fact that people do get curious about strangers surrounding them, as the tunA study confirmed, the fear of disclosing personal details suggest that new applications based only on personal profiling, such as the new 'Nokia Sensor', might not encourage new social connections. Nevertheless, new mobile devices also allow people to share a wide range of personal resources without having to disclose their identity. Music, as we have explored through the tunA project, seems to be a good incentive to create new social links. Music constitutes a general and broadly appreciated content, but it can also reflect aspects of users' personality. Finding music interests in common can establish a connection among unacquainted people, and encourage users to progressively make their digital resources available to others nearby.

> "Well its something (music) that people may have in common, depending on anything in their life, they could be the worst enemies, but they still could like the same music, and that just establish a common ground for people to kind of talk and whatever. Yeah I think music is a good form of communication. Anyway its something that most of the people appreciate."

Apart from music, personal data stored in new handheld devices could constitute the main incentive that people have to communicate with each other through PBSTs and share their resources. These data will constitute the virtual identity that people project to their surroundings, without having to be explicit about who they really are.

The sharing of personal digital resources or knowledge could constitute the first step of a communication process that brings unacquainted people to develop deeper social ties. A connection established on a virtual level between people in physical proximity could then encourage new face-to-face interactions.

Bus Jamming

tunA was evaluated within a college environment. Despite the fact that some of the participants of the trial did not know each other before, the characteristics of the context where interactions took place made people comfortable about using the application. A different scenario to envision for tunA usage could be an urban situation where strangers have to share the same physical space for a certain amount of time, such as in a public transport.

Studies describe how commuters in cities have a reciprocal attitude of 'civil inattention', which implies the attempt to conceal as much personal information as possible from people nearby, and to show respect, but disinterest in each other [Bassoli et al. 2003]. In those situation the portable music player constitutes, for instance, a way for people to create a personal space, to protect their privacy and to disconnect from the surroundings [Levine et al. 1973].

> "I just use it anytime (portable music player), pretty much everyday if I'm travelling on a bus. Its because I love music [...] and also because I love to get into my own little world, chilling out with my tunes."

In this context, tunA represents a provocative attempt to turn the traditional isolating role of portable music player into a means for socialization, which could in the long-term affect the habit of civil inattention. Interactions could happen in such a way that people could still pretend to ignore the others they are exchanging messages and music with, but the physical proximity could influence the way these interactions develop. What happens, in fact, if I discover that I am listening to the song of the person who is sitting behind me? Do I pretend I am not aware of it or do I engage in a face-to-face conversation? What could be the consequence of discovering that some people in the bus are moving their heads to follow the rhythm of my song?

Conclusions

As an example of PBSTs, tunA suggests that new mobile technologies are providing a wide range of social opportunities which were non-existent before, and that, after allowing people to communicate and share resources over distance, they can now provide tools for becoming more aware of the physical surroundings. Simply having a Bluetooth or WiFi-enabled mobile phone will imply to be potentially part of a dynamic and local network of acquainted and unacquainted people who can communicate and share resources, virtually without having to disclose their identity, or even look at each other in the eyes. Despite the fact that this scenario might look almost worrying, an optimistic view on PBSTs highlights the possibility that such interactions, starting at a virtual level, would eventually lead people to become more engaged with others they share a space with, and decrease their level of civil inattention. In public spaces PBSTs will have in fact the opposite effect that mobile phones and portable music players now have, which can be described as an 'absent presence' and implies the act of being 'physically present but [...] absorbed by a technologically mediated world of elsewhere' [Bull 2000].

It is still necessary that researchers and ICT practitioners keep investigating what could be the right balance between a state of openness and closeness that a PBST should allow to users, in order to protect their privacy and security, but also to ensure that they can be able to exploit opportunities for serendipitous and interesting encounters and personal data exchanges. Moreover, contextual motivation as to why people should connect to each other, need to be researched in order to make PBSTs successful and useful products.

Acknowledgements

The tunA project was developed at Media Lab Europe by Arianna Bassoli, Julian Moore and Stefan Agamanolis. I would like to thank the people who have supported this research and the participants of our user study.

Bibliography

Bassoli, A., et al. "TunA: a Mobile Music Experience to Foster Local Interactions." *Adjunct proceedings of UbiComp'03*, 2003.

Bull, M. "Sounding out the city—personal stereos and the management of everyday life." Oxford, UK: Berg, 2000.

Eagle, N., and A. Pentland. "Reality Mining: Sensing Complex Social Systems, Personal and Ubiquitous Computing." 2005.

Gergen, K. "The challenge of absent presence." *Perpetual Contact: Mobile Communication, Private Talk, Public Performance*. Ed. J.E. Katz and M. Aakhus. Cambridge: Cambridge University Press, 2002.

Levine, J., A. Vinson, and D. Wood. "Subway behavior." *People in places: the sociology of the familiar*. Ed. A. Birenbaum and E. Sagarin. London, UK : Nelson, 1973.

Paulos, E., and E. Goodman. "The familiar stranger: anxiety, comfort, and play in public places." *Proceedings of ACM CHI 2004* , 223-230.

Engagement with the Everyday

Margot Jacobs
(Interactive Institute, Gothenburg, Sweden)

The Play in the Everyday: A Past Focused on Use and Work

Our everyday world is overflowing with a sea of artefacts embedded within information technology, offering the potential of seamless connectivity and unprecedented access to information and communication. With the advent of the Internet, mobile and ubiquitous technologies, and the subsequent ongoing miniaturization of electronics, commercialisation of bits and the growing domination of software over materialized forms, a new digital public realm has emerged in both the physical and virtual spheres. Whether one lives in the city, its suburbs or in a small town, the changes affecting everyday environments and their public spaces are vast and significant to our lives. In the light of this, the usefulness of analyzing and understanding this phenomenon becomes quite evident. It is imperative that we explore how this is affecting our everyday lives now and in the future.

> 'We should shift attention from the acts of consumption, or use and focus not on authorship, but on reading, not on urban design but on walking through the city, not on theories of language but on provisional and improvisational aspects of conversation…' [Ainley, 2001; Blauvelt, 2003]

The key question is no longer of how to use technology to support work-oriented tasks or to be more efficient. Instead, the focus has shifted to how technology can support what lies beyond utilitarian need, exploring. for instance, emotionally-driven human needs such as supportive social systems, sustainability and even desires (such as expression, communication, and reflection). In this era, we should forge new design philosophies, combining information technologies with values of a more aesthetic nature, subtlety, and personal meaning that support happy accidents, serendipitous interactions, and the ebbs and flows and of our evolving lifestyles. In sum: "play in the everyday."

Combining aesthetic and even the poetic aspects of everyday life with digital information calls for an exploration of new materials and surfaces that better reflect new qualities and interactions. Perhaps more importantly it calls for new approaches to design practices, actively working to create methods of collaboration, innovation and action to engage multiple voices in the design of technology for our public lives and local spaces.

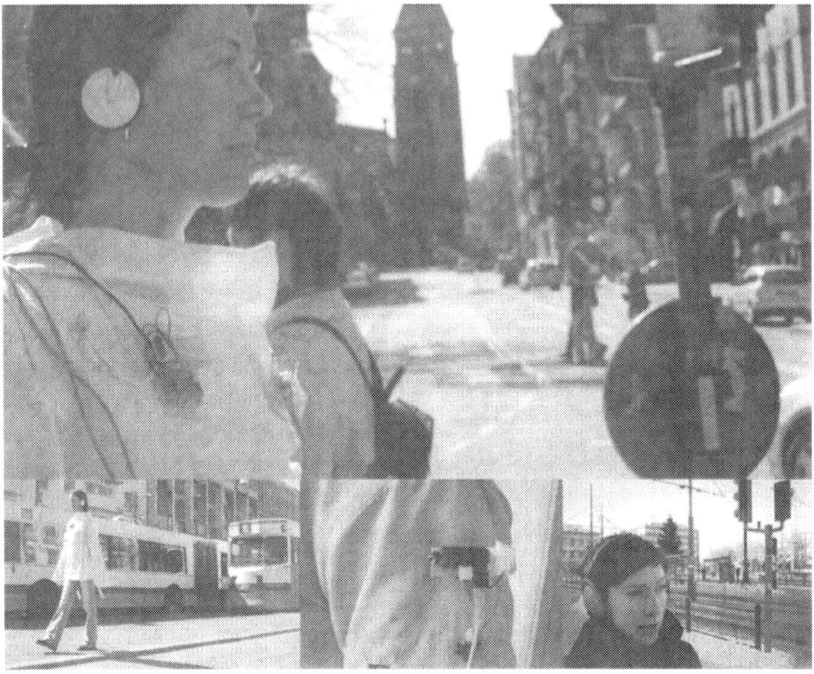

Figure 48: Sonic City

Public Play Spaces

> 'If there is an ambition to make public spaces that raises the stakes for the unknown, for the unexpected, for the moments of status-free intimacy, then the strategies and methods of working in that space have continually to exceed the normal, the expected, the conventional.' [Andersen, Jacobs & Polazzi 2003; Ainley, 2001]

The 'Public Play Spaces' platform at the Interactive Institute was initiated as a means of investigating the surge of computing technologies in the public arena and what the implications are for our society and culture. Essentially, 'Public Play Spaces' is a platform for creative work exploring the playful, emotional and appropriate incorporation of technology into everyday public life, focusing on developing both innovative design methods and experimental prototypes for social interventions in public space. Within this framework, we have taken the opportunity to reflect on, question and reexamine places, relationships and qualities for the design of technology in the public sphere. This requires that we ask different questions, apply new methods and try alternative means of prototyping possibilities [Andreotti and Costa, 1996].[1]

Projects and approaches we have investigated inside the 'Public Play Spaces' platform draw on our combined background and interests in art, architecture, game and interaction design. For instance, we use games as a metaphor for transcending locality and involving people early on in the design process. Inspired by Situationists' ideas of intervention, such as détournement, collective action and methods of agitation for promoting free use and the transformation of the urban environment and Fluxus's notions of experiences including Events and Fluxkits [Bell, Blythe & Sengers, 2005; Andreotti and Costa, 1996; Higgins, 2002], we also explored tactics as to how technologies might offer critique, breaking down current accepted technological practices and challenging people to reflect, as well as to add new layers for expression and participation. The following will illustrate these tactics for intervention and play as well as expand upon our notion of games and game-based methods.

Games: A Framework for Participation

Since 'Public Play Spaces' are based on creating technologies in the public for the public, we believe that the public, or rather people, should have the opportunity to actively participate in the design process. As a result we are committed to exploring methods for engaging people throughout the design process. In addition to applying traditional methods noted for their involvement with people such as performance, participatory design and co-design practices [Andersen, Jacobs and Polazzi, 2003; Brandt and Grunnet, 2000; Buchenau and Suri, 2000; Burns, Dishman, Verplank & Lassiter,1994; Ehn & Sjögren, 1991; Salvador & Howells,1998; Seago and Dunne, 1999], we apply a game-based methodology, drawing from work done in these areas and other artistic practices [Andreotti and Costa, 1996; Gooding,1995; Higgins, 2002] in order to gain perspective about potential use and users.

Games have been successful as a basis for several of the projects completed within 'Public Play Spaces' both as a metaphor for how we approach projects, broadening the scope of participation and for providing concrete tools for negotiating design ideas. As one of our primary objectives for the public realm is to encourage the aesthetic and poetic practices exemplified in play, or to stimulate voluntary actions, which extend imagination in time and space, it was logical to employ games as a format for play to take place 5/16. In addition, it should be

[1] Interactive Institute *Public Play Spaces* project: <http://www.tii.se/pps>

noted that there are numerous parallels between design and gaming with respect to the use of constraints or rules and roles for players (or designers) [Ainley, 2001].[2]

Games have the ability to create worlds, separated from ordinary life that other people can enter into. They can transcend locality and extend their selves into visions of what they would like to be [Brandt & Grunnet, 2000; Andersen, Jacobs, & Polazzi, 2003] We wanted to take advantage of this notion in order to mediate new experiences and discover new extensions of the self, as well as building upon the notion that collaborative design with games improves idea generation and communication between stakeholders [Brandt & Messeter, 2003].

We extended the metaphor: envisioning people as **players**, prototypes, models and mock-ups as **props** and **game pieces**, and experiential factors such as time, context, and rituals of completion or use as **rules** and **boundaries** [Buchenau & Suri, 2000; Mazé & Jacobs, 2003]. This gave us the room to successfully deploy existing creative techniques such as role-play, simulation, mapping, cultural probes and expressive writing all the while re-inventing and modifying these techniques for our own purposes. This also provided us with the freedom to explore the artefact during different stages of the design process. For instance, concepts and prototypes are in some cases deliberately under-designed to prevent premature decision-making while other 'props' have been exaggerated in order to focus on critical issues [Burns, Dishman, Verplank, and Lassiter, 1994; Brandt & Messeter, 2004]. In many cases, it is precisely this incompleteness or indeterminancy that has proven to be the most successful feature of the artefact.

Ultimately, our game-based methodology is a successful framework for increasing the level of player participation and motivating and engaging the whole person, whether before, after or during the design process in the projects we have completed. In addition, with games and through game play we have learned more about design as a social activity, generating experiential learning and influencing the tactics we have chosen to use in order to progress the game.

Tactics for Play

Although the game-based methodology is not necessarily one based on winning or losing, it has still been necessary to employ tactics for implementing our projects, successfully intervening and engaging people, and advancing the game. As we were aiming towards creating technological artifacts that placed the power in the hands of the public as opposed to overarching structures generally inflicted by corporate or political organizations, we chose a tactical rather than a strategic approach.

---

[2] Interactive Institute *Public Play Spaces* project: <http://www.tii.se/pps>

Tactical approaches are defensive, nonconformist, and opportunistic in nature. De Certeau describes tactics as the victories of weak over strong that manifest themselves in the everyday [de Certeau, 1984]. Generally, tactics are 'used in more limited ways and seized momentarily within spaces, both physical and psychological, produced and governed by more powerful strategic relations' [Dunne, 1999; Blauvelt, 2003]. We chose to employ tactics as means for constructing alternatives. In some ways we further extended the game metaphor by engaging tactics in the place of design criteria. In order to ground the discussion, projects will be presented from within 'Public Play Spaces' in order to illustrate tactics that we have incorporated for engaging people up until now. These research and design projects have been developed with expertise drawn from multiple disciplines and are a culmination of work from diverse collaborators and stakeholders (see Acknowledgements section).

Transformation and Identity: Revealing the Extraordinary in the Everyday

Underdogs and Superheroes was one of the first projects initiated within 'Public Play Spaces', exploring design interventions for public places. Rather than starting from a concept or business strategy, *Underdogs and Superheroes* began with people and their desires. The project was primarily about involving people or players experientially, creatively and personally throughout the design process to help create stories about potential uses and users. This project explicitly uses game-based methods in every stage. Overall six games were played throughout the course of the project incorporating paper worksheets, automatic and public mapping, scenarios, invention, imagination props, group treasure hunts, role playing and enactment with the key theme being that of transformation of both identity and place, from the mundane and powerless to the empowered superhero in a magical setting [Ehn and Sjögren,1991; Mazé and Jacobs, 2003].

In this case, the use of games structured participation and interaction among participants and allowed for ideas that enabled inventive and embodied ideation. By tactically transforming the everyday and giving it back to the players, we were able to reinforce transitions and allow players greater space for changing identity and embracing extra sensory abilities [Gaver, Beaver and Benford, 2003; Blauvelt, 2003]. We were able to establish needs and opportunities and develop concepts together with the players. In the end, several design concepts emerged, some of which were continued in other projects. Perhaps more importantly, *Underdogs and Superheroes* provides a conceptual framework and palette of techniques to support the suspension of disbelief during design sessions. The result was that participants could immerse themselves fully in the possibilities in the design space that continues to work as a basis for a major part of the projects within 'Public Play Spaces'.

The project *Reach* also incorporates the tactics of transformation. *Reach* combined aesthetics, pattern exploration, and interactive qualities to create a new language for wearable expression, allowing participants to transform their appearance and perhaps even change their identity. In *Reach*, the props or prototypes were 'wearable sketches' that tested both material and interactive qualities. The wearable sketches included everyday worn items such as hats, bags, scarves, and skirts that react or interact with the environment or persons within the environment. In addition, they explored both additive and subtractive pattern-making processes where patterns grow or are revealed in response to changes in one's personal, social or environmental space. *Reach* areas of investigation included person-to-person communication, collective game play, proximity, reaction to environments and weather stimulus [Gooding, 1995; Jacobs and Worbin, 2005]. This project offered a wide range of props and would-be solutions that enable people to step outside of the everyday and become someone else or perform an act that might not otherwise do.

Generally the tactics presented went hand-in-hand. Transformation of identity and or spaces can also be viewed as a more overarching tactic to some of the others listed below. Largely based on principles borrowed from contemporary art including de-contextualization, de-familiarization, and fragmentation [Higgins, 2002; Bell, Blythe, Sengers, 2005; Blauvelt, 2003; Dunne,1999; Gaver, Beaver, and Benford, 2003], it is through these processes that players became more accepting of alternate realities and willing to engage in new use scenarios.

(Dis)Placement and Deployment

Several projects also examine the tactics of placement and deployment, which can contribute towards transformation and vice-versa. Choosing where to place our design interventions varied from project to project. For the most part we are interested in looking at social corners such as bus stops or city squares versus quiet zones of reflection such as parks or waterfronts. In some cases, such as in Underdogs and Superheroes, Reach and Tejp our goal was to re-envision city spaces and not only affect place, but also displace artefacts in order to reveal new interactions and support discovery. This was accomplished for example by locating objects where they might not otherwise be, along walking paths or in the case of Reach on the body. Deployment techniques include what and if to re-design artifacts, choosing whether to parasite off of existing structures and deciding how and when to intervene. Timing is an important factor of deployment and can affect the outcome of the interventions.

Tejp examined the various possibilities for overlaying and accessing personal traces on public spaces through direct physical interaction. The goal with this project was to connect local communities by providing a space and sounding board for existing social relationships. A series of low-tech prototypes include Audio Tags, simple devices for recording and listening to messages as an open platform for expressing presence and Glitch, which reflects public cell phone usage or digital presence back to local communities through sound. Both placement and deployment were integral to the success of Tejp. Drawing inspiration from existing urban practices, Tejp prototypes were deployed in a similar manner in real settings with the general public, incorporating parasitic methods and placed in clusters along walking paths, at social corners and in more quiet settings. Tejp succeeded in these environments supporting surprising and unfamiliar interactions, transforming spectators into players and encouraging playful ways to personalize territory in the public realm. Ultimately it assisted us to derive design implications for location-based information systems that make full use of the physicality of public spaces [Huizinga, 1950; Buchenau and Suri, 2000].

Exaggeration and Transposition of Scale

Another tactic closely tied to the process of transformation, even placement, and deployment involves scale. Making the small large and the large small can alter one's understanding of the everyday and draw attention to what is normally hidden or imperceptible [Blauvelt, 2003].[3] This change in scale can also be described in terms of exaggeration. In other words: one way to change perceptions of scale was to over-represent and augment something that might otherwise go unnoticed [Jacobs, Gaye & Holmquist, 2003; Gaver, Beaver, & Benford, 2003].

Listening In, a project expanding upon one of the *Underdogs and Superheroes* concepts, supported both active and passive participation intensifying sonic experiences through physical listening. The installation itself consists of a series of three *Listening In* modules placed at three different locations in the public environment. Each module allows people to 'listen in' to other parts of the city by simply approaching two of the three sides of the module. Through physical interaction, the listener triggered sound by placing their ear upon one of the surfaces of the structure. The third side of each module captured and transmitted sounds to the other locations.

---

[3] Interactive Institute Public Play Spaces project: <http://www.tii.se/pps>

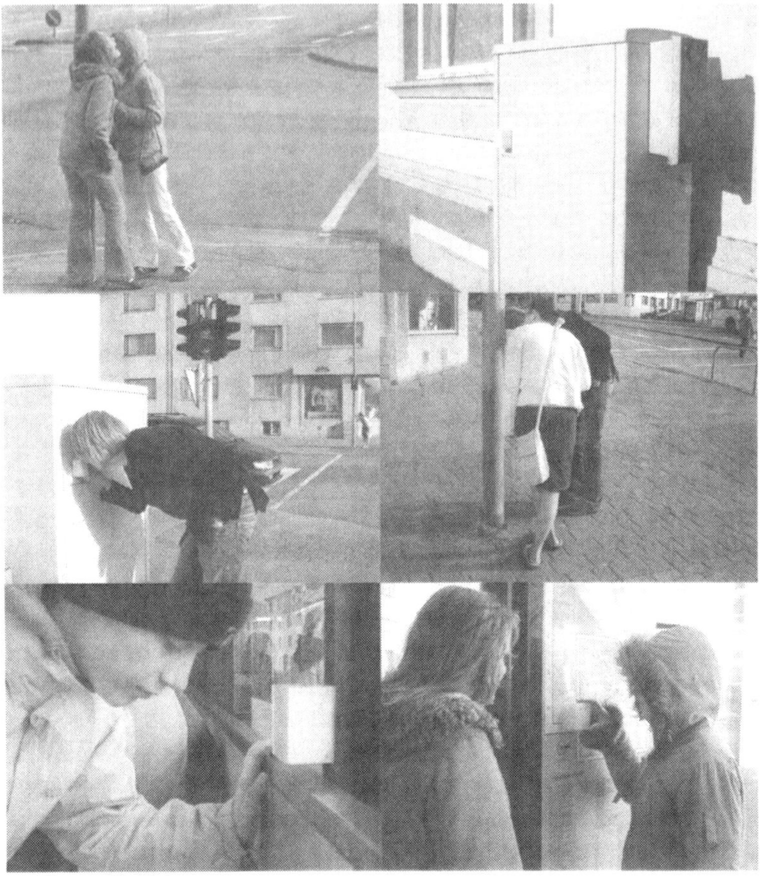

Figure 49: Tejp

The transposition of the scale of the sound, or the raising and lowering of the level, was the basis for this project. Background sounds were brought to the forefront, while local events become broadcasts. As a result, randomness and resolution subtly changed meanings of aural events. This interplay of urban sounds reinterpreted the city; unpredictably combining the colours of different places into a new sound collage; the sound events highlighting the actions of everyday life, enhancing relationships, altering perceptions, collapsing borders, and nurturing intercultural exchanges.

Transparency & Adaptability

In an effort to create modifiable and open platforms it was important to consider the levels of both transparency and adaptability. Adaptability or multi-functionality places choice in the hands of people and allows for greater flexibility, more room for interpretation, transformation and increased mobility and even a greater chance for achieving efficiency in use. Functions are available for different situations. Transparency provides people with 'easy' interfaces that are both accessible and manageable. Combining these two is perhaps the strongest basis for creating open artefacts, environments, and systems.

Sonic City was a project exploring mobile interaction and wearable technology for everyday music creation. We designed, implemented and evaluated a system that created electronic music based on sensing bodily and environmental factors. Mapping these to the real-time processing of concrete sounds, *Sonic City* generated a personal soundscape, co-produced by physical movement, local activity, and urban ambiance. Encounters, events, architecture, (mis)behaviours—all became means of interacting with or 'playing the city' [Jacobs, & Löfgren, 2005; Mazé & Jacobs, 2003].

While dealing with transformation of an otherwise mundane activity into an act of musical composition, as well as altering the scale of gestural interaction to include in some senses the gestures of the city and even choosing its placement on the body, *Sonic City* also extended its scope of what is available through a 'plug-and-play' platform that was both transparent and adaptable. Components of the system are visibly housed in a wearable garment in order to easily 'try on' and rapidly iterate interaction, system and aesthetic variables.

It is through this open platform, which might be viewed as imprecise or incomplete, that we are able to achieve a level of 'digital sfumato' [Jacobs and Worbin, 2005; Gaver, Beaver and Benford, 2003] , increasing the level of ambiguity and blurring the relationship between form and function. *Sonic City* and other projects such as *Tejp* and *Free Energy* (see below) embraced these notions and used them to their advantage. It is precisely this ambiguity, openness, transparency or adaptability that supports user choice and portability, allowing for appropriation or engagement to occur seamlessly.

Extending the Cycle

Besides engaging methods that allow people the choice and freedom to explore alternatives, we were also interested in the life-cycle of design and how this could be used as a tool to further involve people, as well as alter perceptions regarding materials and production processes. In this endeavour, we believed that design continued in the hands of our players, through use. Additionally, we found it important to investigate and understand how to embrace sustainable design procedures. Waste and material use should be addressed by designers [de Certeau, 1984] and should be offered up to people through a heightened degree of awareness and accountability.

The project, *Free Energy* within the Static! project platform at PLAY, increased energy awareness in the public realm by providing access to energy in public space. Energy was harnessed and accessed throughout the city at a series of stations. All of the stations dispensed energy that is driven by either forces of nature (i.e. solar, wind, etc) or through human effort (i.e. mechanical energy). The approach took several parallel paths investigating the placement of stations, their deployment, transparency, as well as what behaviours emerged through use and what impact they might have. In some senses, *Free Energy* was primarily a project about social debate, trying out and experiencing alternative choices in existing situations and generating new ideas together with possible users [Mazé & Jacobs, 2003; Jacobs & Löfgren, 2005].

Our aim was to extend the cycle on two fronts both by making energy more visible, tangible and accessible to people and providing a space for exposing existing energy behaviours or what occurred in the hands of the people. Again, within several of our projects that we were interested in the lifecycle of design and what actually happens when people held our designed artefacts in their hands. *Tejp*, *Sonic City*, *Underdogs and Superheroes*, all employed this tactic. However, it was with *Free Energy* that we consciously looked at sustainable resources and how they could be built into our research practices.

Free Energy not only made energy consumption transparent to users but also investigated the design of interactions with products that enabled choice in energy use, discovery, distribution, and generation—with the ultimate aim of moving beyond 'on' and 'off' and towards an active and participatory experience for people interacting with energy in their everyday life.

Concluding Thoughts

The tactics for play and successfully staging interventions that engage people as introduced above, provide a valuable guide for exploration. However these tactics are not foolproof and are, as yet, unfinished. Our ongoing challenge with 'Public Play Spaces' is to continue to stage people as performers and prototypes as props to discover and uncover emerging behaviours and tactics. As we consider iterations in the designs themselves, we will continue to explore the aforementioned tactics (whether used alone or in combination with one another) as well as any other tactics that might surface and explore whether they can lead to casting doubt, shedding light, stimulating surprise, evoking reflection or curiosity, or even generating creativity. We will also continue to offer new opportunities for outdoor communication and serendipitous exchange, focusing less on accessing information and more on enabling spontaneity, connecting the social and cultural infrastructure between different groups and generations, and further engaging people with one another.

Acknowledgements

Underdogs and Superheroes is a project by Margot Jacobs and Ramia Mazé. The Sonic City project team includes Lalya Gaye from the Future Applications Lab, Viktoria Institute, Margot Jacobs and Ramia Mazé from the Interactive Institute, and Daniel Skoglund from 8Tunnel2. *Tejp* is a project by Margot Jacobs and Lalya Gaye. Both *Sonic City* and *Tejp,* were funded by the Swedish Foundation for Strategic Research through the Mobile Services project and the Interactive Institute. *Reach* is a project by Margot Jacobs and Linda Worbin within the 'IT+Textiles' platform at the Interactive Institute Play Studio. *Listening In* was initiated by Margot Jacobs, in collaboration with Röda Sten and the Göteborg City Museum. The Interactive Institute Göteborg studio has received funding for 'IT+Textiles' from VINNOVA (the Swedish Agency for Innovation Systems) and is a collaboration between Chalmers University, the Interactive Institute, Ludvig Svensson, Newmad Technologies, and the Swedish School of Textiles at the University College of Borås. The Free Energy project team includes Margot Jacobs, Ulrika Löfgren, Sara Danielsson and Anders Ernevi and is funded by the Swedish Energy Authority (STEM). I would like to thank Sara Backlund and Johan Redström and the rest of PLAY for their support and feedback, Anders Ernevi for his constant good spirit and help in constructing some of the electronic prototypes, as well as a special thanks to Ramia Mazé who helped formulate 'Public Play Spaces' together with me and has served as a great inspiration.

Bibliography

Ainley, R., ed. *This is What We Do: A Muf Manual*, London: Ellipsis, 2001.

Andersen, K., M. Jacobs, and L. Polazzi. "Playing Games in the Emotional Space." *Funology: from usability to enjoyment.* Ed. M.A. Blythe et al. Dordrecht: Kluwer, 2003.

Andreotti, L., and X. Costa. *Theory of the Dérive and other Situationist Writings on the City.* Barcelona: Museu d'Art Contemporani de Barcelona, 1996.

Bell, G., M. Blythe, and P. Sengers. "Making by Making Strange: Defamiliarization and the design of Domestic Technologies." *Transactions on Computer Human Interaction: Special Issue on Social Issues and HCI.* ACM Press, 2005.

Blauvelt, A., ed. *Strangely Familiar: Design and Everyday Life.* Minneapolis: Walker Art Centre, 2003.

Brandt, E., and C. Grunnet. “Evoking the Future: Drama and Props in user centered design.” *Proceedings of PDC 2000.* New York: CPSR, 2000.

– – –, and J. Messeter. “Facilitating Collaboration through Design Games.” *Proceedings of PDC 2004.* Toronto, Ontario, Canada, 2004.

Buchenau, M., and J. Suri. “Experience Prototyping.” *Proceedings of Designing Interactive Sytems (DIS) 2000.*

Burns, C., et al. “Actors, hair-dos and videotape: Informance design.” *Proceedings of CHI ’94*, ACM Press, 1994.

de Certeau, M. *The Practice of Everyday Life.* Berkeley: University of California Press, 1984.

Dunne, A. *Hertzian Tales; Electronic Products, Aesthetic Experience and Critical Design.* London: RCA CRD Research Publications, 1999.

Ehn, P., and D. Sjögren. “From System Descriptions to Scripts for Action” *Design at Work.* Ed. J. Greenbaum and M. Kyng. Lawrence Erlbaum Associates, 1991.

Gaver, W., J. Beaver, and S. Benford.*Ambiguity as a Resource for Design. Proceedings of CHI*, ACM Press, 2003. 233-40.

Gooding, M., ed. *A Book of Surrealist Games.* Boston, MA: Shambhala Redstone Editions, 1995.

Higgins, H. *Fluxus Experience.* Berkeley: University of California Press, 2002.

Huizinga, J. *Homo Ludens: A Study of Play Elements in Culture.* Boston: Beacon Press, 1950.

Interactive Institute Public Play Spaces project: http://www.tii.se/pps

Jacobs M., L. Gaye, and L.E. Holmquist. “Tejp: Ubiquitous Computing as Expressive Means of Personalizing Public Space.” *Proceedings of Ubiquitous Computing(UbiComp).* Seattle, USA, October 12-15, 2003.

– – –, and U. Löfgren. “Promoting Energy Awareness through Interventions in Public Space.” *Proceedings of ‘In the Making’ the First Nordic Conference on Design Research*, Copenhagen, Denmark, May 29-June 1, 2005.

– – –, and L. Worbin. “Reach: Dynamic Textile Patterns for Communication and Social Expression.” *Computer Human Interaction (CHI)* 2005 conference, Portland, Oregon USA, April 2-7, 2005.

Mazé R., and M. Jacobs. “Underdogs and Superheroes: Designing for New Players in Public Space.” *Proceedings of User Aspects of ICTs conference (CAST).* University of Art and Design, Finland, 2003.

– – –. “Sonic City: Prototyping a Wearable Experience.” *Proceedings the 7th IEEE International Symposium on Wearable Computing (ISWC).* October 21-23, 2003.

Salvador, T., and K. Howells. "Focus Troupe: Using drama to create common context for new product concept end-user evaluations." *Proceedings of CHI '98*, ACM Press, 1998.

Seago, A., and A. Dunne. "New Methodologies in Art and Design Research: The Object as Discourse." *Design Issues* 12.2 (1999).

Between Improvisation and Publication: Supporting the Creative Metamorphosis with Technology

Cati Vaucelle

(MIT Media Lab, Interactive Cinema/Media Fabrics, USA
Trinity College University, Centre for Research in IT in Education, Ireland)

From an early age, we play, learn and exchange ideas about our identity using stories; we test our hypothesis about the world using toys, telling stories and acting in the world. Today communications technology, with its spectrum of media potential, expands the resources we have at our disposal for exploring and sharing our reflections on the environment we live in. With mobile technology, we are moving into a creative and collaborative world in which images and sounds can mix with local language. Shared movie-making devices can engage people in exploring multi-dimensional approaches to expressing and exchanging point of view on their environment. People can create movies about how their lives interact with everyday means of interaction such as commonly used technological objects; for example, computer game consoles, cell phones, and video cameras, and these objects can themselves serve as vehicles for manipulating personal media to co-construct video games, movies, and songs. Therefore, my technical implementations address the following theoretical questions: How might creative people engage debates with their spectators about the technological environment we live in?

How can the spectators respond and participate in the creation during this process? What role does this dynamic play in the process of creation and reflection? To support this dynamic over the past height years I design objects, mainly visual and tangible media tools, to bridge the gap between improvisation of an idea and its publication.

Part 1—Design of computational video tools to create a collaborative report on the technological environment

The Photographer is a witness, a witness of his time.
[Pierre Mac Orlan]

Visual media is now used everyday and in many forms: from traditional photography to recreation of reality using 3D animation. Therefore, the popular belief that visual media is a mirror on reality has been completely revised. More specifically, an image is not different from the represented object, but over the years its authenticity has changed; the information presented through visual means can be twisted by using specific semantic codes to attract the attention of individuals, therefore its credibility is questioned.

The subject is not that important, any photographer is a reporter; he/she can be "good" or "bad," knowing where to look or passing by without noticing anything; but what has been captured in this mirror of memories has its origins in a scene that has existed at a specific time. The power of photography is immense: we can watch documentaries, read newspapers—by looking mainly at the pictures—and still think we are informed. Today, visual media is the favoured instrument for the development of information. The quantity of information cannot decrease. If the production of events is independent of the journalist, the everyday media consumption of the audience is a constant. Journalists look for shocking visuals in a reality that did not deserve such treatment and this reality then takes on abnormal importance [Vaucelle 1998]. Visual documentary can still be informative through a shape, an object, a costume, and because it witnesses a period, it is possible to date a visual by its subject, to detect a nationality, and even its author. It can help us understand a country or a social group, for instance. Understanding what visual media does tell us and what it does not is already a way to appropriate the medium. At its emergence, photography was conceived as a direct representation of reality and the human factor was not considered in its creation, therefore a photographer was not acknowledged as an artist. In the 20's, the poet, novelist, and photography critic Pierre Mac Orlan considered photographers the best investigators of social reality, because they were the ones capturing the subtle reality of the "life of the streets." Over the years, photographers have captured visuals, looking for important information through the camera, expressing and sharing perspectives on our environment.

We have gone from information to emotion, where the status of image has transcended the limits between public and privacy. Showing visuals of a private event makes this event become public. Now because of the lack of credibility in modern visuals, journalists use emotions to make spectators believe in the truth of the scene. According to the cultural theorist and philosopher Jean Baudrillard, the Gulf War disinformation made not the events, but the information become the scandal [Baudrillard 1992]. We now look for the credibility of the truth. Until very recently I thought that we were moving towards the end of the documentary as a mirror of events, when its credibility was again made possible by a new phenomenon. Take a recent example: during the terrorist attacks in London,[1] individuals captured and shared media with journalists. By using media made by witnesses using their photo-phones, journalists gained a credibility—justified or not—in their presentation of the event. Journalists could use these media as a 'credibility tool' and reinforce their own reporting. The documentary has also evolved into a collaborative report; in fact, any individual who has a photo-phone is potentially a reporter, who can capture key events and by combining these captured visuals, create a popular visual narrative presenting the "truth" of an event.

This demonstrates the shift from *traditional* documentary to *collaborative* documentary around an event. It is within this context that I have mainly used the visual medium through my research as a way to express and share perspectives on our technological space [Vaucelle, Africano, Davenport, Wiberg, and Fjellstrom, 2005; Jehan, Overholt, Solís and Vaucelle, 2004; Vaucelle, Gorman, Clancy and Tangney, 2005; Vaucelle and Davenport, 2004; Nemirovsky, Vaucelle, and Quattoni, 2004]. The public wants to be surprised by information and activist artists respond to social issues and offer by "entertaining," implicit debates, and question the contradictions they find in our society, e.g. the *Korbel Trilogy* by Dell'Arte. (The "Korbel" series was created by the painful and threatening. company over several years and is a kind of Humboldt County soap opera with recurring characters. The themes mirror the changes happening in the area once known for logging and fishing, now a depressed economy increasingly dependent on tourism and dwindling resources.[2]

With the emergence of mobile technology in which visual-capture is a possible everyday action, sharing visuals to collaboratively create videos could be a means to share values in the form of symbolic meetings. Group, as well as personal production of media can be shared over media spaces as a means of understanding relationships with others, their communities, their rules, their habits, and their references to the world. But does this process lead to unexpected experiences and shifts in perspective by exchanging knowledge between creators?

Media creation can provide a means of exploring self-identity, both through individual sharing and group construction of media spaces. Creators can participate in this process by "constructing" their media tools using anything from their video camera to their cell phone; the production and rearrangement of video products could then come directly from them and not be introduced and diffused by third parties. Constructionists such as Seymour Papert have suggested that young users could benefit from systems that support self-expression rather than create content

---

[1] <http://en.wikipedia.org/wiki/2005_London_transport_explosions>(2007)

[2]<http://www.communityarts.net/readingroom/archive/perfcomm/dellarte/dellarteexcerpt.php> (2007)

for the user [Papert 1980]. One main idea in constructionism is that young users "learn by making" [Papert 1991]. Within this context, I analyze which functions of an interface engage people in thinking through making, supporting them in expressing what they think is meaningful in their technological environment, and understanding to what extend it prompts their creativity in challenging their spectators.

Part 2- Design of technologies that support creators in debates with their spectators

Ultimately, the artist alone does not perform the creative act; the spectator brings the work into contact with the external world by deciphering and interpreting its inner qualifications and thus adds his contribution to the creative act [Sanouillet and Peterson].

In 1958, Simondon introduced a new approach to technology [Simondon 1989]. He argued about the necessity for individuals to defend themselves against the technical object in order to appropriate its aesthetic dimension. In fact, while the existence of human reality in the technical object is being denied, only the aesthetic object seems to transmit human values. For the popular culture, the technical object has a function, but does not express a concept. More specifically, individuals protect themselves against the technical object by reducing it to the status of being useful, and at the same time, paradoxically, they mystify the technological object by wanting it to be evil, powerful and dangerous e.g. Fritz Lang's female robot character in *Metropolis*, 1927. It seems that this artificial being that humans create, they fear, and either accuse it of destroying their lives or of only being a useful object without any aesthetic characteristics. As much as Roland Barthes [Barthes 1981] tells us that the status of photography has changed from being purely technical, to being perceived as art, to finally modifying the notion of art itself into that of a concept, e.g. Marcel Duchamp's art works: it seems there is still an unjustified hierarchy among technical objects depending on their more or less common alignment with the artistic sphere. Simondon has denounced this unjustified imbalance between the technical and aesthetic in the sphere of meaning. However, popular culture seems to work in oppositions: how could it justify a possible insertion of the technical into the sphere of meaning, while it is through this lack of knowledge that it can justify the 'raison d'être' of the aesthetic sphere?

A study of the technical and its interconnections to the aesthetic sphere is necessary in order to understand the technological environment in which we live. We are indeed surrounded by technology and even if we chose to deny its human characteristics, we still could understand its implications. The cultural theorist Paul Virilio has analyzed technology both as a threat to collective memory and as the ghost of an "integral accident." He explains how a city can suffer from its conditioning by media, stating that as technology becoming more and more 'mediatic', it can destroy real space to the benefit of *real-time*, provoking a sense of the body lost, the body being de-localised. The evolution of technology is not so much an issue, but rather the fact that humans tend to deny the loss it generates [Barthes, 1981]. Pierre Levy [Levy, 1994] on the other hand, is more positive in his attitude over the impact of cyber space on our lives, and proposes that we should not think of issues in terms of the impact of technology on society, but in terms of understanding it as a collective project, by acknowledging that we all are implicated in it.

I contend that technology can bring people together, and people from different social groups, even those who have never met before, and that belonging to multi-social groups, with different values and life styles, could be a "freedom factor." This freedom would come in the sense that this multi-enrolment could propel us toward universal thinking, away from the tendency toward the natural believe that one's values are the only valid ones, and that those of others are mistakes or anomalies. One archetypal example is the conflict between monotheist religions, each of them affirming that they know the truth and it is the only possible truth.

One of the first questions in my research has then been: whether young adults, by improvising visual stories using technology they are comfortable with, and sharing those across borders, gain a deeper sense of others' values? Besides, identity encompasses many aspects of our everyday life and environment: landscape, sights, sounds, dance style, food, games and a broad range of other activities. Like many other concepts, e.g. *freedom* and *joy*, one could maybe understand others' values better when one confronted by difference. Finally, where being able to explain a trauma with words can relieve its pain, similarly giving reading cues for reunderstanding technological space through visual language in cinema/documentary/video could encourage individuals to change their point of view.

> "We are all designers—and have to be. Professional designers can make things that are attractive and that work well. They can make beauty, create products we fall in love with at first sight. They can create products that fulfill our needs, that are easy to understand, easy to use, and that work just the way we want them to. Pleasurable to behold, pleasurable to use. But they cannot make something personal, make something we bond to. Nobody can do that for us: we must do it for ourselves." [Norman]

Technology is integrated to our everyday lives. I assume that its qualities reside in the way users make sense of it. Designing an electronic object is interrelated to designing for people's lives, and implies reflection on its critical and aesthetic roles

[Dunne]. Some researchers propose that designers develop sensitivity to, and control of aesthetics, for instance, by putting constraints on communications media [Gaver 475-493]. It is within this context that I usually design interfaces with one function, where constraints can be easily integrated for people to adapt to individual needs, and they can feel empowered by re-designing its original concept. These 'transformable' interfaces can enable creators to express their perspective on the influence of technology on their private and public space, by actively responding to it, and receiving critical feedback from other creators. As in the work of the artist and engineer Natalie Jeremijenko,[3] abusing popular references is one of my research positions, chosen to make people feel comfortable in expressing themselves using objects they are familiar with, e.g. joysticks, computer game consoles, cell phones, and still cameras. The technology I have developed is meant to enable a transparent process between improvisation, e.g. capturing a video, creation of a final piece, e.g. digitalizing and editing, and publication, e.g. exhibition. Using this system, they can debate and provoke others over the contradictions in the technological space. These pieces typically take the form of video games, animated stories, interactive art, and collaborative movies.

Part 3- Design of technologies facilitating process between improvisation/publication

My prior research has been concerned with the underlying mechanisms regarding the improvisation of narratives by children and the way they integrate their beliefs about the world during this process [Sam Ryokai, K., Vaucelle, C., Cassell, J. (2003); Vaucelle, 2002; Vaucelle, 2000; Vaucelle, 1999]: the modifications made by computational customized toys on these narratives in order to prompt reflection on what has been improvised [Ryokai, Vaucelle, and Cassell, 2003; Vaucelle, 2000]; the reflections that occur while they share their narratives [Vaucelle 2000; Vaucelle, 1999], or create movies in real time according to their stories, or play games [Vaucelle and Davenport, 2004]; and the perspective transformations of their beliefs during this process (see *Figure 50*). I have also been researching technologies that drive the students to learn that content and creative ideas could drive self-expression by developing a mobile, tangible, easy-to-use, loop-based music mixing interface for children in which students consider issues of composition, musicianship, and improvisation in modern music; build sampled music (without requiring any technical and musical skills); and understand how mathematical concepts apply to music (Emailer [Vaucelle,1999] Story Generator [Vaucelle, 2000] and Dolltalk [Vaucelle and Jehan, 2002]).[4]

---

[3] Natalie Jeremijenko's project hub : <http://xdesign.ucsd.edu/>

[4] Droppin' science <http://www.brillomag.net/Learning/workshop.html>

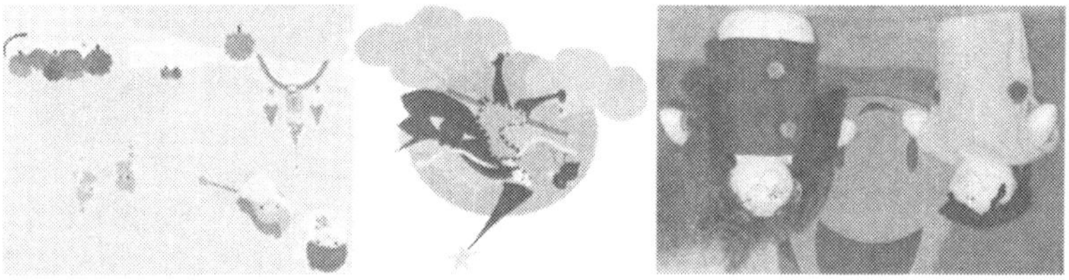

Figure 50: Visuals of the narrative systems Emailer'99, StoryGenerator'00 and Dolltalk'02

My past research has also been concerned with the transformation of space using technology. I have explored communication and aesthetic channels such as devices that enhance remote telecommunication. The system designed provides a sense of the presence and activities of remote users through a ball floating in the air [Hover Maynes-Aminzade et al, 2002]; designed a video background behind virtual agents to create a sense of shared space between the agent and the user [Sam Ryokai, K., Vaucelle and Cassell, 2003; Cassell et al., 195-208]; created a physical interface to interact with the agent in a "natural" and "transparent" way [Cassell et al, 2002]; inspired by the experimental cinema works of Kenneth Anger[5] in which the rhythm between music and images is synchronized: the soundtracks being mapped onto the video, I have explored the concept of the relationship between audio and video co-controlled by an audience in which the video is controlled by a soundtrack [Jehan, Lew and Vaucelle, 2003]; explored how to enhance social interactions in developing a shared video space co-controlled by physical objects in which participants wield magic wands and practice wizardry [Kanis, Agamanolis, Vaucelle and Davenport, 2003]; created an online 'adaptable' space for children to integrate the activities of play and video editing into a never-ending process of mixing and 'subversion' of each other's material, encouraging playful collaboration in an exchange network of unique media artefacts [Nemirovsky, Vaucelle and Quattoni, 2004] (see Figure 51). I have worked on transforming urban spaces like bus stops or train stations through the medium of cell phones [Jehan, Overholt, Solís, Garcia, and Vaucelle, 2004.]. Finally I have designed a system for visitors to contribute to the museum exhibition space [Vaucelle, Gorman, Clancy and Tangney, 2005].

[5] ANGER, K. 1953. *Eaux d'Artifice*, 12 min, USA.

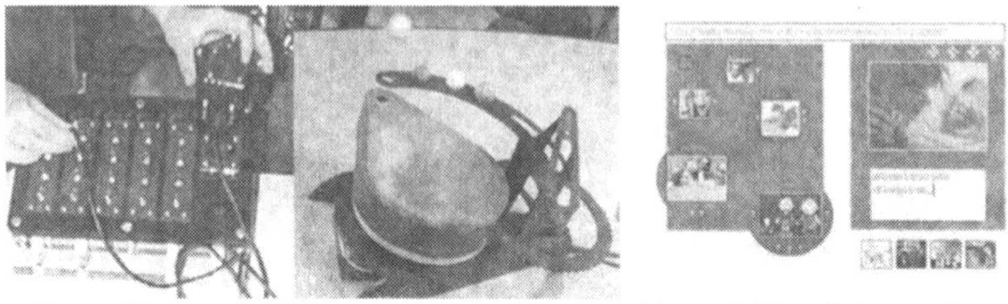

Figure 51: Visuals of Droppin Science'01, Hover'01 and Mixer Subverter'03

Improvising movies with a personal video database: Textable Movie

I have always been fascinated by Marcel Proust's experience of having tea and biscuits, i.e. his famous Madeleine, that brought back emotional memories. When I created the first version of *Textable Movie*, I wanted to recreate this same phenomenon. The scenario I envisioned was being comfortably installed on my sofa, telling a story and instantly being projected visually into my past, by watching corresponding videos and sounds. This would have inspired me for more storytelling to orally drive the video projection. The original concept was that by immersion in their own memories, storytellers could become engaged into telling rich, and passionate stories, based on their past experience (see *Figure 52*).

The media tools we have developed have, for the most part, emphasized the convention of a finished form with a beginning, middle and end, much like a letter from the past. For instance, there is currently no moviemaking tool that allows us to drive visuals by verbal input, therefore I conjectured that such real-time movie-making device would have potential in personal exploration, enabling investigation of multi-dimensional approaches to express and exchange perspectives on a memory. To this end, I designed a system and implemented a software system called *Textable Movie*. Its graphical interface takes text as input and allows users to improvise a movie in real-time, based on the content of their writing. Media segments are selected according to how the user labels their personal audio and video database. As the user types in a story, the media segments appear on the screen, connecting writers to their past experiences and inviting further story-telling. Using language as an interface seems to allow a natural approach to story building and play for both the person who is trying to understand the world and for the person who engages in improvisational story-telling with someone else using the system [Vaucelle and Davenport, 2004]. *Textable Movie* is open-ended allowing anyone to become a "video-jockey" by mixing, applying effects, and rearranging video samples in real time (see *Figure 52*).

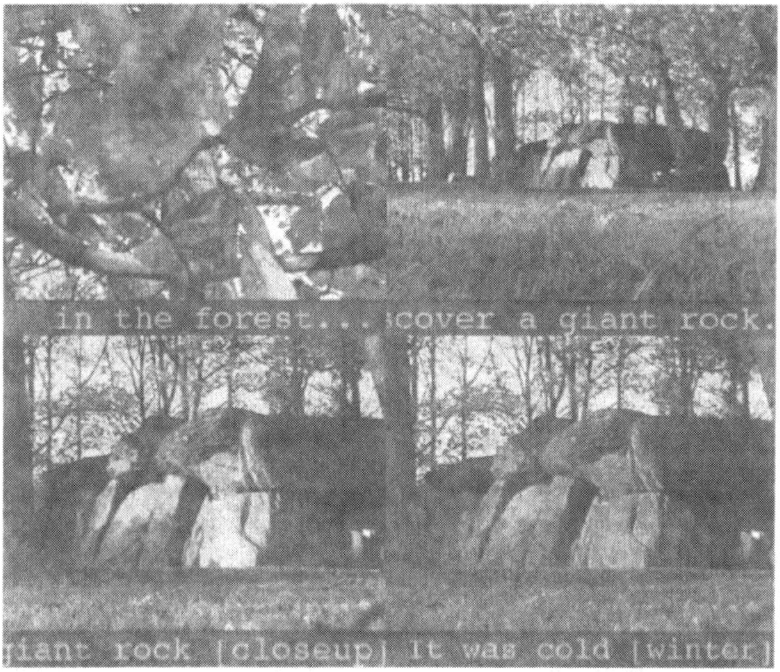

Figure 52: Example of a narrative driving visuals and sounds in Textable Movie'02.

In the framework of computational storytelling, *Textable Movie* promotes the idea of maker-controlled media and can be contrasted to automatic presentation systems. By improvising movie-stories created from their personal video database and by suddenly being projected into someone else's video database during the same story, users can be surprised as they visualize video elements corresponding to a story that they could not have expected. With *Textable Movie*, users make their own inference from these discoveries rather than in using artificial systems that make the inference for the users.[6] They can then create a new interaction, e.g. mapping keywords to videos, and incorporate new video clips and sound samples in their database.

---

[6] LIEBERMAN H., AND LIU H. 2002. Adaptive Linking between Text and Photos Using Common Sense Reasoning, Conference on Adaptive Hypermedia and Adaptive Web Systems.

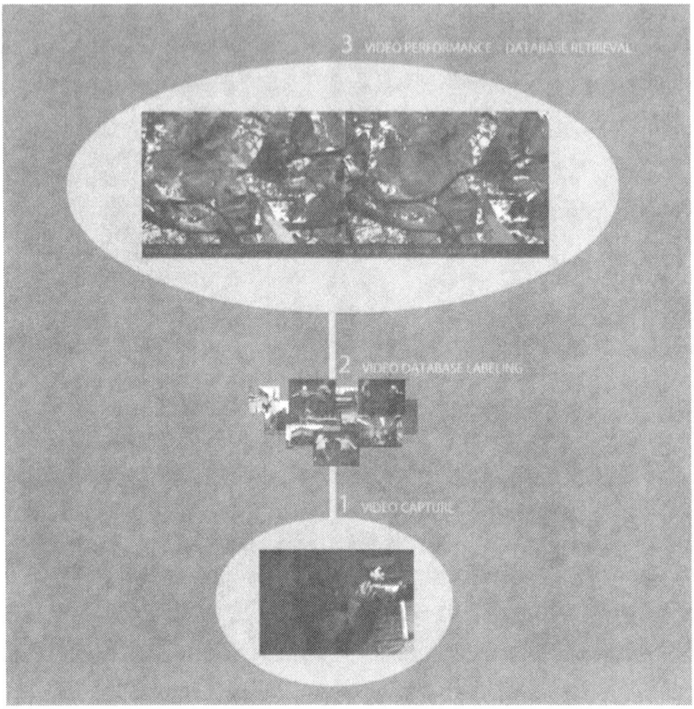

Figure 53: Textable Movie General Scenario

From the beginning, one of my goals for this tool was to encourage teenagers internationally to share views of their communities with each other. However I had to be cautious with the concept of *community*. Indeed, there can be more common points between teenagers from a same social class from two different countries than between teenagers from the same country, but from different social classes. Therefore I positioned this work through a shared view of the teenagers' mental model of their environments [Lynch]. They can then drive a discussion around the notion of living in the city, versus living in the countryside and therefore understand the implications of technology in their life. During the user studies, when teenagers were asked to share their opinions on living in different locations, they were extremely motivated, and decided to talk about their 'city' or to another extend 'their country' by grabbing some video clips of key elements in their city, e.g. a cell phone conversation (see *Figure 54*) and sometimes performed their own movie and acted what were perceived as interactions in their city, to finally create a movie out of their collaborative video database.

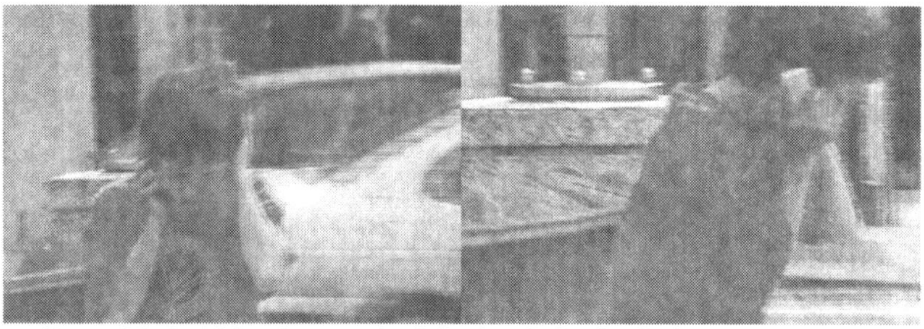

Figure 54: A child acting out the following: 'a loud cell phone conversation due to the noise in the city.'

During user studies, young teenagers were engaged in the process of capturing of videos, however, they were easily distracted in digitalizing and editing, i.e. cutting videos into video segments. Finally, they enjoyed the stage of improvising a movie-story in real-time using their collective video database, and sometimes wanted to re-create pieces of video that they considered unsatisfactory. However, because of this editing phase most of their original vision was not followed and was dropped from the final published piece. Reviewing these results, a need for more fusion between creating an idea and producing it, seemed necessary. I then hypothesized that mobile technology, e.g. cell-phone, ipaqs etc, could be useful for this purpose. However, a technology materialising their ideas, allowing them to easily retrieve and reflect on them had to be implemented. Hence, tangible objects as metaphors of the captured elements sounded like a solution. Based on the work of Hiroshi Ishii[18], I explored the concept of tangibility of digital data as a way of gathering and capturing digital data around the city for later retrieval. I researched the potentiality of another device using mobile technology combined to tangible objects called *Moving Pictures: Looking Out/Looking In.*

Textable Games

The power and flexibility of *Textable Movie* as a transformable interface can be evaluated in part through applications that I have created, which I did not initially envision. The immediate response from the system showed it comparable to a video game. Then, we created *Textable Game* that extends the concept to the realm of video games (see *Figure 55*).

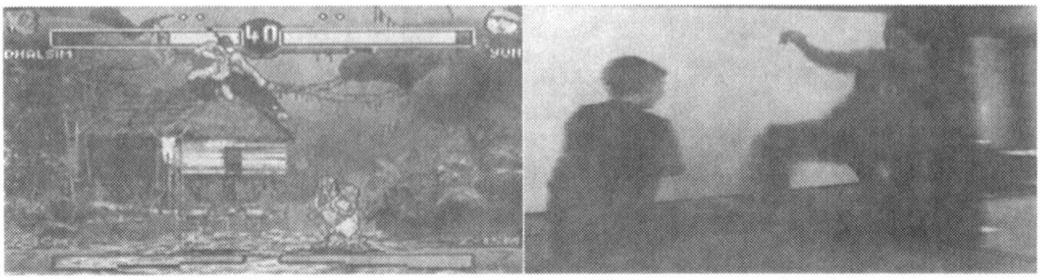

Figure 55: A traditional action game vs the children personal action game

This application aimed to engage teenagers in building their own games, e.g. action games, exploration games, mystery games, using their own footage, e.g. their activities in the city, and sounds, and allowing them to create their own rules and scenarios. They created footage about themselves doing sport in the city in order to integrate it to the *Textable Game* system (see Figure 56).

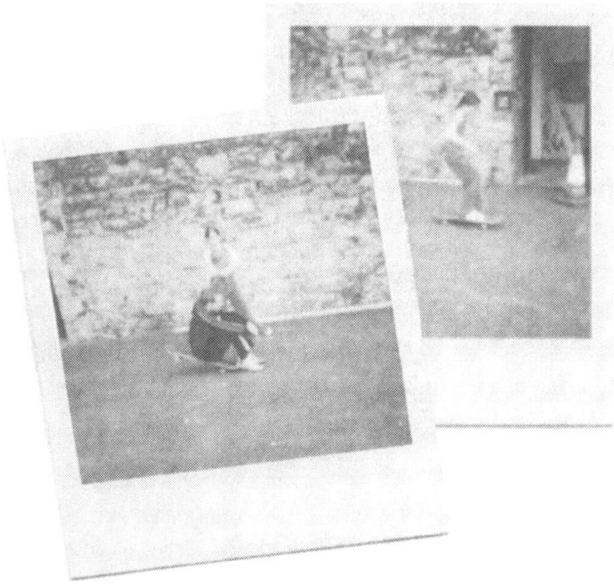

Figure 56: Children transform the traditional action game

The children create a palette of actions for a set of keywords, e.g. one user uses the letters 'a, s, z, x' to control one action, and the other one uses the numbers '1, 2, 3, 4' (see *Figure 57*).

The goal is to invite teenagers to be their own video game producers. In Dublin, a workshop for 10 to 14 year-old children was conducted with *Textable Game* in order to examine how they created strategies of interaction for a video game they have built based on the *Textable Movie* engine; looking *out* to capture the world, looking *in* to reflect onto it.

Figure 57: User A kicks user B by typing '1' vs User B responds to user A by typing 'a'

Moving Pictures

I have also been researching on new ways to materialise the creation of space that are digital media combined with tangible interfaces. Tangible interfaces combine operations on physical objects with digital data. In these interfaces, digital data can be overlaid onto physical objects in a display space or physical objects can act as handles into the digital space.

Investing some research on tangible interfaces for movie making, I invited Diana Africano[7] and Oskar Fjellström[8] at the Umeå Institute of Design to join my efforts in designing a computational and tangible version of *Textable Movie*. From my observations of teenagers using video in workshops, I had hypothesized that *Textable Movie* would be even more powerful if I could construct a physical movie-making device that would allow teenagers to easily understand and make videos using traditional cinematic language. *Moving Pictures: Looking Out/Looking In* [9] allowed children to gather outside and look around in their environment collecting visual clips, e.g. in the city centre, in parks, and then capture their short videos using their video cameras, and come back inside to their video editing station to reflect on their media collection (see *Figure 58*).

[7] Diana Africano is a researcher at the Umeå Institute of Design: <http://www.dh.umu.se/default.asp?sida=236>

[8] Oskar Fjellström is a researcher at the Umeå Institute of Design: <http://www.dh.umu.se/default.asp?sida=240>

[9] Moving Pictures web site : <http://web.media.mit.edu/~cati/movingpictures.html>

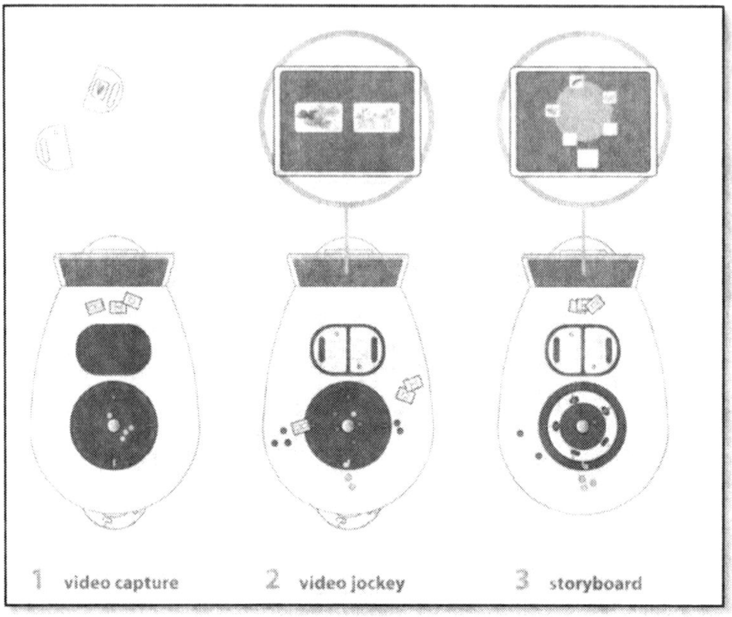

Figure 58: General Scenario of Moving Pictures: Looking Out/Looking In

Different types of input tools appropriate for young users were explored before finally proposing an interface that invited several users to collaborate in the creation and sharing of media stories. Thus:

> "Moving Pictures is a multi-user station that invites young users to create, explore, manipulate and share video content with others. A video station containing a set of two cameras, tokens, a screen and an interactive platform, enables a spontaneous and sociable approach to video creation, selection and sequencing. The station supports multiple input devices and group interaction, encouraging collaborative creation." [Vaucelle, Africano, Davenport, Wiberg and Fjellstrom, 2005]

Integrating spatial components in the design, such as two video cameras, gave the option to take two perspectives in order to give a sense of space, (e.g. close-up and a wide-angle view of a scene, at the same time. With this transformable interface, I wanted to realize my original vision, which allows users to establish and exchange through the use of visuals and sound, where capturing and editing is made very transparent for the user, and in which the process of improvising a story is accessible as a way to create a final movie (see *Figure 59*).

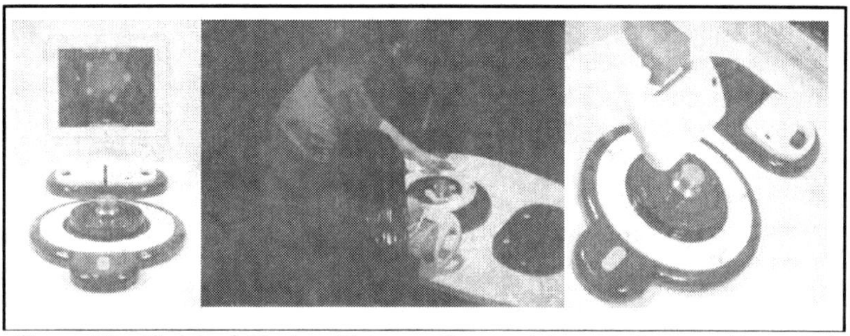

Figure 59: Model of Moving Pictures, Group picture during our User study, closeup onto Moving Pictures

With the use of digital cameras, the technical barriers of producing a final picture have been minimized. The possibility of taking risks and experimenting had been encouraged, so enabling more expressive thought in the use of visual media. Bridging the gap between innovation and production in the creative process is important, as it makes it easier to express an idea. However for a majority of the population that do not master the conventions of visual media, some scaffolding, reflection, context and constraints could be necessary. Intentionally in *Moving Pictures* the user is limited on the length of the captured movie. Indeed, the tangible metaphor of a token symbolising a single shot had to be consistent and refers to the different shots in a storyboard (see *Figure 60*).

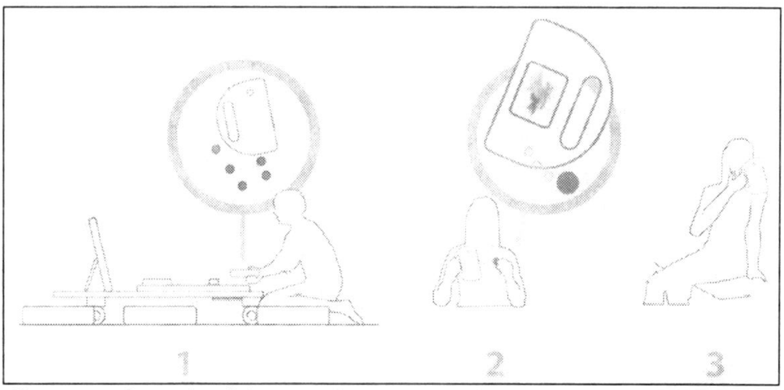

Figure 60: Scenario of the use of the wireless Camera and the tokens.

During user studies,[10] young adults took this physical metaphor very seriously and accordingly used it to capture their video-story. They were very careful of the length of the movie clips captured. It enabled them to possess some rules in standard video editing, without being too conscious of it that could have broken the creative flow (see *Figure 61*).

Finally, the Irish movie director David Keating, mentioned his strong interest in *Moving Pictures* and the physical device to "test" cinematic ideas, and storyboard a movie. Indeed, *Moving Pictures* allows a movie director to easily test his/her movie scenario by capturing a few shots and immediately assemble a quick movie draft of it. Improvisational theatre sessions will also be organized using the wireless video camera as an expressive tool using the *video jockey* table as an instant review display of the improvisation, to later assemble a collaborative video.

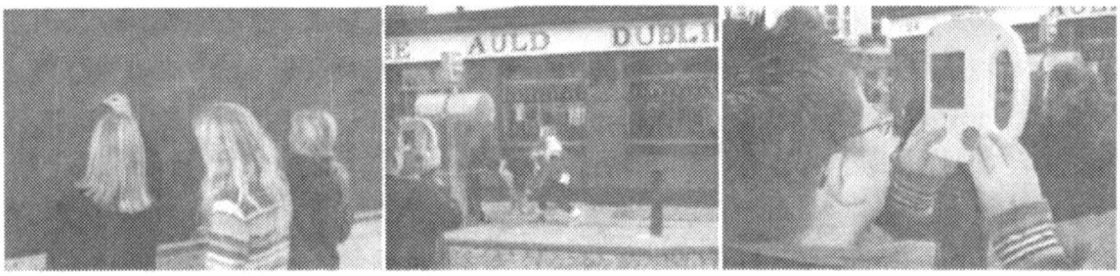

Figure 61: Children storyboarding, acting out their scene in the city, and capturing their video clips

Passing Glances/Texting Glances

With colleagues at the University of Trinity College Dublin, I have explored a similar interface concept to Textable Movie described above: *Passing Glances/Texting Glances* [Jehan, Overholt, Solís, Garcia and Vaucelle, 2004]. Imagined for installation in an urban space like a bus stop or train station, *Passing Glances* invites people to use SMS messaging on their cell phones to drive the selection of images to recreate their space through media usage; with a photo-phone, they would also be able to contribute to a video database. This works by allowing cell phones to send SMS messages which are then displayed publicly along with images related to the content of those messages (see *Figure 62*).

Digital public boards could also be installed for leaving secret messages, videos, and pictures. Connected to a server, each user's cell phone could browse inside their personal computer to display videos, pictures, and sound files. This would be done anonymously in the sense that no one would understand the message except

---

[10] A video of children interacting with the Moving Pictures system can be downloaded at : <http://web.media.mit.edu/~cati/movingpicmedia/new/MPSIGGRAPH05_HIGHRES.mov>

the sender or the recipient, thus respecting the secrecy of the authors. Because these messages are unpredictable, they create an atmosphere of surprise and excitement; after all, why do people carve their names on the sidewalk or spray-paint arcane messages? It is through this unique combination of secrecy and public display that people express themselves and make personal statements. In this way, I explored the "subversive" acts and the ongoing exchange of media and ideas coming from the participants.

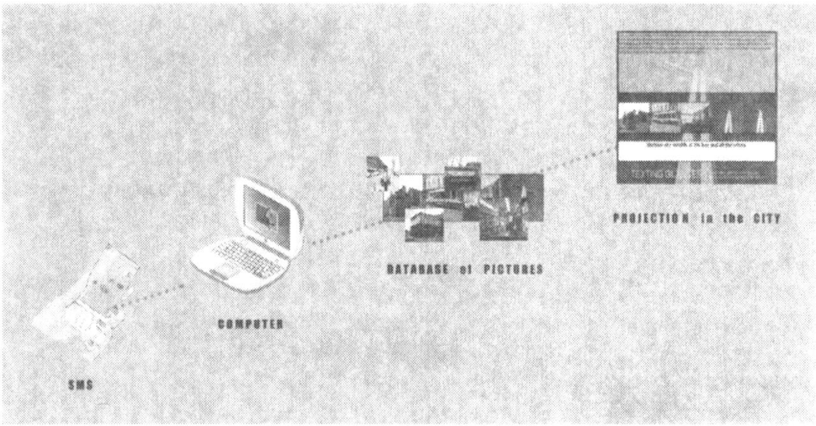

Figure 62: Passing Glances General Scenario

Participating in the exhibition space brought technological landscapes and robotic art into the scene to within the context of a three-month exhibition on the history of robots at The Ark: a Cultural Center for Children,[11] in Dublin, Ireland. Michael John Gorman, curator of the exhibition, asked me to design *Terraria*, a system for visitors of the exhibition to make their movies taking the simplest idea of children's play: playing with toys, adding voices, and turning toys into characters, and allowing children to capture the play on video, and turns movie-making into a natural extension of play. Previous work on immersive storytelling for children showed that children became engaged in the physical space when they could control the physical elements of their story [Bobick et al. 1999]. I then proposed a system for visitors to author the content of the exhibition space by creating their own movies and designed a technology that would enable visitors to experience movie-making using popular devices, e.g. a joystick to control the movie-making process. I also wanted to re-think real-time videomaking for the museum exhibition space. Indeed, within

[11] See < http://www.ark.ie>

the context of an interactive exhibition space, dividing the tasks of recording and editing of digital media between production and post-production can be disruptive of the visitors' experience. The challenge of making a finished video is compounded by the challenge of engaging exhibition visitors to make their own thematic visual story [Vaucelle, Gorman, Clancy, and Tangney, 2005]. *Terraria*,[12] is then a graphical and tangible interface which synthesizes performance and editing into a simultaneous act. *Terraria* consists of four landscapes with robots to play with, four video camera, four joysticks and five wireless networked computers. *Terraria* captures movie segments, voice over, and offers real-time video and robotic sound effects. The user records sounds and visuals, and applies real time effects using a joystick. The joystick commands the movie creation, and consequently, a novel movie gets created and generated in a very transparent, and easy manner (see *Figure 63*).

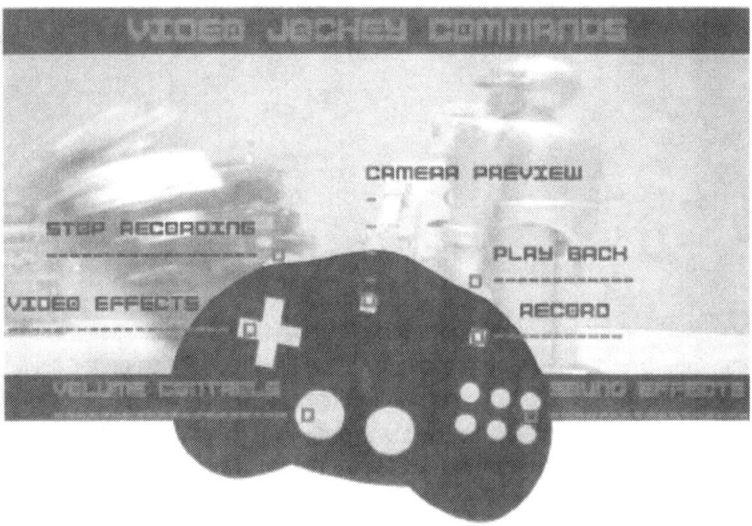

Figure 63: The joystick 'commands' to control the movie making

Within the short time-frame allowed, and given the low attentional focus that can be expected of visitors, capturing and projecting a video would certainly be satisfying, however I wanted to allow more novelty in the interaction and provide a minimum of editing features such as voice over, soundtracks, and video effects. Finally, the simplicity of use, and immediacy of response aimed to engage visitors in

---

[12] Terraria's web site : <https://www.cs.tcd.ie/crite/projects/terraria.html>

movie creation. Video capture, editing and publication of the final video were optimized (see *Figure 64*).

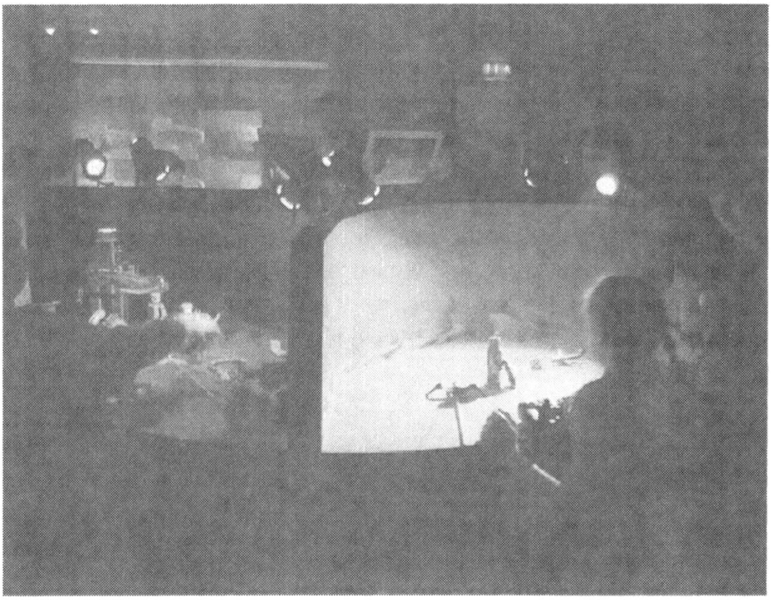

Figure 64: The Terraria landscapes and the projection screen during the exhibition.

Terraria is a direct continuation of my previous research on bridging the gap between video capture and edition within the context of an exhibition. While usual editing tools such as iMovie™ facilitate the editing process, they require a longer time-frame of interaction than the one offered in an exhibition space and are mainly suitable for home use. There is also no transparency between capturing a video, digitalizing it and visualizing it. For this reason, digital jockey systems offer a paradigm for simultaneous real-time performance and editing of sound for a show space, and could be applied to video. My past research [4] and other researchers [Lew 144-149] extended the music concept to video performance of pre-recorded footage.

Maintaining their learning function, 'Science and technology' museums of the last half-century afforded visitors an entirely new experience [Bell]. Within the context of a three-month exhibition on the history of robots, *Terraria* is a system to perform movies using a joystick, editing the movie while capturing, and exporting a final piece instantly to contribute to the existing exhibition space. Indeed, after having created a short movie clip, the user has the option to project it in the exhibition space, and to contribute to a Diaporama of the visitors' movies.

With this interface, I have realised my original goal to not only "unmake" the sacrosanct space of the museum, (e.g. the spectators can touch), but to also allow spectators to transform and contribute to the exhibition space. Constructionists tell us that people learn by constructing their knowledge. People are also engaged in the process because they have fun, because they are challenged, but also perhaps because they anticipate spectators for their creation. The extension of the constructionist field I am exploring comes from the Community Art notion of 'challenging' the audience by the object created. I am also exploring how people can be empowered as artists, in the sense of involving them in creating pieces to challenge their recipient about their assumptions. These assumptions can be about their environment, their identities, and also their culture. The core issue of my research has been to focus on video-stories in order to bring to light contradictions and resolve them.

Bibliography

Barthes, R. *Camera Lucida.* Trans. Richard Howard. New York: Hill and Wang, 1981.

Baudrillard, J. *L'Illusion de la Fin.* Paris: Éditions Galilée, 1992.

Bell, G. *Making Sense of Museums: The Museum as "Cultural Ecology."* Intel Labs, 2002.

Bobick, A., et al. "The KidsRoom: A Perceptually-Based Interactive and Immersive Story Environment." *Presence: Teleoperators and Virtual Environments* 8.4, Conference on Adaptive Hypermedia and Adaptive Web Systems, 1999.

Dunne, A. *Hertzian Tales: Electronic Products, Aesthetic Experience and Critical Design.* London: RCA Press, 1999.

Gaver, B. "Provocative Awareness." *Computer Supported Cooperative Work* 11.3 (2002).

Levy, P. *L'intelligence collective. Pour une anthropologie du cyberespace.* Paris, 1994.

Lew, M. "Live Cinema: Designing an Instrument for Cinema Editing as a Live Performance." NIME 2004.

Lieberman, H., and H. Liu. "Adaptive Linking between Text and Photos Using Common Sense Reasoning." 2002.

Lynch, K. *The Image of the City.* Cambridge: MIT Press, 1960.

Norman, D. A. "Epilogue: we are all designers." *Emotional Design: Why we love (or hate) everyday things*. New York. Basic Books, 2004.

Papert, S. *Mindstorms; Children, Computers and Powerful Ideas*. Cambridge: MIT Press, 1980.

Papert, S. "Situating Constructionism." *Constructionism*. Ed. I. Harel and S. Papert. Norwood, New Jersey, Ablex Publishing Corporation, 1991.

Sanouillet, M., and E. Peterson, eds. *Salt Seller: The Writings of Marcel Duchamp*. New York: Oxford University Press, 1973.

Simondon, G. *Du mode d1'existence des objets techniques*. 1958. Paris: Editions Aubier, 1989.

Vaucelle, C. "La Madone Algerienne by Hocine." Paper for Andre Rouillé 'du document a l'expression: le document' course, UFR Art, Philosophie, et Esthétique, Paris VIII University, France. 1998.

Virilio, P. *Cybermonde, la politique du pire*. Editions Textuel, Collection Conversations pour demain, 1996.

Works Bibliography

Moving Pictures: Looking Out / Looking In

Vaucelle, C., et al. "Educators Program Moving pictures: Looking Out/Looking In." *Proceedings of SIGGRAPH'05*, Los Angeles, California, USA, 31 July—4 August 2005.
Web site: <http://web.media.mit.edu/~cati/movingpictures.html>
Video: <http://web.media.mit.edu/~cati/movingpicmedia/new/MPSIGGRAPH05_HIGHRES.mov>

Passing Glances

Jehan, T., et al. "Real-time Distributed Media Applications in LANs with OSC." *Proceedings of the 1st Open Sound Control Conference*, Berkeley, California, USA, July 2004.

Vaucelle, C., et al. "Ambient Urban Interludes: Passing Glances." *Proceedings of Conference on Human Factors in Computing Systems CHI '04*, Vienna, Austria, 24-29 April 2004.

– – –. "Texting Glances: ambient interludes from the Dublin cityscape." *eNarrative 5, hypertext. narrative. art. tech*. Boston, Massachusetts, May 2003. <http://www.mee.tcd.ie/~moriwaki/passing/>

Terraria

Vaucelle, C., et al. "Re-thinking real time video making for the museum exhibition space." *Proceedings of SIGGRAPH'05*, Los Angeles, California, USA, 31 July—4 August 2005. <https://www.cs.tcd.ie/crite/projects/terraria.html>

Textable Movie

Vaucelle, C., and G. Davenport, G. "An open-ended tool to compose movies for cross-cultural digital storytelling: Textable Movie." *Proceedings of ICHIM 04 'Digital Culture & Heritage'*, Berlin, Germany, August 30-Sept 2004.

– – –. "A system to Compose Movies for Cross-cultural Storytelling: Textable Movie." *Lecture Notes in Computer Science*, 2nd International Conference on Technologies for Interactive Digital Storytelling and Entertainment TIDSE'04 v3105, pp. 126-131, ZGDV e.V, Darmstadt, Germany, June 24-26 2004.

Mixer Subverter

Nemirovsky, P., C. Vaucelle, and A. Quattoni. "Mixer-Subverter: an Online Improvisational Video System." Accepted for AINA 2004 International Conference on Advanced Information Networking and Applications, Fukuoka Institute of Technology (FIT), Fukuoka, Japan, March 29—March 31, 2004.

Sam

Ryokai, K., C. Vaucelle, and J. Cassell. "Virtual Peers as Partners in Storytelling and Literacy Learning." *Journal of Computer Assisted Learning* 19.2 (2003): 195-208.

– – –. "Literacy Learning by Storytelling with a Virtual Peer." *Proceedings of Computer Support for Collaborative Learning CSCL'02.* January 7-11, 2002, Boulder, CO, pp. 352-360

Dolltalk

Vaucelle, C. "Dolltalk: a computational toy to enhance narrative perspective-taking." Master in Media Arts and Sciences, Massachusetts Institute of Technology, Cambridge, USA, September 2002.

Vaucelle, C. "Dolltalk: A computational toy to enhance children's creativity." *Proceedings of the Conference on Human Factors in Computing Systems CHI '02*, Minneapolis, USA, April 20-25, 2002

Story Generator

Vaucelle, C. "Interactive Storytelling for Children." D.E.A. Diplome d'étude approfondies, esthetiques, technologies et créations artistiques, Paris VIII University, France, June 2000.
<http://web.media.mit.edu/~cati/design.html>

Emailer

Vaucelle, C. "L'appropriation active de l'espace virtuel par l'enfant." M.S.T. Maitrise des Sciences Techniques, spécialité photographie et multimedia spécialisation édition électronique, Paris VIII University, June1999.
Video : <http://web.media.mit.edu/~cati/video/Emailer.mov>

Droppin' science

<http://www.brillomag.net/Learning/workshop.html>

Hover

Maynes-Aminzade, D., et al. *Hover: conveying remote. Conference Abstracts and Applications of SIGGRAPH '02*, San Antonio, Texas, July 21-26, 2002, ACM Press.

Mack

Cassell, J., et al. "MACK: Media lab Autonomous Conversational Kiosk." *Proceedings of Imagina '02*, Monte Carlo, February 15, 2002

Cati Dance

Jehan, T., M. Lew, and C. Vaucelle. "Cati dance: self-edited, self-synchronized music." *Proceedings of SIGGRAPH'03 Conference Abstracts and Applications*. San Diego, California, USA, 2003, ACM Press.

Wanderful Alcove

Kanis, M., et al. "The WANDerful. Encouraging constructive social interaction with a socially transforming interface." *Proceedings of Interact Federal Institute of Technology (ETH)*, Zürich, Switzerland, 1-5 September 2003. IOS Press, pp. 966-969.

Developing Creative Audience Interaction: Four Projects by Squidsoup.

Anthony Rowe

Introduction

This paper discusses the processes of creative development in four projects by Squidsoup: *Altzero* (1999-2003), *Come Closer* (2004-5), *Freq2* (2006) and *Driftnet* (2006-7). Each piece presented users with a structured audiovisual composition that could be manipulated, probed and explored in different ways. Together they illustrate a search for novel, intuitive and subtle forms of interaction where the processes of interaction are creative, rewarding and (as far as possible) understood by those taking part. This is an investigation not only into direct user interaction, but also into the participant's interpretation of their experience; what it makes them feel and think, and what meanings they ascribe to their experience and the work.

At a conceptual and structural level, the sonic component in all these pieces is of greater importance than the visual one. Though in all cases sound and visuals are intertwined and feed off each other, the pieces are conceived primarily as interactive musical compositions. The visuals are abstract and highlight movement and structure, complementing the inherently abstract nature of the music and sound in the pieces.

The visualisations tend to have a fluid materiality—every component of the space affects the rest of the space, and can be manipulated by user interaction. They exhibit little predefined temporal change outside of user interaction; they react to events but do not define them. The user creates the context and narrative during, and after, the period of interaction. But they are definitely *spaces*; the feeling of depth, volume and spatial presence is important enough in the pieces to warrant the use of anaglyphic[1] glasses in *Come Closer* and *Driftnet*.

Vital to the user experience is a lack of technological clutter. Apart from *Come Closer*, where hats containing small ultrasonic sensors were used, the projects utilised no worn devices: pure body position and movement in space defined user interaction. The four pieces have been shown in public venues at various times, and user responses have been recorded through observation and questionnaires.

---

[1] *Anaglyph images* are used to provide a stereoscopic 3D effect, when viewed with 2-colour glasses (each lens a different color). Images are made up of two color layers, superimposed, but offset with respect to each other to produce a depth effect (Wikipedia 2007)

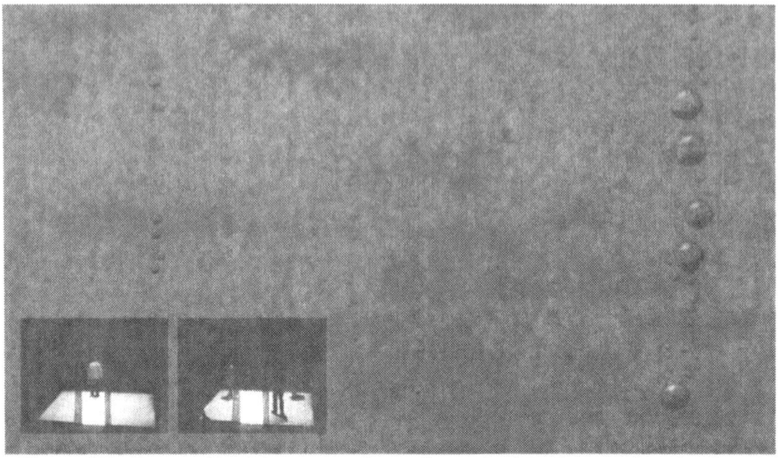

Figure 65: Altzero, Participants navigate through virtual space by altering their physical position within an active 5m x 4m area, as if a piece of physical space is transplanted into an immersive virtual audiovisual environment. A projection shows a first person view of the space; speakers play the sounds heard at that location.

Altzero—Music in space

Altzero allows people in an art gallery or online to move through the visualization of a spatialised musical composition. Musicians (Icarus, Scott Arford, Cyan) were asked to compose a piece of music existing primarily in space rather than over time, and their compositions were then placed in virtual spaces, accessible both online and in gallery settings, for 'active listeners' to immerse themselves, and explore. The idea links games technologies (where navigable soundscapes are now normal—but not the focus) and musical composition. In a sense, the project envisages a gaming environment where everything is stripped away save the sound and the ability to move. The visuals then become referents for the sounds—another means to help in deconstructing and understanding the structure of that which is being listened to by the audience.

Several navigational methods and metaphors were used in the various incarnations of the project, ranging from mouse and keyboard (mandatory for online access), through joysticks to a multi-user camera-based gallery setup. The version shown at the ICA (London, 2001) used a portion of the theatre as an active physical space within the virtual composition that was controlled by body position within an active area. Participants stood at the front of the space to tip down, on the right to bank in that direction and so on.

Gallery-based participants were joined in virtual space by visitors to the website. Each could see the others and, by following other visitors' trails, could hear their experience of the piece. One aim of the ICA installation was to create a truly multi-user experience, not only in terms of the virtual space, but also for the physical gallery space. There were no restrictions on how many people could enter the physical space at one time; if more than one person entered, the system would take the average position of all people in the area and use that to control movement through virtual space. Technically, this approach worked well, but from a user perspective the experience became somewhat confusing and unresponsive when several people took part simultaneously (see below).

The project also grappled with issues of authorial control. In all but the first version of the project, the positions of the sounds was regarded as part of the composition and therefore defined by the composer rather than the listener. This made for an aesthetically pleasing end result, but one that lacked in terms of user control or 'creative' interaction. Participants could determine to an extent what they heard but could not generate, control or compose the sounds themselves. This issue was partly addressed in the final iteration of the project, which included a software tool allowing musicians to create their own navigable spatial compositions—the piece then became a hybrid performance/installation, involving performing musicians as well as an active audience.

The issues brought up by *Altzero*, particularly the notions of shared use of physical space, collaborative and 'creative' interaction (where the user is given a greater share of control over the piece), have become central to Squidsoup's work, particularly in the other three projects discussed here.

Come Closer—Music in shared space

Edward Hall, in his book 'The Hidden Dimension' [1963], identified and measured 'personal space'. He showed that we do not, in normal social behavior, encroach upon others' personal space. In the West, the distance of this boundary is around 1.2 metres; any closer is regarded and 'personal' or 'intimate' behavior. *Altzero* used a 4 metre x 5 metre semi-enclosed active space. With two or more people in the space, movement was severely affected by these boundaries, causing them to constantly evade each other. The system responded to average position, so with people in diametrically opposite corners the net result was that the balance point was usually very near central; equivalent to there being nobody in the space. Even with the person in front unaware of anyone behind them, the rear person would still generally move in opposition to the front person, cancelling out their effective presence

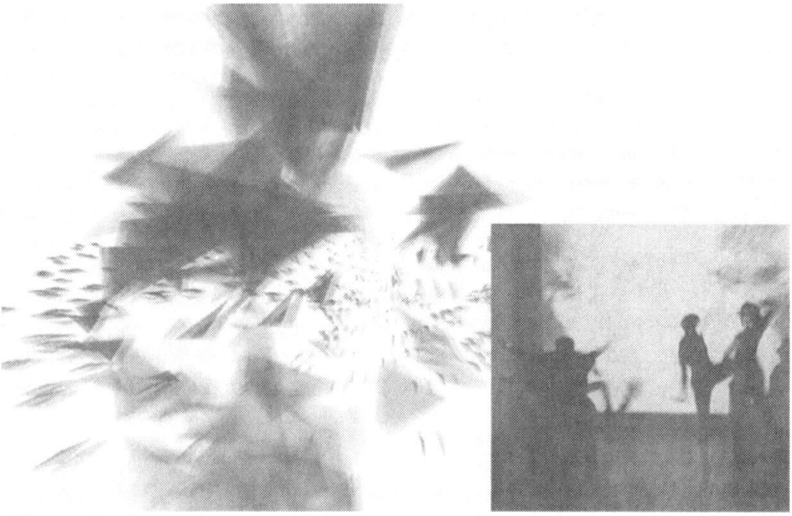

Figure 66: Come closer. Multiple users control the behaviour of a visual and musical structure through their position in physical space. The effects they have on the piece are expressed sonically and through anaglyphic projection, where the projection is an abstracted and enhanced mirror image of the physical space, showing the relationships between participants. The system responses highlight participants' proximity to each other.

This, then, was the premise for *Come Closer*: to probe and investigate the boundaries that define our personal space; to test them and to cross them. The first *Come Closer* experiment recreated the balancing active space system used in *Altzero*, but using far more accurate ultrasonic position detection, a different visualization, and implemented in a larger, 8 metre x 8 metre, active area. The aim was to see if the problem was reproduced, or whether higher system sensitivity and larger space in which to manoeuvre would make the processes of interaction more transparent. Increased lighting was also employed to heighten awareness of others' physical presence.

However, these changes had no real effect. The relationships between individual movement, group movement and the onscreen response were still not generally understood; at least until they were explained. Once an explanation of the processes was given, participants took far greater active control, but this proved that the mechanisms were not sufficiently intuitive for the purposes of the artwork.

It was apparent that any system where individuals are expected to cooperate and work together in a shared experience either requires prior explanation, or each individual needs to be able to see the results of their own interactions as well as the group effects—a direct one-to-one connection between multiple users and a shared audiovisual output. Another approach was needed.

We opted to use the projection as a *virtual abstracted mirror,* with each person visually represented in the virtual space—i.e. a static virtual camera in a moving space, rather than the converse. Two implementations of this system were produced. User tests were immediately conclusive: the relationships and processes were much clearer to all participants. Participants were visible in the virtual space, and were clearly aware of each others' presence and proximity in both physical and virtual spaces. This, then, allowed us to start encouraging participants to go beyond Hall's boundaries for social behavior, and to move into a more personal space—to literally *Come Closer* together.

The final *Come Closer* piece uses position (both absolute and relative to the other participants) as the main variable of interaction to control sounds and visuals. The visuals consist of several thousand exploded polygons, affected as if by a form of 'fluid magnetism' by the positions of participants. As people near each other, their joint magnetism increases, resulting in increased visual stimulus. Similarly, piano couplets and triplets are triggered only when two or more people come within the 1.2 metre boundary. The piece encouraged participants to behave in a more intimate manner than they would normally, by suggesting that interesting consequences would result from their increased proximity. The more people are in this personal grouping, the higher the climax achieved in the audiovisual experience.

User behavior in *Come Closer* was recorded and analysed in some depth [Randall and Rowe], concluding that such an experience can begin to break down Hall's boundaries, albeit temporarily.

Freq2—Space as a musical instrument

Freq was in part inspired by a sound installation by LaMonte Young at The Dreamhouse, New York.[2] Young's piece, a series of sine waves played loudly in a heavily dampened room, has the effect of producing very different audible sounds depending on how these sound waves enhance or cancel each other out. Physical position determines what is heard; as one moves around the space, so the various sine waves (coming from two large speakers) progressively go in and out of phase,

---

[2] LaMonte Young "The Base 9:7:4 Symmetry in Prime Time When Centered above and below The Lowest Term Primes in The Range 288 to 224 with The Addition of 279 and 261 in Which The Half of The Symmetric Division Mapped above and Including 288 Consists of The Powers of 2 Multiplied by The Primes within The Ranges of 144 to 128, 72 to 64 and 36 to 32 Which Are Symmetrical to Those Primes in Lowest Terms in The Half of The Symmetric Division Mapped below and Including 224 within The Ranges 126 to 112, 63 to 56 and 31.5 to 28 with The Addition of 119." Documented online at <http://www.melafoundation.org/dream02.htm>

changing their perceived volume. It certainly seemed to me that when standing perfectly still, small movements of an arm or even a finger had clearly distinguishable sonic consequences.

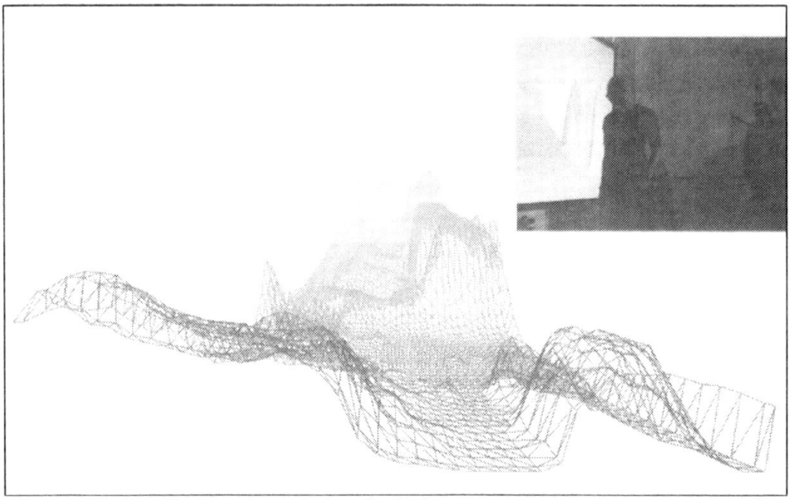

Figure 67: Freq2. A projected virtual wireframe structure reacts to passers-by, tracing with its leading edge the outline of their silhouette. Simultaneously, sounds begin to emerge, derived directly from these shapes.

The idea of being able to manipulate sound directly through whole body interaction; using space as a musical instrument, was intriguing, and also plugged into a rich lineage including such work as David Rokeby's *Very Nervous System* and Golan Levin's *Scribble*; projects that respectively explore the use of physical space as sonic space, and real-time interactive sound generation.

Freq began as a series of experiments into plastic audio generation, and resulted in answering the question "what does your shadow sound like?" by converting the outline shape of participants' silhouettes directly into an audible sound wave. Any sound can be defined as a two-dimensional line; for example a sine wave or square wave. The shape of the line defines the timbre of the sound, and its base pitch or *frequency* is defined by how quickly it is repeated over time.

A camera detects the outline shape of anything between it and a light source, and traces its outline in real time. Thus, if there is nothing in front of the camera, a straight horizontal line will be produced—as a waveform this is silence—but as soon as there is a physical presence in front of the camera, shapes begin to emerge and sounds are generated. Even tiny physical movements have immediate and distinct sonic results.

The piece works with multiple simultaneous users and has no worn equipment, allowing for unhindered interaction. It can also be perceived as an interpretation of a space; creating a feedback loop by causing users to react to its response to their presence. *Freq2* (2006) placed this idea into a more complex and composed musical structure. Using the waveforms created by the *Freq* engine as the sole sound source, a musical composition was built that would utilize these sounds in a more structured way, allowing participants to play space as an instrument within a score. The score uses the sound waves at a range of frequencies and volumes, and also includes a series of time loops, giving the piece a sense of 'memory'; this too is reflected in the visuals.

Due to its simplicity and immediate responsiveness, *Freq2* works well with audiences with little explanation. The exact relationship between interaction and sonic output was often not accurately understood, but this had little effect on the perceived experience, at least for non-specialist users; people do generally and quite readily take active control of the piece.

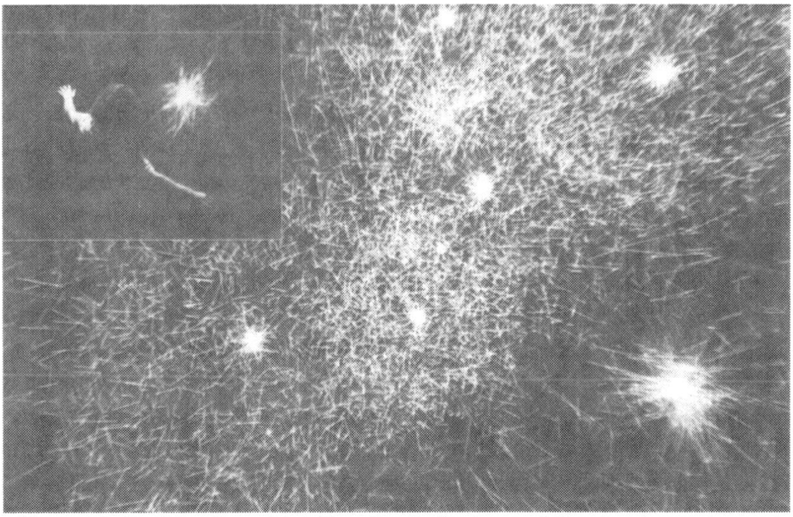

Figure 68: By flying like a bird (flapping their wings, gliding, turning and so on), participants can fly through a series of dynamic virtual spaces. An anaglyphic stereo projection shows what their virtual self sees, and spatialised sound allows them to listen to their virtual surroundings.

Driftnet—Flying in dynamic musical space

Driftnet returned to the notion of navigable musical composition but, unlike *Altzero*, the composition is dynamic, constantly changing and reacting to the participants' presence. As such, it was to become much more of an interactive experience than one of pure navigation. It was also felt that the experience needed, above all, to be immersive, intuitive, and musical.

The *Driftnet* virtual space is populated by autonomous reactive agents designed to respond (both aurally and visually) to external factors—principally other agents, sound input and the navigating participant's presence. These agents are audiovisual entities with their own programmed behaviors, which together form the sonic and visual components of the space. By moving through the space, visitors affect the agents' behaviors and therefore become a part controller of events as well as a listener/explorer.

Visually, the piece continues the sparse and highly abstract aesthetic explorations of *Come Closer* and *Freq2.* The space has depth (again using anaglyphic stereoscopic vision) and uses conventional perspective to achieve a high level of immersion, yet the components of the space consist entirely of separate and unadorned polygons—the basic building block of virtual space; in a sense a form of "truth to material." There are three connected virtual spaces in Driftnet, the feel of each defined by the sounds and behaviors of the agents in each space. The central space is a surface, upon which sounds skitter and polygons dance. Above the surface, everything is clear, crystalline and light; whereas below it the spaces are dark, muffled and slow moving. Away from the surface (above or below) are ostensibly limitless spaces. These spaces have no up or down, and are filled with active sonic agents that move freely around, reacting to each other and to the participant's presence.

These are dynamic spaces that need to be actively *flown* through. The project needed a navigation interface that was simple, intuitive and flexible. Flight seemed to be an honest metaphor; bird flight seemed particularly intuitive: ask a child to 'fly like a bird', and they will gladly show you. Tilting and flapping arms/wings is a generic gesture set. Navigation in virtual space is still an area of active research. Numerous systems and metaphors are used, yet the generic ones are restrictive, non-intuitive and simply not fun. Many interesting interfaces have been built; e.g. the breathing system used by Char Davies (in *Osmose* and *Ephemere*, a jacket that detects breathing controls rate of ascent/descent, much as a diver) and Brenda Laurel's *Placeholder* project where the navigator assumes an animal character—fish, spider, snake or crow. The crow is functionally similar to 'Fly like a bird', but its technical conception, and its entire feel, is very different. *Placeholder* was a VR experience, using the full paraphernalia of VR: headset, earphones, body motion-tracking sensors and so on. *Driftnet* needed a system with no worn components, and that could be shown in public without any individual calibration or setup required.

Our experiences using a camera to recognize whole body interactions for *Freq* were exciting—cameras are reliable, flexible and non-intrusive (beyond their well documented use in intrusive surveillance). There are no physical or worn components—making them very easy to use in public spaces, and allowing participants completely free and uninhibited movement. They also allow for an analogue-style form of interaction; areas with levels of grey rather than on/off pixels.

Twenty people, with a range of backgrounds and levels of computer literacy, were asked how they would perform the standard set of navigation tasks if they were imitating a bird in flight. The responses were not as unanimous as expected. There was general agreement on accelerating (flap your arms). When asked to turn (left or right), most tilted their arms up and down, but some rotated to the left or right. Turning up and down was fairly uniform (tilt backwards/forwards or look up/down), though some people stood on tiptoes or crouched. Decelerating, stopping and strafing (sidestepping—or moving in a direction other than "forwards"), caused confusion and no consensus whatsoever. These differences of opinion led to a simplification of the mechanics of bird flight to the point of abstraction, mainly to avoid asking participants to perform tasks with ambiguous gestures attached to them, and also out of an artistic desire for users to focus on the experience rather than the process of flying.

Function	Associated action/gesture
Accelerate	Flap arms like bird's wings
Turn down	Head forwards
Turn up	Head back
Bank to left/right	Rotate arms outstretched about the vertical axis OR Tilt arms about forward vector
(Strafe)	No action: happens automatically when turning
(Stop/decelerate)	No action: happens automatically if not accelerating.

Table 5: gesture/function set for Driftnet

The final gesture set used (see *Table 5*) required image analysis in 3 dimensions—a single camera cannot give accurate distancing information other than in very carefully defined laboratory conditions; but two cameras, or a stereo camera, can be used to derive real-time distance information. A Point Grey Bumblebee2 stereo camera was tested and produced depth map information of sufficient accuracy to be used as the basis for interaction detection in 3 dimensions. Although in theory the camera could be mounted almost anywhere, it was ceiling mounted, as this produced better depth map imagery, and is also beyond the reach of the public.

Like any non-standard interface, there is a learning curve to flying in *Driftnet*. User tests showed that after reading the very simple graphical instructions (a process that typically takes less than 10 seconds—(see Fig. 76), people would be able to control the system as soon as they were in the active area—but interesting alternatives (swimming, wading, pointing) are attempted if the instructions are not read. User interactions during public trials at Shunt (London Bridge, London, June 2007) are documented online.[3]

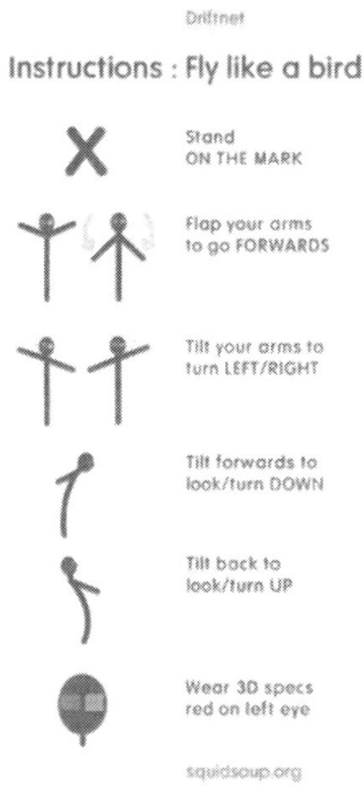

Figure 69: User instructions for Driftnet

---

[3] Online documentation for the Squidsoup projects: Altzero: <http://www.squidsoup.org/altzero>; Come Closer: <http://www.squidsoup.org/comecloser>; Freq2: <http://www.squidsoup.org/freq2>; Driftnet: <http://www.squidsoup.org/driftnet>

Conclusions

These projects each deal with one aspect of a larger picture; creating interactive experiences in public places that allow for subtle and creative forms of interaction between individuals and an artwork. The works are all about relationships; illustrating the connections and synergies between participants, between real and virtual space, between sound and visuals. They also explore the various problems associated with installing interactive work in open public spaces without resorting to very simplistic forms of interaction.

User feedback on all of four projects shows that people are also keen, in the right circumstances, to actively engage in the creative processes involved in assigning meaning, both during and after an interactive experience. The potential for this process is left as open as possible through the use of generally non-referential and abstract visuals and sound within controlled structures. Comments left from a single day showing an early version of *Come Closer* (Watershed, Bristol, 2005) include: "deeply immersed in the soup; lovely dream-world—fairies I think and two different kinds of fairyland; loved making waves; like being in a superman film—white and graceful and spacey; I like the hover flowers; I liked the feathery things; felt removed from myself; like pulling a dog along on a lead and getting bigger so I could catch it; liked the wind chime; liked the electronic rain"

Acknowledgements

Altzero: music by Cyan Collier, Scott Arford and Icarus. Altzero[4] commissioned by Cornerhouse with additional funds from Arts Council England. *Freq2:* commissioned and part funded by Cybersonica 06. *Come Closer:* developed with funds from the Clark Digital Bursary, Arts Council England (Grants for the Arts) and Watershed Media Centre. Development was supported by Watershed Media Centre, Mobile Bristol, JA Clark Charitable Trust and The University of the West of England. Positioning technologies developed by Cliff Randell[5] (University of Bristol) *Driftnet*: supported by Future of Sound. All projects by Squidsoup: Anthony Rowe and Gaz Bushell. Additional work on *Altzero* by James Lane and Chris Bennewith. Additional work on *Driftnet* by Ollie Brown.

---

[4] See <http://www.squidsoup.org/altzero>
[5] See <http://www.cs.bris.ac.uk/~cliff/>

Bibliography

Hall, Edward T. *The Hidden Dimension.* New York: Bantam, 1963.

Randall, C., and A. Rowe. "Come Closer: Encouraging Collaborative Behaviour in a Multimedia Environment." *Interactive Technology and Sociotechnical Systems.* 12th International Conference, VSMM (ACM).

3.3 Wearable Computing

The Emotional Wardrobe

Lisa Stead[A], Petar Goulev[B], Caroline Evans[A], Ebrahim Mamdani[B]

[A]School of Fashion and Textile Design, Central Saint Martins University of the Arts
[B]Department of Electrical and Electronic Engineering, Imperial College London

Introduction

With the advent of new 'wearable' electronic and material technologies an innovative union between fashion and technology offers potential to satisfy the fashion industry's continual search for novelty. However, few fashion designers are experimenting with its application, although notable exceptions are designers such as Hussein Chalayan [Quinn 26-32, 50-55] and Elise Dee Co.[1] We have used this quote to highlight the problem,

> "What we are talking about here is a…revolution—one which will require the electronics industry to 'think emotionally'…we can not expect the fashion industry to adapt itself to technology. Rather the technology industry will have to learn how to deal with fashion."
>
> [Stefano Marzano; Phillips, 7]

It is the fields of electronics and technology that are embracing technology for the body. The majority of 'wearable' technology has utilised clothing as a carrier of entertainment and communication systems, adding increased functionality. Less research has focused on the exploitation of technology for aesthetic purpose and the social and interactive potential that this could offer.

This research aims to investigate a garment that responds to the wearer, translating an inferred change in emotional state with a change in the garment aesthetics. By using the AffectiveWare platform to mediate between the wearer and the garment

---

[1] Co, E.D. (1998). Computation and Technology as Expressive Elements in Fashion. Massachusetts Institute of Technology. <http://acg.media.mit.edu/people/elise/thesis/> Viewed 12/04/04.

display, a degree of autonomy in design may be possible, allowing the wearer to actively change the aesthetics of the garment. This use of the AffectiveWare platform as a tool provides a human-centric context to the technology, to encourage and facilitate human-human interaction between the wearer and society.

It is proposed that a fusion of fashion and technology will enhance fashions interactive and expressive possibilities to create a personalised vision of identity and a 'poetic' use of technology. Through emotional design, it suggests that aesthetic communication can be used to provoke paradoxical responses, by creating a narrative between the body, garment and society.

The second section will introduce the design context and inspiration and its reflection of attitudes within the design and technology communities. The third section will explain the design concept and how it could be materialised through technology enablers. In the fourth section a method for conducting experiments to provide data for the AffectiveWare platform is discussed. In the fifth and final section the technology and design elements are brought together to create the overall vision.

Context—The Age of Emotion?

A survey of design, science and technology was conducted to look at how other disciplines integrate technology within design, and a predominant interest in emotion and emotional interaction with technology was discovered. There are publications that analyse the nature of emotional design, such as Don Norman's 'Emotional Design' [2004] and organisations that specifically address these concerns, such as the Design and Emotion society.[1] Within the design of technology and computing more human orientated solutions are being sought through research fields such as interaction design. The area of affective computing is endeavouring to give computers the ability to feel, recognise and react to emotions [Picard 1997]. Emotional music cues are being researched to enhance the emotive quality of computer-generated music [Bresin and Friberg, 2000]. There is also neuroscience research into the origins of emotion in the brain [LeDoux 1999] and debate about biotechnology's ability to control them [Fukuyama 2002]. Why such interest in emotions?

It seems appropriate given that we live in an age that caters for our functional requirements that it is now time to address the essence of what makes us human

---

[1] See <http://www.designandemotion.org/> (Viewed 12/04/04)

and individual; our emotional and spiritual needs. There is a realisation that technology needs to become personalised and human-centric, adapting to the user as the user adapts to the technology and that user interaction can be made meaningful by looking to our social and cultural world. Cass, Goulden and Kozlov [218-223] write,

> "These needs are not functional or physiological but psychological and cultural. They include the need to represent roots and heritage, to create a sense of belonging and connectedness and to demonstrate personal identity and achievement."

Fashion is an emotional discipline that fulfils all of these requirements; it functions in contemporary culture as an interface between the individual and society. It is a visual display of choice and a powerful cultural communicator of self. Norman proposes three categorises to accomplish an emotional product; visceral, behavioural and reflective. He suggests visceral design is about the look, feel and sound of a design. Behavioural design focuses on function and usability and reflective design is about meaning and how product messages inform our self-image. Fashion can be seen as a merger of these three elements. It is visceral, as aesthetics and sensory perception are integral to fashions appeal and it is the communication of aesthetics and their meaning [Barnard,1996] that provide its reflective qualities. Fashion can have practical functions but function can be more than utilitarian protection or comfort; it is also psychological. Fashion is able to bring us psychological benefit through our emotional response to what we wear, how we look and how we are perceived [Hollander,1980] reinforcing its reflective qualities.

Fashion has an emotional context and through the integration of responsive material technology and computing capabilities these elements could be enhanced. But should the emotions that it evokes be pleasant? Desmet [8-12] comments,

> "It may be interesting for designers to investigate the possibilities of designing paradoxical emotions because this may result in products that are unique, innovative and rich, and more challenging or appealing than those that elicit only pleasant emotions."

Contemporary design practitioners that could be considered 'critical' in their approach, such as Dunne and Raby [Dunne & Raby 2001; Dunne, A. 1999] offer a paradoxical perspective on design. In their work they look at darker human needs and the role that electrical products play in everyday life. A key theme is, 'Making

Visible the Invisible'. The work engages the user by 'visualising' electromagnetic energy in various forms, which could be seen as worrying or playful, inspiring a range of emotions. They aim to create poetic experiences, which involve the user in a narrative with the product rather than dictating a generic use.

By placing fashion and technology within this paradoxical and emotive framework, fashion could act as a mediator of poetic experience through provocative aesthetics. A garment that changes its aesthetics in accordance to inferred changes in emotional state could stimulate a myriad of responses. For the individual, it could be used as a form of self-therapy, aiding self-awareness. It could become a measure of, or route to wellbeing by highlighting your emotional state so that you can adapt your behaviour accordingly. There are times when it is not social acceptable to show ones emotions and this can become stressful. For example, you have been queue jumped but are too polite to say anything, or you overhear a conversation on the tube that makes you angry. By creating a playful form of non-verbal communication, the wearer could visually comment on a situation without others knowing. Those around you may find it disturbing to see changes in garment patterns and wish to translate the hidden messages. In comparison, you may want loved ones to decode your mood through familiarity to your aesthetics. How often do we really say or show what we mean? *The Emotional Wardrobe* could also create a democracy of design by implementing personalisation through the integration of interactive material technology. In this way the *Emotional Wardrobe* could be a tool to enhance and explore social interaction and self-expression.

Technology Enablers : AffectiveWare

By merging fashion with responsive material technology and an existing real-time platform termed 'AffectiveWare' we aim to create clothing that is personalised by the emotions of the individual; an Emotional Wardrobe. The AffectiveWare platform has been developed at Imperial College [Goia et al, 2004] inspired on the principles of affective computing as documented by Rosalind Picard [Picard,1997].

Affective computing aims to improve human computer interaction by simulating the emotional interaction experienced between human beings. It is based on the premise that psychological significance can be inferred from physiological signals, as proposed by Cacioppo and Tassinary [16-28]. Human computer interaction is explored through computers that recognise, represent and respond appropriately to our emotional states. This is done through the measurement of physiological signals via sensors and the assessment of body language facial expression and gesture.

It has been suggested by Rees and Cass [2003][2] that such computing could become a 'robot poet', a machine that reads and simulates human response but is less interesting as an interface to stimulate 'poetic' dialogue with the user. They propose that within 'Poetic Interfaces', 'the computer remains the tool and people the poets, the ones sharing emotional experience'. However, can not a 'reader' also become a 'mediator' of emotions? When the AffectiveWare platform relays changes in affect back to the garment, new social interactions may be created through a new 'poetic', aesthetic vocabulary. Much of affective computing is centred on human computer interaction but by addressing this area from the notion of 'fashion as social communication' the emphasis has shifted from human computer interaction to human human interaction. The computer, sensors and display technology become the interface between the body and clothing and the wearer and society.

Emotional Wardrobe

The Emotional Wardrobe includes a set of sensors, AffectiveWare platform, a Garment server and a display controller. The system is configured from an undergarment and overgarment. The undergarment contains the *AffectiveSensors* and the *AffectiveCoder*. The overgarment has inbuilt electro-luminescent display panels.

The physiological sensors send data to the *AffectiveCoder*, which translates the analogue information into digital signals. They are transmitted wirelessly to the *AffectiveWare* Platform, which estimates the mood of wearer and passes the information to the garment server. To personalise the Emotional Wardrobe we have defined a set of rules, which control the way in which the garment display will respond. The wearer can select a subset to match the desired aesthetic. This choice is facilitated by a Personal Digital Assistant connected via Bluetooth or GPS. Following the wearers personal setup, the garment server sends this XML commands to the Bluetooth enabled display controller which is circuitry inbuilt within the over garment. The controller reads the instructions from the server and changes the display to create a changing aesthetic.

---

[2] Rees, F. and Cass, J. (2003).Poetic Interfaces. In E. Aarts and S. Marzano (Eds), *The NewEveryday, Views on Ambient Intelligence* (pp.224-229). Rotterdam: 010 Publishers

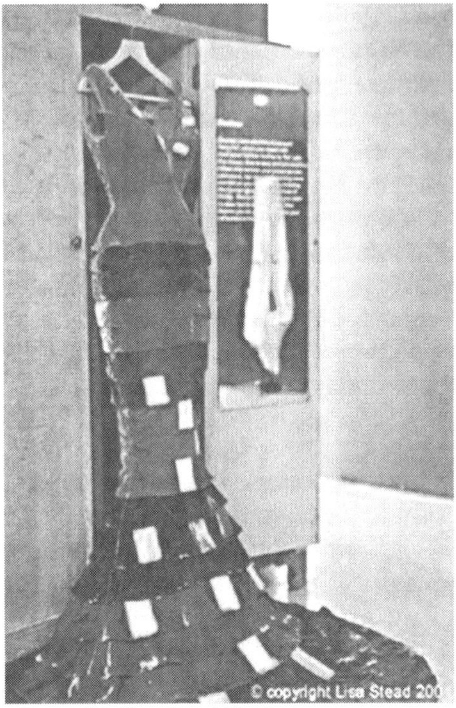

Figure 70: Emotional Wardrobe

Personal Sensor Network

The *AffectiveSensors* are built into a close fitting undergarment to provide maximum comfort for the wearer and to ensure unrestricted movement. The *AffectiveCoder* is located in the small of the back to maximise comfort and to ensure that the aesthetics of the garment are not disrupted.

In most cases the heart is positioned on the left hand side of the body. Therefore, the *AffectiveHeart* sensors are located accordingly on the front of the undergarment. The *AffectiveRings*, are also positioned on the left hand so that the technology is located in a focused area, minimizing wiring. This organisation of the Personal Sensor Network ensures that the technology is evenly distributed. The copper cables are covered with silicon insulation and are embedded within the seams of the garment to provide extra comfort and strength during washing. During washing the AffectiveCoder can be detached from its holder in the garment. The undergarment is made from a breathable and stretchable fabric to allow the skin to breathe normally and to avoid stimulation of sweat glands, which could interfere with the Affective Readings.

Display technology

The Emotional Wardrobe is concerned with expressing the inside on the outside, making the invisible, visible. In order to visualise the inferred emotions a responsive material was needed to create the display. Colour-change technology was chosen, as it is the most accessible display technology. Colour-change can be produced through thermochromic inks, fibre optics or other light sources such as light emitting diodes or electroluminescence. Electroluminescent panels were selected for the display as they are thin and flexible and provide a uniform light. The intensity of light and colour change can be manipulated quickly, by altering the voltage or/and frequency. Therefore, Electroluminescence provides a controllable colour/ light source that is supple enough to be embedded in panels within a garment.

The display system comprises a series of electroluminescent panels, which are housed within the top garment. They can change their colour or/and intensity depending on the mood and the display rules employed. The display controller changes the aesthetics of the light in order to express inferred emotion to the self or others.

Method for Experiments

To provide physiological data for the *AffectiveWare* platform [Goia et al, 2004] and the garment a group of actresses will evoke 8 emotions based on Plutchik's model [Plutchik, 1962]. These emotions are anger, sadness, surprise, joy, fear, acceptance, disgust and anticipation. From this group one will be selected as an emotional model for the garment, to which the garment will be, tailored both bodily and emotionally. The remaining actresses will act as a control group with which to cross-reference the data.

The physiological states strongly connected with human emotion will be measured in real-time using Affective Sensors. We have begun experiments using a galvanic skin response sensor. It measures the conductance of electricity across the surface of the skin [Goulev and Mamdani, 2003]. We are currently developing heart rate and temperature sensors that will be integrated into the AffectiveWare platform. The actresses will also be video recorded so that body language, facial gesture and speech intonation can be assessed. This information will then be analysed by the AffectiveWare platform, to enable us to determine the emotion expressed by the corresponding physiological signal and this information will then be expressed through the garment display to create the emotional aesthetics.

At present, we have conducted initial experiments with 6 actresses. All participants were asked to wash their hands before the experiments were conducted to remove any cosmetic lotions that may interfere with the readings. The actresses were recorded as they evoked each emotion, repeatedly for a minute and a half; followed by a calm period of the same duration. This enabled us to obtain clear demarcations between the readings.

Of the 6 actresses, 2 stood and used very physical body language to evoke emotions, 3 sat down to create internalised emotions and 1 used both methods. From the group of 6 participants, 2 of the actresses' readings were too extreme to be interpreted. It could be assumed that a demonstrative display of emotion would cause this to happen but interestingly, of the 2 actresses, 1 actress used standing body movement and one remained seated. This highlights that galvanic skin response differs widely between individuals regardless of the type of emotive activation. We are currently analysing the results to select the actresses for further testing.

Garment Design: Incorporating display technology with fashion aesthetics

A series of experiments were conducted to determine the use of the electroluminescent panels within the garment, working with and against its flexible properties. The chief concern was to make the technology tactile and aesthetic rather than sterile and cold, remembering why we enjoy wearing fabrics and fashion. In this way a series of tradition fashion techniques were explored as methods to embed the technology within the felt fabric of the garment. One idea was to weave strips of electroluminescent panels into the fabric of the garment so that the panels became incorporated. However, the panels emit noise when bent and the polyester casing is still very visible. Further ideas were developed to use the panels as a flat light source behind the felt fabric. These were employed to work with the qualities of the felt, as it is able to be punctured without fraying and can also be moulded into 3D shapes through the application of steam.

Cutouts were made in the fabric and diaphanous fabric inserts were applied but the natures of the panels were still evident. Next, the felt was punctured to create areas for the light to emit. It was noticed that a double layer of cut-outs added dimension and colour change to the emitted light, through shadowing. Finally, the ability of electroluminescence to be reflected and to cast light onto the surface of the fabric was utilised. By creating contours with the felt, the panels can be hidden at a 45° angle within the garment, so that only their light can be perceived. Through experiment it was realised that there would need to be a reflective and undulated surface to carry the light out from the contour and onto the garment. This gives a subtle effect of light seeping from the garment contours. Various methods were tried and eventually a pearlised print binder was printed onto the felt and the felt then steamed into the desired shape. After experimentation, the electroluminescent panel encased within the garment contours was chosen as the most successful solution.

Although, the technology has provided restrictions to its use, it had also provided answers to its integration. The contours enable the panels to remain hidden, retaining the tactile nature of the garment and also produce a housing for the chip inverter, which alternates the current for the electroluminescent panels and controls the illumination. The effect is more subtle and poetic than if the panels had have been used in a flat format, attached directly to the garment. Once the technique of incorporating the panels within the garment had been determined, a display program, or display rules were constructed.

Display Rules

The display rules are influenced by research into music and emotion to offer a more ambiguous model of aesthetic communication than common colour/ mood associations, for instance. The eight rules match the eight emotions in the Plutchik model. The inferred moods are fed from the AffectiveWare platform to the garment server, which contains the display rules. Based on the inferred emotion, the corresponding display commands are selected and sent to the display controller within the garment. The electroluminescent panels are illuminated in sequence, transforming light intensity and colour to create the changing aesthetics of the garment. The display rules are based on the analogy of music and emotion, as there are close relationships between the body and music. Music is a very emotive medium and many of the rhythmic patterns in the body such as breathing and heartbeat are mirrored in popular rhythmic patterns. Also, parallels can be made between emotional speech intonation and musical cues. For instance, a sad speech pattern is slow and dips in tone and intensity at the end of the sentence. These configurations can be found in the phrasing and tempo of a sad piece of music. It has been shown by Bresin and Freiberg [44-63] that emotional meanings in music performance can be communicated through musical devices such as: tempo, sound level and note articulation. It is these devices that inform the illumination of the panels to create emotional light patterns. The light from the panels is pulsed to reflect tempo and note articulation. Colour and light intensity are used to portray music volume. Green and blue electroluminescent panels are used for the display. The luminosity of the green is perceived as brighter than the blue; therefore blue is used to signify a dip in tone or visual intensity.

The placement of the panels in the garment is informed by the use of body language [Givens, 2003]. The body is sectioned into generalised areas of expression such as face, shoulders, arms, hands and wrists, torso and legs to provide the locations for the display panels. The panels are sequenced to reflect the use and posture of the body when expressing emotions.

Acceptance

Acceptance is represented by long, constant blue pulses at a low intensity of light. This arrangement is shown on all panels at the same time to reflect an even tempo and body language.

Fear

Low intensity blue pulses at irregular tempo, expressed in mainly staccato articulation, visualize fear. The phrase ends with a long, low intensity pulse. The panels are sequenced to move into the centre of the body and remain stationary, reflecting a crouching motion followed by a frozen stance.

Surprise

Mixed rhythms of long and short durations, correspond to low and high levels of light intensity. These are interspersed with short, 'staccato' light pulses in the highest intensity of green. The sequencing appears on all panels to represent a body 'shock'.

Sadness

Sadness becomes long, slow, green pulsations representative of a slow tempo and 'legato' articulation. The panel is changed to blue at a low light intensity at the end of the phrase, to express a dip in tone and volume. The panels pulse in a descending display to mirror the slumped posture of the body when expressing sadness.

Disgust

Descending levels of light intensity, moving through green to blue, represent disgust. This is reversed on the panels, which ascend the body from hand, torso to head, reflecting a hand to mouth movement.

Anger

One second, green light pulses at the brightest intensity setting represent anger. These articulate a rapid tempo, loud sound level, and non-legato articulation. The display sequence moves up the torso and remains at the neck flickering, reflecting an upright torso and a jerking of the head. At the wrist, the panels light up from the outside, moving inwards to the hip panels, conveying a palm down expression and hand on hip body motion associated with anger.

Anticipation
The light pulses for anticipation begin long and low intensity, shortening and building in brightness through the blue range. A short, high intensity pulse in green follows this phrase. The display sequence mirrors this crescendo as it moves from the bottom to the top of the body.

Joy
Short, fast pulses in alternating blue and green light at a medium intensity express happiness. The panels are illuminated in random arrangements across the garment to mirror 'joyful' body movements.

The final aspect of the Emotional Wardrobe is the garment design, which incorporates the display system within its aesthetic.

Dress design

A major concern of the design was to conceal the *AffectiveWare* sensors, coder and electroluminescent panels. A dress was chosen for the Emotional Wardrobe as the shape can be used to disguise the technology. The dress brings together the various design issues discussed earlier. It incorporates the contour feature and the electroluminescent panels, which are positioned to correspond to body language zones and to add to the overall aesthetic. The outside of the garment is made from a light felt, as it is tactile and able to retain the contour shapes. It is also dense enough to aid the concealment of the technology.

The Emotional Wardrobe research presents an affective garment that is concerned with extending the aesthetic possibilities of a fashion and technology merger. This concept endeavours to unite fashion with technology in a provocative way, to create a garment that can be personalized by the wearer and used to stimulate social and personal communication. Through the translation of inferred emotion into aesthetics, *AffectiveWare* has been transformed into Computer Aided Emotional Fashion (CAEF).

Conclusions and further development

By integrating traditional sensory and aesthetic elements of fashion with the functionality of the *AffectiveWare* platform and the controllability of the electroluminescent display, the paradigm of the Emotional Wardrobe is realised. Through the development of the Emotional Wardrobe, we show that a merger of fashion and technology is able to extend the interactive and expressive qualities of fashion. It is a model that is able to offer a more personal vision of visual identity that utilises a dialogue between the body and the garment to create the emotional aesthetics of the wearer. The wearer is able to co-author a unique, active and adaptive identity, which can playfully communicate with it and others, providing a provocative narrative between the wearer and society. By using the computer as a poetic interface the Emotional Wardrobe contributes to a wider design and technology context that endeavours to make computers more human-centric and technology more emotive.

The wearability of this design could be further improved by the development of small-scale and soft electronics, embedded within the fabric of the garment. This would allow greater flexibility for design rather than the design being determined by the technology. Also, the under and over garments could also be combined to create an integrated system.

This study provides the foundation for continuing research, which will involve the development of additional sensors, investigations and evaluations. These will inform further design concepts, which will aim to expand the aesthetic and emotional potential of fashion and technology.

Acknowledgements

The fashion aspect of this study forms part of an AHRB funded project entitled "Fashion and Modernity." The AffectiveWare hardware and the electronics inbuilt into the garment have been constructed in the laboratories at the Department of Electrical and Electronic Engineering in Imperial College under the supervision of Neil Todd. The research of the hardware and software has been kindly supported from two IST projects—SAFIRA(IST 1999-11683) and ISIS(IST 2002-37253).

Bibliography

Barnard, M. *Fashion as Communication*. New York and London: Routledge, 1996.

Bresin, R., and A. Friberg. "Emotional Coloring of Computer-Controlled Music Performances." *Computer Music Journal* 22:4 (Winter 2000): 44-63.

Cacioppo, J. T., and L. G. Tassinary. "Inferring psychological significance from physiological signals." *American Psychologist* 1.45 (1990): 16-28.

Cass, J., L. Goulden, and S. Kozlov. "Intimate Media." *The New Everyday: Views on Ambient Intelligence*. Ed. E. Aarts and S. Marzano. Rotterdam: 010 Publishers, 2003. 218-23.

Desmet, P.M.A. "From disgust to desire: how products elicit emotions." *Design and emotion: the experience of everyday things*. Ed. D. McDonagh et al. London: Taylor & Francis, 2004. 8-12.

Dunne, A., and F. Raby. *Design Noir: The Secret Life of Electronic Objects*. Basel/Boston/Berlin: Birkhauser, 2001.

Dunne, A. *Hertzian Tales*. London: RCA CRD Research, 1999.

Fukuyama, F. "They may live for ever, but will they still be human? How science is destroying us." *The Guardian* 13 May 2002, 2-5.

Givens, D. B. *Nonverbal dictionary of gestures, signs & body language cues*. Spokane, 2003.

Goia, P., et al. ISIS: Intelligent scalability for interoperable services, 1st European Conference on Visual Media Production (CVMP), London 15-16 March 2004.

Goulev, P., and E. Mamdani. "Investigation of human affect during electronic commerce activities." IEE Eurowearable'03, 4-5 September 2003, The University of Birmingham, UK, Wrightsons 2003.

Hollander, A. *Seeing Through Clothes*. New York: Avon Books, 1980.

LeDoux, J. E. *The emotional brain : the mysterious underpinnings of emotional life*. London: Phoenix, 1999.

Norman, D. A. *Emotional design: why we love (or hate) everyday things*. New York: Basic Books, 2004.

Philips. *New Nomads: An exploration of wearable electronics by Philips*. Rotterdam: 010 Publishers, 2000.

Picard, R. W. *Affective computing*. Cambridge and London: MIT Press, 1997.

Plutchik, R. *The emotions: Facts, theories, and a new model*. New York: Random House, 1962.

Quinn, B. *Techno Fashion*. Oxford/New York: Berg, 2002.

Rees, F., and J. Cass. "Poetic Interfaces." *The New Everyday: Views on Ambient Intelligence*. Ed. E. Aarts and S. Marzano. Rotterdam: 010 Publishers, 2003. 224-29. <http//w.w.w.members.aol.com/nonverbal2/diction1.htm 2003>

Social Fashioning and Active Conduits

Katherine Moriwaki

(Disruptive Design Team, Networking and Telecommunications Research Group, Dept. of Electronic and Electrical Engineering, Trinity College, Dublin)

1. Introduction

In the early days of the Internet the dematerialization of place allowed for the development of a digitally embodied aesthetic, where "virtual" equivalents of physical objects were believed to be capable of replacing their material counterparts. The physical world, with its spatial and temporal limitations fell out of vogue, unable to compete with the chimerical seduction of the electronic agora [Mitchell 1995]. However, as ubiquitous computing technologies have emerged in the "third wave" [Weiser 2004] of the digital revolution, the importance of the physical and the local has acquired renewed relevance as emerging technologies are increasingly incorporated into the everyday environment. New communications infrastructures stress decentralized and temporary networks, which support mobile and distributed computing needs. More importantly, these emerging network communications technologies have the capability to reinforce and supplement existing social and behavioural structures allowing for deeper resonances between the "virtual" and "physical." This paper will present examples of *socially fashioned networks*, or networks which use social behaviour and human mobility as the means for establishing a network infrastructure between co-located individuals. The various projects discussed represent social fashioning at different degrees of behavioural and technical resolution. They function as lenses through which spatial-temporal constraints are focused and applied in order to create interrogative disruptions in everyday experience. Through this process individual and group relationships in public and urban space are challenged and re-imagined using everyday objects and activities.

2. Social Fashioning and Network Infrastructure

In 1995 Negroponte announced the end of geographical limitation in the post-information age. The promise was, through working with digital information, individuals could connect, and communicate anytime and anywhere, irrespective of geographical location. From the telegraph to the internet, new communications technology has provided a third space for interaction, linking people from across the globe, cutting across national and cultural borders. While such technologies have undoubtedly brought their advantages, speeding the pace and reach of human interaction, the specific opportunities created through situated proximity were often left unexplored by early evangelists of the medium. Meanwhile insights into the resonances and discrepancies between virtual and physical space led to a greater understanding of the ability of online forms of interaction to provoke new types of social presentation and behaviour, often spilling over into "real" life. This relationship between the virtual and the physical, when moved beyond the simplistic paradigm of substitution (with information space replacing physical space) provided the groundwork for recent developments in mobile and ubiquitous computing and embodied interaction [Dourish 2001]. Through starting with the assertion that social beings are the basic unit of design [Chalmers et al] everyday activity in the physical world becomes the idealized setting for digital technologies coupled with spatially situated interaction and locally relevant information. Like the minutiae of non-verbal exchange invested in simple acts, digital media has developed conventions for both near and remote interaction. Researchers in Computer Supported Cooperative Work (CSCW) have looked extensively into augmentation of face to face and distance communication between individuals [Greif 1988], but have so far looked less at how new conceptions of social and psychological constructions of space are developed, and more at assisting workflow or business meetings. However, while work plays an important role in most people's lives, it is not the sole focus, and a good deal of people's lived experience is oriented not towards supporting work, but subverting it. It is in this spirit that interactions, in physical space and augmented through digital technologies provide a window for viewing new relationships between individuals and others.

In the process of socially fashioning the works presented, situated experience in urban space and subversive uses of disruption and constraints provide the environment in which to provoke social interaction and awareness. Through changes in the material and metaphoric significance of existing everyday artefacts, and the augmentation of accepted behaviour and use of these objects, a greater individual and group awareness can emerge. The projects presented all use common accessories as garments as their interface, capitalizing on the notion of fashion as a "second skin" and "final layer" [Derycke & Van De Veire]. Each presents a return to situated interaction, where boundary conditions surrounding everyday performative actions are "made strange" and transitory or ad-hoc connections to socially formed networks are promoted. Each project functions at a distinct degree of social and technological resolution, providing illustrative case studies of different communicative infrastructures.

3. Recoil

"*RECOIL*" is a set of garments (Figure 71), which contain small yet very strong embedded magnets, which attract and attach to other people and objects in the cityscape without the wearer's permission.

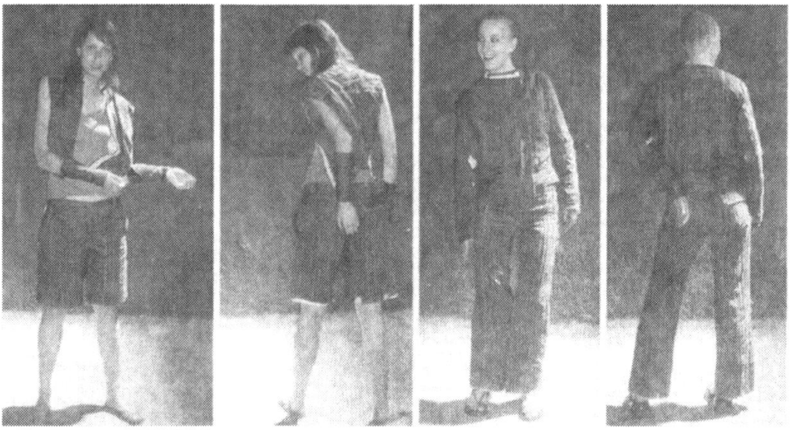

Figure 71: "RECOIL" garments

The project addresses the social infrastructure of high-density urban spaces, and the effects of overcrowding and public anxiety. Due to the attractive force of the magnets embedded into the clothes, "*RECOIL*" garments attach themselves to other people and objects in the environment (through latching onto another person's metal belt, or a turnstile, for example.) The wearer of the garments are unable to control what the clothing attaches to, and often do not know what elements in the environment will be magnetically attractive, creating an environment where an everyday stroll through familiar streets in the city can become a strange and awkwardly playful experience. Within crowded urban conditions, as the proximity to others increases, the likelihood of unexpected, unwanted, and sometimes-inappropriate physical contact, rises. Both Canetti [1973] and Hall [1969] point to the aversion individuals have to unwanted and accidental touching by strangers. When physical space is scarce, psychological barriers are erected to compensate. This desire to demarcate territory is highlighted by the "Hertzian Space" [1999] identified by Anthony Dunne, which is saturated with the emissions of electronic devices and encroaches upon human psychological space through neuroses and irrational fear of the invisible spectrum. "*RECOIL*" was a way of drawing individuals out beyond their social and physical comfort zone, while at the same time investigating possible ad-hoc peer-to-peer physical re-configurations of the body, which could create changes in the subjective experience of the individual and group.

While using no digital technologies, "*RECOIL*" presents a novel way in which to think about socially fashioning a network. By creating disruptive breaks in the wearer's everyday experience through clothing, "*RECOIL*" challenges ordinary means of social, psychological, and physical spatial navigation. The project could easily be extrapolated to include digital communications devices, and future work in this area is planned.

4. Inside/Outside

Like "*RECOIL*," the next project, "*Inside/Outside*" looks at ad-hoc peer-to-peer connections, but rather than creating involuntary connections between individuals, "*Inside/Outside*" encourages the exchange of information.

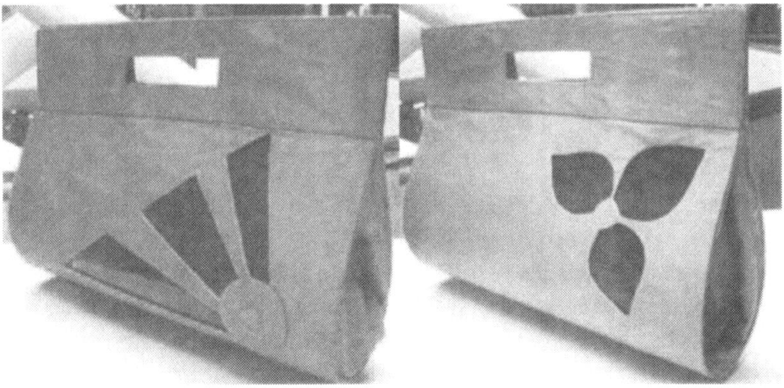

Figure 72: "Inside/Outside" bags

"*Inside/Outside*" is a set of handbags (Figure 72) that monitor environmental irritants and share maps of individual pollution levels based on physical co-location. An "*Inside/Outside*" bag monitors ambient air quality or noise pollution levels and displays the data along two axes. The first is an ambient display, which acts as a "mirror," reflecting real-time exposure levels. The second is a data "diary" which stores environmental exposure levels over time. The data can then be reviewed at a later date through mapping visualizations. The bags are equipped with 802.11b wireless communications, and when in radio range of each other, form a network and swap environmental data, creating *necessarily incomplete* maps of the city's pollution. The size and reach of the map depends on the number of individuals the wearer has passed throughout the day (Figure 73). The emergent diagrams are locally specific and contextually relevant views of pollution information in the city, one that is coupled tightly with the movements of each bag wearer, and others who cross their path [Moriwaki, Doyle & O'Mahony].

The bags were designed as means of creating personally invested information access, using data that is not commonly available to regular individuals at the granular level, and allowing information exchange based on situated co-location. Through instrumented use of an everyday object (the handbag) external information (environmental irritants) would be integrated into the awareness of an individual, resulting in an expanded mental map of the city. In aggregate the information collected by multiple individuals in the city could be used by the inhabitants of a community to police and monitor their own neighborhoods and public spaces. By introducing an augmented everyday object into the social infrastructure of people's everyday lives, personal expression, function, and wireless communications form a socially fashioned network, utilizing "local knowledge" and connection with the environment. "*Inside/Outside*" provides alternative views of the city and urban life, through connecting unseen social networks (that of shared urban space between individuals) embedded in daily activities. As a succession of new connections are temporarily made, the system retains the data "snapshot," building on the forgetfulness of the "*RECOIL*" garment, and instead filing away each instance of data sharing. Rather than altering the bodily configuration of the wearer, "*Inside/Outside*" creates reflective displays of impact, allowing the bag to serve as proxy for the physicality of the individual.

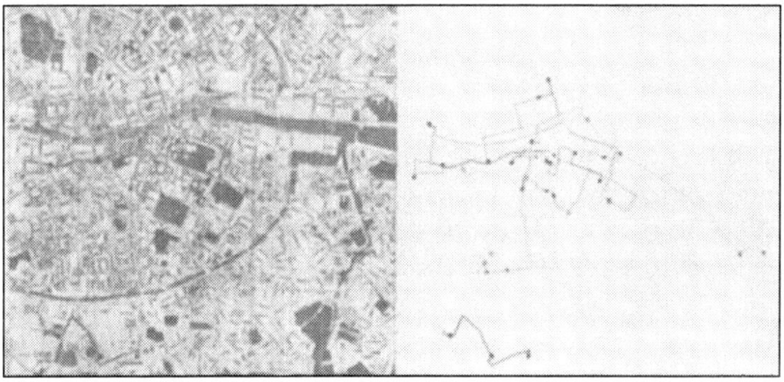

Figure 73: Collaborative Mappings

5. Circumstantial Networks

Whereas "*RECOIL*" and "*Inside/Outside*" examined ad-hoc peer-to-peer relationships, the following projects, "*Oscillating Windows*" and "*Umbrella.net*" interrogate multi-hop ad-hoc networks (MANETs). MANETs are typically described as temporary associations of mobile nodes, which do not depend upon fixed support infrastructure [Perkins 2001]. As a result such networks are non-hierarchical, distributed, and dependent on the proximity and cooperation of other nodes. In the previous projects, ad-hoc *connections* were used to swap data and create physical bodily configurations. "*Oscillating Windows*" and "*Umbrella.net*" use mobile ad-hoc networks as the infrastructure for socially fashioned networks.

The network topology of an ad-hoc network may change rapidly and unpredictably according to available resources and shifting infrastructure requirements. Though typically reliant on wireless radio communication, the organizing principles (decentralization of control, re-configurability, mobility, and physical proximity) of ad-hoc networks are their distinguishing characteristics rather than the physical method of transmission. It is these organizing principles, which provide compelling corollaries to the physical world of movement and social relations, as wireless radio signal range mimics the territorial spatial rings [Hall] surrounding the body, first demarcating space through wireless signal propagation, then integrating a constantly changing and complex configuration of mobile nodes into a network body. This tightly coupled relationship between the topology and behaviour of social and ad-hoc network infrastructures creates a rich area of emerging exploration where wireless and social networks are interleaved.

6. Oscillating Windows

The project "*Oscillating Windows*" presents an application for close-proximity network communication where physical co-location, proximity, and group interaction move a digital image to and from opposite sides of a space. Through short-range ad-hoc wireless communications worn on individuals a signal is passed through a crowd from one person to another. By introducing consequential rewards (the changing of the visual properties of the environment) the project attempts to influence and change the social behavior of individuals in groups. Based on the number of people and their physical proximity and orientation, the ad-hoc network forms and allows the transmission of data.

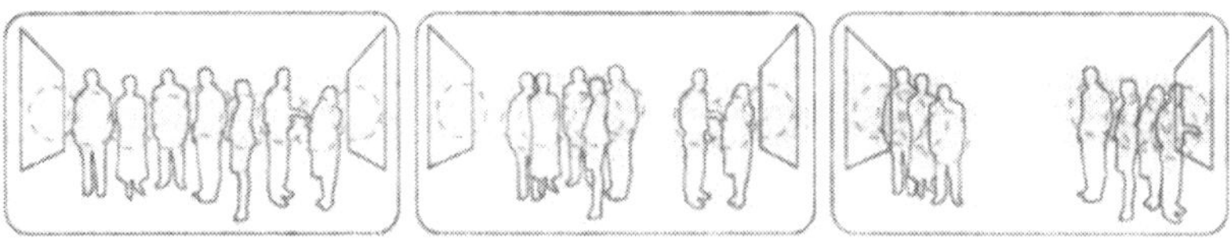

Figure 74: Critical mass determines the data pathway

Without a critical mass of people in the space the digital image in the "*Oscillating Windows*" application cannot move locations, but as critical mass is reached a pathway for the data is created, and bits begin to flow. Likewise, if critical mass cannot be achieved or there is an uneven distribution of people within the space, there is no data pathway, and data transmission does not occur (Figure 74). The existence of the network is entirely dependent on the individuals and whether or not they organize around particular configurations. Typically this has been seen as a weakness of ad-hoc networks as it means communication between network nodes is not always possible; however, this "weakness" has been leveraged into the design and implementation of the project into a feature of the application. The goal is not to maintain a continuous network, but to see how changes wrought in the environment as the result of aggregate human presence and the presence of the network might create greater awareness amongst individuals in a co-located environment. Over time perhaps repeated visitors to a place running "*Oscillating Windows*" may slowly realize their actions and behaviours change the environment. Certain behaviour or actions may be "learned" in order to produce particular environmental effects, or serve as an ambient indicator of the activity and movement of crowds. Unlike "*RECOIL*" or "*Inside/Outside*," the ad-hoc network links entire groups of people together in a chain of connections, which make their impact not only on the individual, but also on the environment. The changing perceptual elements of the space, which depend on the proximity and presence of individuals is an example of a circumstantial network, where specific conditions must be met before the network can instantiate. When such networks require enforced cooperation, a new type of social play might form. "*Oscillating Windows*" indicates how ad-hoc networking may lead to new questions as to how to indicate network presence and activity.

7. Umbrella.net

Similarly, "*Umbrella.net*" takes circumstantial action and co-location to form a network, which acts as a *visual footprint* of network activity in a public space. The project consists of a wireless ad-hoc network, which is deployed when multiple people individually open their umbrellas in the rain. The reach of the network grows and shrinks proportionally to the number of people who have opened or closed their umbrellas. Contained in each umbrella is a wireless communication device, which initiates the network (Figure 75).

In the same way that "*Inside/Outside*" used a handbag to deploy a wireless network, "*Umbrella.net*" uses another everyday object, the umbrella, as the starting point for a socially fashioned network. The "*Umbrella.net*" system works with a hardware and software component that is integrated into the object (an umbrella) without damaging its original use value. The umbrella still works in the way it normally does, yet has additional functionality. The prototype includes a handheld PocketPC (iPaq) computer that interfaces to the umbrella and only communicates with the other umbrella devices when the need exists: i.e. when rain is present and other nodes exist in close proximity. When the umbrella is closed the device is off.

When the umbrella is open, the device engages and begins looking for other wireless nodes. To visualize the network to passerby, the umbrella illuminates in a pulsing red pattern to indicate that it is not connected to, but looking for the ad-hoc network, and turns blue when the network has been found. The software includes a chat messaging application allows fellow umbrella users to communicate with each other, and see how many hops (nodes) they are away from each other. A "sociability" range is provided which determines from how many hops away users can to receive messages. "Umbrella.net" creates a *visual footprint* of network activity, similar in effect of the visual sync of blinking of fireflies [Lefebvre 1996].

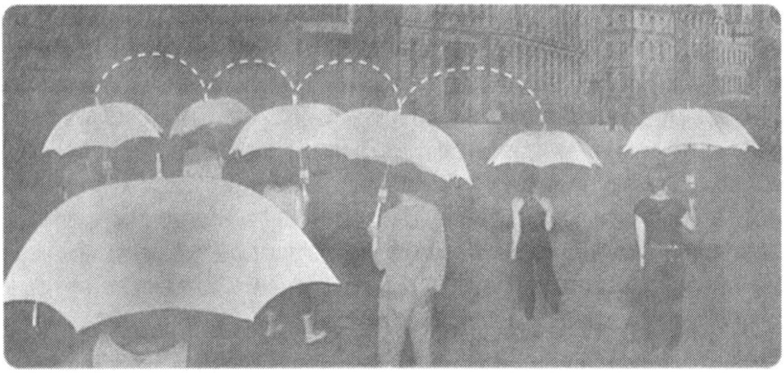

Figure 75: "Umbrella.net"

In terms of social fashioning, the "*Umbrella.net*" project takes its inspiration from the contextually specific environment of the crowded city street, where small individual actions have larger aggregate results. The condition of being caught in the rain in a public place is something, which cuts across social and economic lines, as well as various goals and intent. Street vendors, tourists, and pickpockets all occupy the same public space with very different purposes, and for varying amounts of time. These diverse renderings of time and activity can be caught in a "point of intersection" between flows, highlighting emergent coordination in a polyrhythmic society [Strogatz]. Like "*Oscillating Windows*," "*Umbrella.net*" is a circumstantial network, which requires specific conditions to be met in order to exist. However, instead of using the environment to visualize network presence and activity, "*Umbrella.net*" uses the people that make up the network to display its *visual footprint*. The effect is a poetic disruption, borne of the apprehension of sudden, striking, and unexpected connections between people in public space. As with all previous projects presented, "*Umbrella.net*" is driven by proximity and co-location, but also highlights a point of intersection between urban flows, harnessing individual yet coincidental action into an aggregate and social phenomenon.

Conclusion

This paper has presented examples of *socially fashioned* networks, which operate along varied technical and behavioral axes. While "*RECOIL*" and "*Inside/Outside*" examine ad-hoc connections, "*Oscillating Windows*" and "*Umbrella.net*" interrogate mobile ad-hoc networks.

"*RECOIL*" takes a "no-tech" approach to ad-hoc behavior, using physical constraint and social anxiety to disrupt boundaries between strangers and the environment, whereas "*Inside/Outside*" creates enlarged mental models for its users, by developing new relationships between the individual and the environment. Rather than sensing the environment, "*Oscillating Windows*" enforces cooperation amongst users in order to produce changes in a public space, and "*Umbrella.net*" creates a visual footprint of a wireless ad-hoc network. All of the projects discussed take the urban condition as the prime environment for application deployment, and re-imagine relationships between the individual and the group in ways that utilize common interfaces and everyday objects. By focusing on everyday objects, which are worn or carried, each of these projects generates a performative duality between actual and observed use, allowing clothing and accessories to direct social interaction and self-expression. Through exploiting perceived technical weaknesses as strengths, as well as anticipating and embracing disruption and subversive use, "*RECOIL*," "*Inside/Outside*," "*Oscillating Windows*" and "*Umbrella.net*" provoke discussion regarding the appropriate use of technology in public space, and raise opportunities for beautiful, poetic, and disruptive interrogations of urban life.

Bibliography

Canetti, E. *Crowds and Power*. Trans. Carol Stewart. Harmondsworth: Penguin, 1973.

Chalmers, M., et al. *Seamful Design: Showing the Seams in Wearable Computing*. Eurowearable 2003, University of Birmingham, UK.

Derycke, L., et al, eds. *Belgian Fashion Design*. Ghent- Amsterdam: Ludion, 2000.

Dourish, P. *Where the Action Is: The Foundations of Embodied Interaction*. Cambridge: MIT Press, 2001.

Dunne, A. *Hertzian Tales: Electronic Products, Aesthetic Experience and Critical Design*. London: RCA CRD Research Publications,1999.

Greif, I. *Computer-Supported Cooperative Work : A Book of Readings*. Morgan Kaufmann, 1988.

Hall, Edward T. *The Hidden Dimension: Man's Use of Space in Public and Private*. London: Bodley Head, 1969.

Lefebvre, H. *Writings on Cities*. Trans. Eleonore Kofman and Elizabeth Lebas. Oxford: Blackwell Publishers, 1996.

Mitchell, W. *City of Bits*. Cambridge, MA: MIT Press, 1995.

Moriwaki, K., L. Doyle, and M. O'Mahony. "Inside/Outside: An Everyday Object for Personal Environmental Monitoring." *Adjunct proceedings of UbiComp*, October 2003.

Negroponte, N. *Being Digital*. New York: Vintage Books, 1995.

Perkins, C., ed. *Ad hoc networking*. Boston: Addison-Wesley, 2001.

Strogatz, S. *Sync: Rhythms of Nature, Rhythms of Ourselves*. London: Allen Lane, 2003.

Weiser, M., and J. Seely Brown. "The Coming of Age of Calm Technology." *Ubiquity* 5.4 (2004).

Wunderkammer: Wearables as an Artistic Strategy

Laura Beloff

Introduction

Many of us are currently, and possibly permanently, living in a hybrid space, conditioned by the use of mobile and wireless technologies. Various theorists and researchers have scrutinized this space—which emerges from our technological condition—during recent years. Amongst others, Timo Kopomaa has written about the concept of a *third space*, Anthony Townsend about *phonespace,* and Adriana de Souza e Silva has defined the more general concept of hybrid space which is formed from a merge of physical and virtual spaces.

Our use of technology is generally limited to standard applications and commercial ready-mades, which commonly exhibit a very functional and task-oriented approach to technology. The wider possibilities of hybrid space are often left unnoticed due to the ordinary and pre-defined perspectives offered by mobile devices available for consumers.

The focus of my research is in artistic experiments and works, which investigate and refer to concepts of hybrid space, ubiquitous computing, the individual *third space* and/or *phonespace*. These works enable a person to experience hybrid space in a different manner than from standardized perceptions. They aim at directing the focus away from a functionally oriented approach to hybrid space, towards a more conceptual approach. In my practice I have introduced the Figure of the *hybronaut*; a person coupled with an artistic wearable device. The *hybronaut* exists within and continuously explores hybrid space. By simply focusing on being rather than doing, the *hybronaut* refers to the wider phenomenon of technology within our society and in our future.

WORLD: anecdote

Some time ago in a Finnish newspaper there was an article concerning the life circumstances of a person who has spent the most part of his life in a wheelchair. At the end of the article the journalists asked him about the possible communities formed amongst people in wheelchairs. He commented:

> "When people travelling with caravans meet on the road, they always wave hands and joyfully greet each other. But people travelling in wheelchairs, they don't. Everyone carries his own world with him."

SPACE: space, time, hybrid

Timo Kopomaa writes that the mobile phone itself, as a communication device, can be interpreted as a kind of virtual space parallel to work and home. A mobile phone with its connections forms *a third space* for socializing and meeting friends, as well as functioning as a place to withdraw into privacy and safety within the midst of an urban lifestyle. According to Kopomaa chatting in the *third space* is a primary function and a major way of expressing one's personality and individual self. The meeting of two people in third space is considered completely private in relation to the surrounding environment: it is one's own personal zone of freedom. The mobile phone is firmly linked to sustaining connectedness. As Kopomaa writes, the mobile phone is not solely a device to keep in touch, but a device for being together [Kopomaa, 2002].

Anthony Townsend has observed that time is the most important change occurring with new lifestyle opportunities offered by mobile phones. Time is a commodity, which is bought, sold and traded over the phone. Instead of living within a more traditional schedule based on minutes, hours and weeks, individuals now live within a constant stream of negotiations, reconfigurations, and reschedulings. One can be interrupted and can interrupt others at any time. Townsend has defined this as *phonespace*. According to him, a person living in the *phonespace* cannot let go of it; it is a primary link to the temporally and spatially fragmented world of personal networks of friends, colleagues, as well as offering constant reconfiguration of schedules and meetings. "It has become their new umbilical cord, pulling the information society's digital infrastructure into their very bodies" [Townsend, 2001].

Adriana de Souza e Silva defines a concept of hybrid space in order to re-conceptualize physical spaces through the connectivity of digital mobile media. She writes about hybrid space: "…the mix of social practices that occur simultaneously in digital and in physical spaces, together with mobility, that creates the concept of hybrid reality" [de Souza e Silva, 2006]. She has focused her research on the use of mobile media devices as social interfaces, which reshape space. Obviously, hybrid space is not solely a technological construction. According to de Souza e Silva hybrid space is emerging explicitly as a social space created by social networks and interactions, which are developing simultaneously in physical and digital space. Based on de Souza e Silva's definition, hybrid space inherits properties from both worlds; it grows larger than the sum of its parts.

As mobile phone users we carry individual spaces, as described by Kopomaa and Townsend, with us everywhere. Although these individual spaces are first defined as private, as our personal area of freedom, nevertheless the surrounding physical and public reality is an inherent part of them. Personal memories about situations or meetings with friends, which may have happened in these technologically-enabled private spaces, are often still anchored, as memories, as to their physical location[1]. Physical reality still continues its customary function as a reminder of past events and encounters; the only difference being that now the physical anchor point is no longer shared, but one event may have several "stages" located anywhere in the world.[1]

In any case our perception concerning physical space is obviously under transformation. The amplified importance of individual spaces, as pockets of privacy within public space, is impacting on our developing relationship with our physical surroundings. When large parts of our lives and important events take place within technologically-enabled private spaces, it raises the question as to whether physical surroundings will eventually be converted into a mere backdrop for the alluring individual spaces. Or possibly, as many seem to believe, the era of ubiquitous computing [Weiser & Brown, 1996][2] will eventually remove the rigid separation of physical environment and virtual space.

INVISIBILITY: standard

Without being aware of the fact, the majority of us are already inhabitants of the hybrid spaces defined by de Souza e Silva through the use of mobile phones or other connected mobile devices. Even if our mobile phones are always on and we are constantly connected through them, the concept of hybrid space that we share with other connected bodies, and its wider possibilities, are easily left unnoticed. This could be claimed as the result of our standardized attitudes towards technology. Technological devices are often solely scrutinized as to their functionality and usability. Our practice with them is still very much influenced by instrumental perspectives.[3] Mobile devices are commonly considered mainly as tools for pre-specified tasks, for example: SMS-messaging, phone calls, reading emails, and paying bills. The possibilities offered for the use of technological devices are

---

[1] Jussi Parikka has written about games on mobile devices and how they are "leaking" in a way, such that the player is constantly aware of the surrounding disturbances, especially when playing in public places. This has been used in newer mobile games as a feature; the game is capturing the everyday reality as a part of its concept. PARIKKA, J. (2003) Liike, paikka, havainto - mobiili kuvaruutu ja pelaamisen uudet ulottuvuudet. (Movement, place, perception - mobile screen and the new dimensions of gaming) *Widerscreen 2/2003*

[2] This has in some sense been true throughout the history of the telephone. However, with fixed phones that were always related to a place, the caller usually had an idea or mental image of the environment surrounding the receiver without being present there. With the use of mobile phones, the caller only has an idea about the person, not of the location.

[3] Mark Weiser first articulated Ubiquitous Computing in 1988 at the Computer Science Lab at Xerox PARC.

also often standardized and restricted to specified functions. For example, the personalization of our private phones with unique phone covers, wallpapers, ringing tones and amulets readily creates a false idea of freedom. Though we all can apparently choose from within a wide range of computers, phones and other commodities, these are all still enveloped within larger technological systems.[4] David Nye writes: "It is easier to select among many telephones than it is to do without one" [Nye, 2006]. Likewise in the current situation, hybrid space is only attainable by way of standardized interfaces such as mobile phones with specified functional possibilities. To be able to function or experience hybrid space differently is out of the reach of the average consumer.

ART: hybrid space

The development of wearable, small-footprint technologies, and wireless and mobile networks has impacted on the manifestation of various artistic projects using mobile and wearable devices. Artists have begun to expand the notion of hybrid space and its possibilities by exploring the limits and creating customized situations. These kinds of artistic works experiment with new perspectives that oppose the pre-fabricated standard perceptions, which neither offer possibilities for exploring the shifting notions concerning hybrid space, nor address critical notions concerning the ways we use technology. Yet, many of these artistic projects have a strong goal or task-oriented approach to hybrid space, such as in the form of a game or by sending messages from a mobile phone to a large display. An example of this would be the project Blinkenlights[5] by Chaos Computer Club, in which one could play Pong or send SMS-messages onto a house-sized screen.

My own interest is focused on experiments and artistic works, which, in contrast to the functional approach to technology as a tool, often appear as awkward looking wearable devices and objects offering a more conceptual direction, rather than a strictly useful functionality. They are not designed for, or solely focused on the user's perspective, in which case they would be designed as an aid for or an extension of the user, and would commonly have a sleek or almost invisible appearance. These devices could have an absurd look, and can even be uncomfortable to wear. Their inherent playfulness in relation to technology, and

---

[4] Johan Redström writes in the following way: "...though phenomenological, sociological and other studies have challenged and expanded our understanding of technology, practice still seems to be dominated by an instrumental perspective. Central to our understanding of technology still lies notions of use, the idea that technology is the means for achieving certain ends, often by amplifying the power of our actions." REDSTRÖM, J. (2005) On Technology as Material in Design. IN REDSTRÖM, M.REDSTRÖM, J. & MAZE, R. (Eds.) *IT + Textiles.* Edita Publishing Oy.

[5] See <http://www.blinkenlights.de/>

humor or irony combined with an awkward appearance opposes the standardized views and (restricted) possibilities of commercial mobile devices. These kinds of works function as an essential and critical commentary questioning the direction and the development of mobile and wearable technologies.

For example: German artist Stephan Schulz's "Tin Drum"[6] [2006] is a wearable physical drumming machine, which reads GPS data and translates this into drumming patterns. The artist writes that it is important to combine the virtual layer with a real world intervention in order to create a stronger connection between the two, and to use new technology as a catalyst for social interaction. In this work the user of the device becomes a public performer, functioning as a focus for public attention and as a juncture between the virtual and real layers of public space.

FIGURE: hybronaut

A person, or a body, coupled with a peculiar looking wearable device becomes a kind of space traveler, who is equipped to be able to exist within hybrid spaces. This figure, which I will call the hybronaut, aims to explore hybrid space, and its possibilities by producing a non-standardized perspective on this space, and by pointing to the restricted manners in which we are currently allowed to use hybrid spaces.

Kopomaa and Townsend have articulated the occurrence of individual private spaces, which are formed by a connectedness enabled by mobile devices. These individual spaces become noticeable mainly when we are using our mobile devices to perform specific tasks. At other times, the existence of our technologically connected world is largely imperceptible. In contrast the hybronaut is only able to exist within a continuously connected world; within a hybrid space. His/her equipment is specifically constructed for exploring, experiencing and observing hybrid space rather than existing as a properly functioning tool for performing certain tasks.

This, often bizarre, looking wearable device, which is an essential part of the hybronaut, raises curiosity through its visual appearance. This fosters interaction and communication with the public. We could say that the hybronaut is a user turned into a performer. The hybronaut carries his/her own (connected) world with him/her, pointing out to the public his/her private investigations concerning both connectedness and our shifting notions of space, presence, the real and the virtual.

---

[6] See <http://www.maybevideodoes.de/sites/tindrum.html>

In some way the hybronaut can be compared to a flaneur.[7] The appearance of the flaneur was influenced by the material circumstances of the city, for example the newly developed concept of department stores. In a similar manner the concept of hybrid space and even the Figure of the hybronaut have appeared within the technological development of (commercially enabled) wireless and mobile networks. The flaneur, as well as the hybronaut, is a Figure on the street, strolling around the city without any specific goal or destination, simply observing life; the only clear difference here is that the hybronaut is strolling and observing life within a hybrid space while the flaneur was enthralled by emergent urban life within the physical world.

CONCLUSION: artwork

In my artistic works, the hybronaut's equipment is constructed from standard technological components and already existent possibilities. However, the way these standard parts are linked together differs from that which is usually expected, or possibly even allowed for normal consumers. This method enables the hybronaut to experience hybrid space differently than from within standardized perspectives. These works aim at directing the focus away from the functionally oriented approach to hybrid space, towards a more conceptual approach. By simply focusing on being rather than doing, the hybronaut refers to a wider phenomenon of technology within our society and in our future.

---

[7] See *Other Voices* e-journal: <http://www.othervoices.org/gpeaker/Flaneur.html>

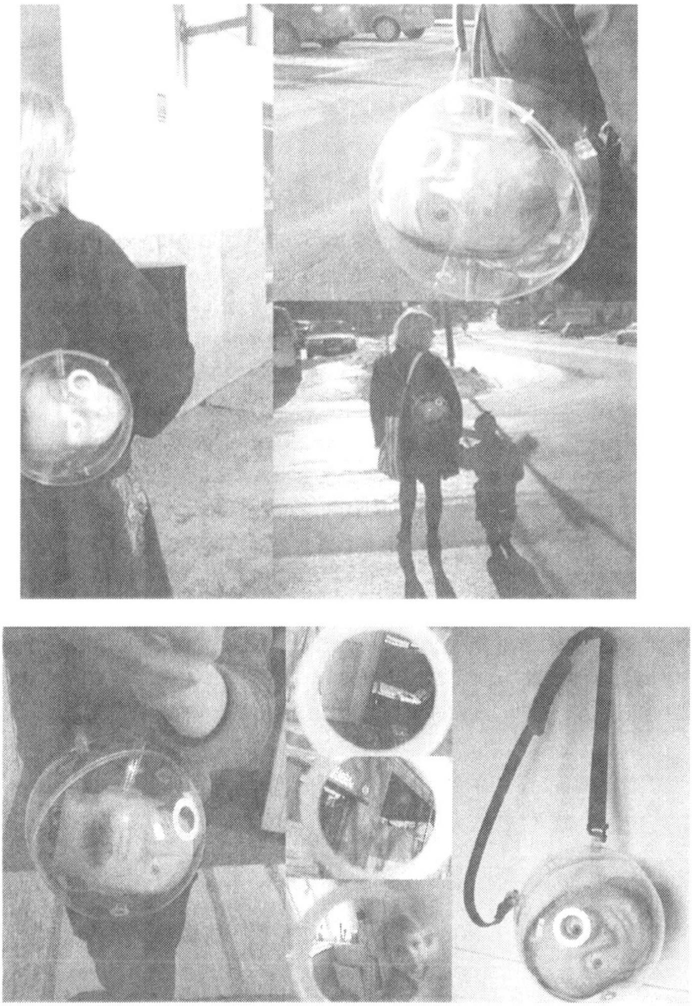

Figure 76: Wearable Head by Beloff—capable of photographing and sharing images by text, Bluetooth and WAP to internet sites.

HEART-DONOR

Heart Donor[8] by Beloff & Berger with Mitrunen (2007), is a wearable art work addressing our life in hybrid space. You can "wear" the hearts of your close network of selected persons on you, and observe the presence of these people in physical and virtual space. (Figure 77)

The work takes its point of departure by rejecting the (common) concept of the differentiation of virtual (digital) and physical ("real") layers of the world. This work is specifically constructed for hybrid space. The work attempts to make the renewed concept of space visible and materially concrete in our everyday lives. The work Heart-Donor is a physical instantiation of a concept concerning personal social networks and life in hybrid space. It is imagined as one's personal apparel (a vest) for long-term everyday use.

One can collect 30 recordings of the heartbeats of friends and family in the HDvest. These heartbeats will be stored in 30 small lamps embedded on the front of the HDvest. The blinking heartbeats function as personal mementos of close people and friends. The heartbeats are combined with another concept related to the technological world. The default colour of a recorded heartbeat is green, but it changes to blink in red when the person (whose heartbeat is stored in the HDvest) goes online (with Skype). The "owner" of the HDvest can follow his/her selected social network of people shifting their presence between the physical and the virtual layers of the world wherever s/he and the people in the network may be located geographically. The HDvest and its wearer reside continuously within hybrid space.

---

[8] See http://www.realitydisfunction.org/heartdonor/

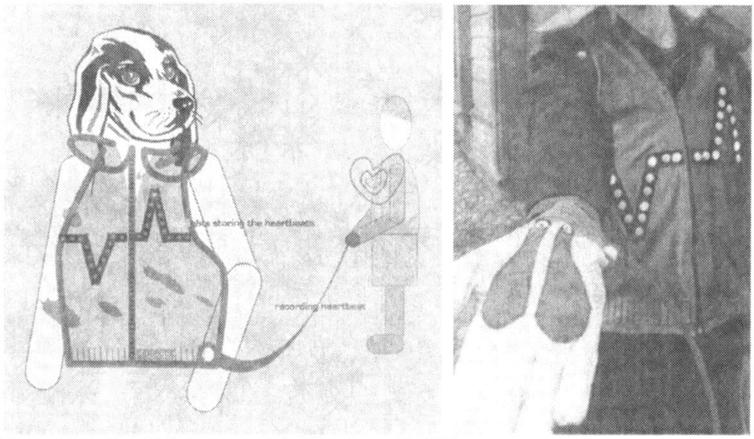

Figure 77: Beloff & Berger with Mitrunen (2007)

The form and design of the HDvest is inspired by the traditional life-vest. This technological version of the life-vest refers to the fragility of life. The heartbeat is used as a sign of physical life and presence, which is combined with another sign for a global presence within a technological sphere of the world. The work is not created as a tool or function aimed at specific tasks. It is created as wearable apparel enabling everyday existence within a hybrid space. One becomes an observer of the hybrid world, the one who simply exists within it.

Bibliography

Barfield, W., and T. Caudell. "Basic Concepts in Wearable Computers and Augmented Reality." *Fundamentals of Wearable Computers and Augmented Reality*. Lawrence Erlbaum Associates, Inc, 2001.

Beloff, L. "When The Cables Leave, The Interfaces Arrive—immaterial networks and material interfaces." *Technoetic Arts: A Journal of Speculative Research* 4.3 (2006).

de Souza e Silva, A. "From Cyber to Hybrid: Mobile Technologies as Interfaces of Hybrid Spaces." *Space and Culture*. Sage Publications, 2006.

– – –. "Re-Conceptualizing the Mobile Phone—From Telephone to Collective Interfaces." *Australian Journal of Emerging Technologies and Society* 2006.

Gleber, A. "The Art of Taking A Walk; Flanerie, Literature, and Film in Weimar Culture". Princeton: Princeton University Press, 1999.

Huhtamo, E. *Pockets of Plenty: An Archaeology of Mobile Media.* 2004.

Kopomaa, T. *Kännykkä-yhteiskunnan synty; Tihentyvä arki, tiivistyvä kaupunki. (The birth of mobile society.)* Helsinki: University Press Finland, 2000.

– – –. "Kännykkä, paikkaan kiinnittyminen ja samanrytmisyys." ("Mobile phone, attachment to place and synchronisation.") 2002. <http//:www.m-cult.net/mediumi/article.html>

Mann, S. "Mediated Reality with implementations for everyday life at presenceconnect.com." *Presence: Teleoperators and Virtual Environments.* 2002.

Nye, D. *Technology Matters: Questions to Live with.* Cambridge: MIT Press, 2006.

Parikka, J. *Liike, paikka, havainto -mobiili kuvaruutu ja pelaamisen uudet ulottuvuudet.* Widerscreen, 2003.

Redström, J. "On Technology as Material in Design." *IT + Textiles.* Ed. M. Redström, J. Redström, and R. Maze. Edita Publishing Oy, 2005.

Townsend, A. "Mobile communications in the 21st century city." *The Wireless World: Social and Interactional Aspects of the Mobile Age.* Ed. B. Brown. Berlin: Springer-Verlag, 2001.

Weiser, M., and J.S. Brown. *The Coming Age of Calm Technology.* Xerox PARC, 1996.

Wodiczko, K. *Designing for the City of Strangers: Critical Vehicles, Writings, Projects, Interviews.* Cambridge: MIT Press, 1997.

Section 4: Artist Interviews

4.1 Locative

Flirt and Mset

Fiona Raby

(Interview with Martin Rieser: Dunne and Raby London studio December 2004)

MR Could you tell me a bit about the work on FLIRT you started in Helsinki in 1999?[1]

FR It was an EU project, where we were brought in initially, but not exactly to make nice graphics or icons, but as designers.

MR They didn't think you could provide ideas?

FR No, we <u>were</u> involved with the concept development. The stages of the project are laid out here. We worked with a group comprising Phillips at Redhill, so the partners were Phillips Research and their software development group, and Consumer Communications who were the ones that made all the handsets and prototypes: they were based in France, and then the Helsinki Telephone company, who are actually called something else now-they've been bought out. They were the ones supplying the network at the time. Infogrames Entertainment, who were doing the transfer, building the software platforms to put it onto the network and developing their own thinking, and then there was us as the design team: so we

---

[1] "1998-2000 FLIRT: Flexible Information and Recreation for mobile users. A European Commission research project under the 'IT for Mobility' theme. The development of digital cellular structures by the mobile communications industry generated a genuine fusion between information space and urban territory. City location, the time, day and date can all begin to shape relationships to information sources. The tight constraints of mobile displays, juxtaposed with the spontaneity, unpredictability and transience of everyday mobility, require a fresh approach to how this relationship might work. <http://www.dunneandraby.co.uk/designing/FLIRT/flirt5/> (accessed July 2005)

FLIRT investigates the potential of location specific information, not only as an information resource, but also as a medium for social interaction and play.

PARTNERS : Philips Research Laboratories (UK), Philips Consumer Communications (Fr), Helsinki Telephone Corporation (SF), Infogrames Entertainment (Fr), and the Royal College of Art (UK).

RCA Research team : Neil Clavin, Paul Farrington, Marcus Gosling, Nick Durrant, Shona Kitchen, Ben Hooker, Arita Patel, Fiona Raby, Niell Sweeny, Brendan Walker."

were a very small part of the whole massive project. At the time there were five of us and none of us had mobile phones, which was quite interesting...

MR Very interesting—was that liberating in a way?

FR In the end the fact that we didn't understand very much about them, because we had never had them gave us a different and fresh way of looking at them and so the first thing we did was go to Helsinki for a trial. It was supposed to be a trial looking at future technologies using WAP. We went to see one of the partners in Helsinki and they gave us a phone. In Helsinki they had cell broadcasts where in each cell you entered they would broadcast something into the cell.

MR You mean a kind of "push" information?

FR Yeah, so every cell you went into you got another little bit of information-all in Finnish. We had no idea what it said; it was all text with no animation or graphic imagery. So we couldn't understand the way the city was working. We had been told that the model for the future would be surfing the Internet on your mobile phone, but we had this tiny black and white screen and we were freezing cold, running around the streets with this big lump and thinking how on earth is anyone going to sit in sub-zero temperatures playing with the internet? So we thought that whole notion of delivering content was ridiculous. The other thing we did that was quite critical, because we didn't understand, we made this map which became quite a strong tool in the whole of the project: helping to understand what was going on, and of course the telephone company didn't tell us where any of the antenna were.

MR They didn't tell you?

FR They couldn't for commercial reasons. So what we did was drawn from the little we knew of how area cell antenna worked. Where there are many buildings they block the signal. So you have to have more cells in built up areas. Over by the harbour you might have only one antenna for the entire space. It really created a different reflection of the city. So we made this map and showed it to Helsinki Telecom and they said it wasn't accurate, but it was realistic to use as a model. So it gave us a basic idea if you are walking through. We also found out that the cells became quite elastic, so when you are in one cell you pulled it with you, until the strength of the next one takes over. So for us this sort of ambiguity became very interesting. This mapping of a virtual space on top of the city became a different way of looking at this project. We then spent a lot of time working out what kinds of services we might provide. There was a point where we felt we had to look at all our partner's needs. At one end you had Infogrames and at the other you had Helsinki Telephone Company, who wanted information such as bus timetables and tourist information, whereas Infogrames wanted to do gaming. We did a whole chart, which provided a sweep of all these needs. What we found was a space in-between, which wasn't information and wasn't gaming. It was this space, which in the end Phillips called "social services." We did a few ideas based on this in between concept starting with *Pixel Kissing*.[2]

MR Yes, please explain that.

FR We had an idea about being close to someone in a cell and because the cells were so variable in size you can't tell which. There is this ambiguity, a fuzziness-what the engineers wanted to do was find a precise way of finding where you were. I don't know if they can do it now with triangulation. We thought it would be a shame to locate in that way. We felt we could use the ambiguity as a medium. In *Pixel Kissing* you sign up to a service-a game in real space. We called it *Love Tectonics* at first because it is like two plates meeting. It was like a dating service and would put your lonely hearts "classified " in a place, somewhere in the city for which you felt an affinity, for example, a coffee house where you liked to hang out. You put your advert in the place.

MR Like a digital post-box?

FR Sort of—so if you were part of this service and were looking for somebody, you might go to a cell and check through all the different adverts and see who was around. Because the cells were elastic you couldn't know if they were actually there, just that they had an association with the place. So you could choose the place as a way of connection. You could only have five potential lovers and when you joined the service you were given a questionnaire of 30 questions over a month. We were interested in a period of time where you got to know someone, rather than an instant relationship. You were asked more and more personal questions as the month wore on. The idea was that if you choose somebody for a date, they become part of your database, so you would have these five people in your database and every day you would get an answer to a question that you had sent, so you would get to know them after a period of time. If you found someone else you would have

---

[2] <http://www.hookerandkitchen.com/flirt/pixel_kissing_01/flirt_window_01.html>

to choose which lovers to keep. Over time you might find you might not like some of them any more.

MR It was all done remotely by the phone—and you didn't meet the people?

FR This is where *Pixel Kissing* comes in-the idea was that if you were anywhere in the city getting on with your daily business and one of your five partners was in the same cell, you would get a message. The idea was that you have a picture in your head of what the person looks like so you might immediately look up and examine the people around you and project on them who you think it might be. They of course might be around the corner or up in an office block or they could be right next to you!

MR So it was a kind of mystery tour?

FR The idea was a break through into reality using a fantasy idea. It is not so fantastical that you will never meet-there is always a possibility. We wondered if it might inspire people to be brave enough to ask strangers if they were the person. The idea was that the story wasn't just in the phone and the content delivered to you. The story was constructed by you yourself and through your own imagination: by the people surrounding you and by your location, which almost becomes a stage. One of the things we considered was the screen not being the narrative, but the narrative being in the city. The screen just triggers the narrative at various places. I think a lot of the things we have done since then were really about constructing the narrative in your head. The content isn't here (pointing to the mobile); it is here (pointing to her head) and is triggered by the things around you.

MR Many successful mobile artworks now do the same, but you were well ahead in the way you were thinking-and much of this isn't known and needs to be put back into the domain, because people are reinventing the wheel all the time.

FR The other thing we thought about this was that if at the end of 30 days people would actually want to meet or whether playing the game was more interesting than actually meeting. It came about through using the map and trying to understand the configuration.

The next project we did was *Lost Cat* in which we were looking at the idea of a narrative embedded in the city over a period of time. This was at the time very high-tech animation-we produce twelve frames and at the time the systems could only support seven! The programme was huge, we have a whole fat book in the cupboard. The idea was we would have a spatial narrative where we would send out to our twenty users a message preparing them for the next step. We would say "The network cat has gone missing-keep an eye out for it" and the next day another message would go out "The cat has been seen in certain parts of the city-up by the train station." At the time we were looking at nodal points like stations and changeover points where people might be waiting, or hanging around. After that we would release the cat onto the network.

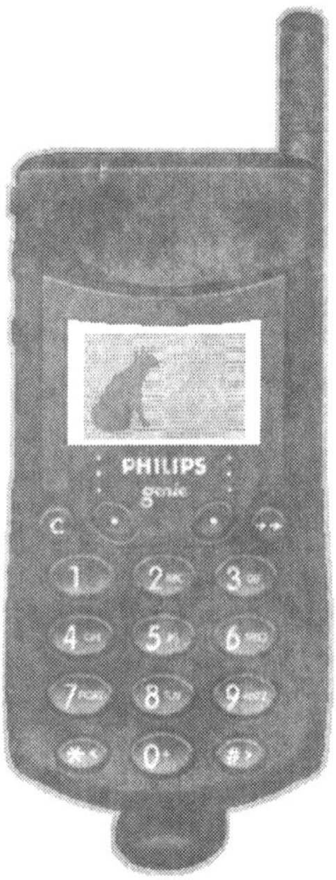

Figure 78: Lost Cat

Apparently the cat walks around until it finds somebody and if there were somebody in that area it would go into his or her phone. So the cat will appear on your phone as a tiny animation (embedded in the phone because of transmission problems). The cat would disappear into your menu and become a menu item. We had an accumulation algorithm, so the first time it would appear it would be on your phone for about an hour and at any time you could go to the menu and show the animation. So the cat was with you and then after an hour, it would walk off and go and be with someone else.

MR Quite a good iconic animal to use because it's got no loyalty!

FR Exactly, it will go off with anyone

MR Anyone who feeds it

FR The idea was that if you visited the same cell again at the same time and place, the cat would come on to your phone again you could keep it for two hours, the next day for three hours, the next four and so on. You know you will get the cat for longer, but you don't know when it's going to leave you. The interaction was not about a Tamagochi pet—we looked at questions such as "do we start to feed it?" etc. But we decided that we didn't want to go that route. The interaction is about you having been special and the idea that you could show it to anybody and that you "own" the cat and it stays with you.

MR I suppose the next project was *Stampede?*

FR Yes-we liked the *Lost Cat* more than *Stampede.* It was just a bit of fun to explore this virtual landscape set on top of the city, to exploit a narrative placed over the city. Again we set up by preparing people. A day before we would say: "In North Helsinki a group of reindeer are starting to gather" and on the second day we would send a message saying: "The reindeer are starting to get agitated"

MR All this in Finnish, presumably

FR Yes, and then on the day of the event we would plan a route through the city which the stampede was going to run through. Then, if you were in one of the cells near the stampede-we made the stampede huge to include as many people as possible-if you were so many minutes away in its path you would get a warning." Run! The animals are stampeding!" Then five minutes later you would get a miniature image of the stampede on your phone. Then a minute later you would get a second closer image and then another. This is the one we loved the most—you just get a huge pixel, then immediately afterwards you get this black pixel. Then a message goes out to the whole community "Sonny was trampled by the stampede." If you were in another part of the city you would get to hear of all the deaths occurring as they were happening. Then, when the stampede has passed through the whole city. You'd get a death toll of all the people-in loving memory of all the people killed.

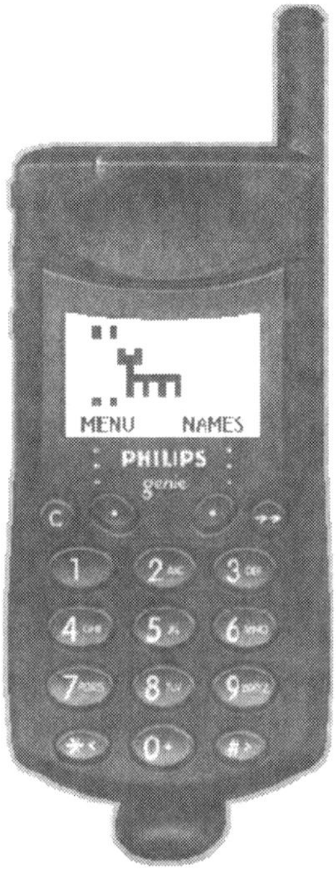

Figure 79: Stampede

MR Fascinating-you know Mobile Bristol has done a located narrative called *Riot,* where they read out a death toll at the end of all those who died in the riot. There are so many parallels with what is going on now. Blast Theory's Uncle Roy also has many parallels.

FR What happened was that halfway through the project-there were fifteen services to put through-we had the usual tourist stuff at one end and at the other Infogrames just wanted to repurpose some existing games; just download them with no location-based adaptation. Very tenuous, but in the end Infogrames were disillusioned with WAP and already halfway through the project were thinking about 3G. I guess in the end they only did the bare minimum to get the project through the EU. We were meant to have 50 users from Helsinki and we ended up with 10 of them, with only a handful of services. We were so disappointed that in the end it came down to the code for *Lost Cat*-the engineers at Phillips had worked it out and it was all ready to go—but they didn't do it. So in the end all of this didn't happen. We were devastated. I have a phone with everything on it-but that's all.

MR It was before its time, I think now it would be taken up with alacrity because the companies are desperate for content.

FR But in the end they were more interested in technology and not in the content

MR One of the problems with 3G is that they have mundane ideas on content-they won't think sideways.

FR Afterwards Phillips were doing a project, making the chips for the 3G phones, so they wanted us to come up with ideas for using 3G. None of the services went anywhere, but it was interesting for us after experiences at the end of the 1990s where other projects didn't come off and we lost interest and started doing other things like the Placebo project. Then the MSET (Multimedia Services and Technologies)[3] project came along. This was done in-house for Phillips with no EU funding (with Ben Hooker and the Redhill group, including Dave Bell the project manager). We came up with 10 ideas, but the agreement was they would keep three secrets that they would work on. We felt strongly that we needed to show the ideas at the end or there would be no point in doing it. The theme was very much entertainment, so we wanted to learn from some of the things we had done in Helsinki.

To start with the idea of having colour and photos freaked us out. We thought we would fall into the trap of being seduced by the screen. We came up with a few things-I will show you my favourites. The buses one was about the idea that if you were regularly travelling on a bus and turned one of the surveillance cameras outside, you could connect into it via you're mobile and see where you were going. The idea was to phone your bus while you were waiting and literally see where it was and the traffic conditions ahead of it. When you got on the bus you could tune into the camera, but when another bus approached you could swap to their camera, while your body continued its actual journey.

MR So you get a virtual journey in parallel?

FR So you jump from bus to bus on your journey, and if you have knowledge of the city you try and work out where you are and where you are going. You might set a

---

[3] Mset: Multimedia Services & Enabling Technology with Phillips Personal Technologies Group, Software and Interactive Systems in Redhill, Surrey, UK (2000-01) Project team: Fiona Raby, David Bell, David Yule, Steffen Reymann, Duncan Mcpherson, Ben Hooker <http://www.dunneandraby.co.uk/>

definite destination or just see where you end up. Then we wanted to add more dynamics, so we were thinking of extending to courier services, which might be more exciting.

MR So you could leapfrog onto different carriers.

FR Taxi companies might also be into that. You could have Nike sponsoring DHL.

MR I could see the commercial potential if it were a game for Taxi drivers doing "the knowledge"

FR Then we took *Pixel Kissing* and turned it into what people thought was a mad idea: "*Making Babies*" We asked what an increase in resolution might mean in terms of being different. We did this work with my students. Extending *Pixel Kissing,* we let people declare if they were up for making virtual babies. The moment you volunteered you had to photograph yourself. In *Pixel Kissing* the idea was you didn't know who the person was; in this project we wanted to disguise the person. If you made a regular journey to work through the same cells everyday that too became part of your 'genetic' data-there might be a pattern that emerged. If you meet another person up for *Making Babies* in a cell, their profile of cell structure and your own are put together. The possibility of meeting becomes the whole engine of the game. If there is a likelihood of sharing a cell with this other person, then the probability is high that it is a good criterion for reproduction. If it is low you don't want to *Make Babies* with them.

With a 90 % chance of cell sharing, what happens immediately is that your eyes are cut from your picture and the other person's eyes are put on your face. The only bits of them you can recognise are their eyes. As with *Pixel Kissing* you were told of a potential partner and try to identify the partner just through the eyes, although you of course might not want to. The first time you meet you make a baby. The baby sits on your phone and has your face and the eyes of your lover. Over the first few times you pass through cells, the baby has to grow a bit and stays on your phone. After five passes it is big enough and you put it into a cell to live. It lives in a virtual place in the city. Hopefully it ends up in a cell with other babies growing there. This is the next stage of dating. If you meet in your cell your baby glows and becomes susceptible to romance. If there were any other glowing babies-they might glow for two hours-they would also get to have a date. After four dates they have a marriage.

MR Does this means anything to you as a user?

FR You don't know anything in relation to your virtual child except you get to know how many times you are meeting your reproductive partner. The whole idea was based on pictures of the family showing off a baby. So you get sent an image. If you have a marriage, you get to see whom you are married to, so you get to show off your baby's partner. But if the baby stays single, we worked out a whole lifecycle or logic tree leading to affairs and divorce etc. So you may end up single again-and you start to travel to other cells to find more babies.

MR This was all planned on paper—did it actually get implemented.

FR No, we used it to work out a whole system as a technical analysis.

MR What was the prognosis of the company that would have to implement it?

FR This one was really just a thought-experiment. We asked if we had better resolution, what could this mechanism be used for? Could it be an engine? We were always looking for concepts where people became the engines to make the content. The content wasn't made by somebody and then delivered. You would make the content. This was about finding that engine. It was a crazy idea.

MR It does seem hugely complicated

FR Hugely complicated and a nightmare for someone, but with a good Alife programme it could have been made ten times better. The network was revisited with *The Garden in your Pocket*. This one was about using video-we thought with those capabilities on the phone people would be taking pictures of stuff around them. We really wanted to do something really pixel based, something distinctly digital. You would be given a garden on your phone with a perfect lawn. The idea was you would go into the menu and there would be all these seeds with Latin names, but not real ones. The idea was that when you are sitting on the bus and bored you plant some of the seeds in the garden into the green lawn. Basically you create these psychedelic flowers and wonderful blooming lawns. It is all based on location. You have your *Garden in your Pocket* and go about your daily business. It may be I go to college in South Kensington and I check my pocket. Some flowers only grow in certain locations under certain conditions. Maybe in South Kensington a huge purple flower would come up, which you hadn't seen before. So wherever you go in the city you check your garden and see what things have grown. Maybe there would be prompts to look after certain plants, which could die because of a cold breeze. Or you might get animals moving into the garden. You might be in King's

Cross and a bear invades the garden. The idea is that you get to narrate these changes to others.

MR So the garden also embeds your physical journey?

FR Or it triggers these other narrative events and you show them to people on your phone. We also had this idea of embodying wildlife videos in certain places.

MR You haven't heard of *Savannah* from Mobile Bristol where children became Lions hunting Wildebeest on a savannah? It was again very similar and you had to rapidly learn how to hunt or your cubs died. It was role acting for kids, but distributed over a real location. It was based on GPS not cells on PDAs.

FR Then we did "Sticks and Stones" which was about fiddling-you know when you commute and fiddle—like we used to sit in the canteen and pick at plastic cups. These little things you would get to form collections—beautiful things like tomato stalks and keys and trivial things like staples. The idea was that the city became a storage cupboard. So the keys might be at Angel and the magic tape is down elsewhere else. Where you went to collect materials would depend on the materials on your phone. Lots of people would have different materials. If you had car keys at Kings Cross lots of people might want to go there to collect them. When you are bored on a bus you might stick them together.

MR Like potato men?

FR Yeah, you make little beings with behaviours-they might start to wiggle or jump.

MR You get an A-life effect?

FR Yeah, they start to get characteristics. The important factor was sharing it with someone else. If you had made a creature you could put your phone together with that person and your creatures could walk off your phone onto theirs and vice-versa and they could interact.

MR You could make little Arks of these creatures?

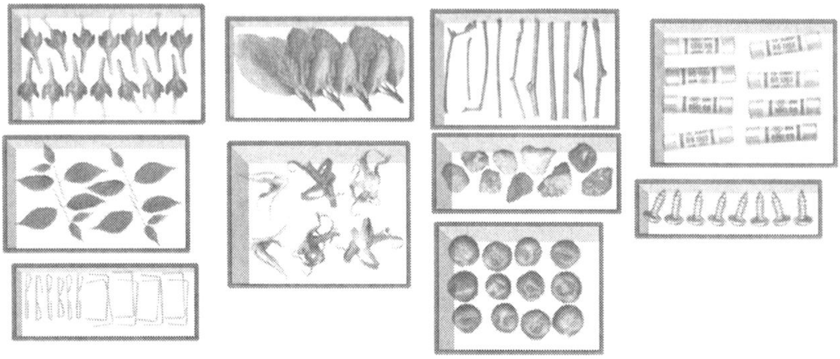

Figure 80: Sticks and Stones

FR Another one I really liked was *Lazy Crow*-similar to *Lost Cat* in that it used location. When we did this one we became so very aware of real crows. We suddenly saw them everywhere. So again the narrative only exists in your imagination. The idea is that you have a pet crow that comes on your phone and knows where its home is. It knows it lives in Clerkenwell or the Barbican. If you went up to Liverpool you could release your crow across the city. You would get reports of the route it was taking. But the crow is lazy and likes to hang out with other crows. So it looks for roosting points in the city. Maybe these could be giant electronic billboards where crows hang out together.

MR This would be displayed as an adjunct to the project? On boards with lots of crows?

FR Or any digital display-we even had a chocolate machine where a tiny crow could appear. They would eventually find their way back to you. The idea was you would have a mental map and would have a shrewd idea of the route it should take.

MR Would you get reports of their progress?

FR It would email back messages-for example crows like shiny things, so you would get messages of the shiny things that it finds. The whole thing was the journey. It might take a week getting back to you-and it relates all the things it finds and the crows it meets and then it might actually appear in a physical location.

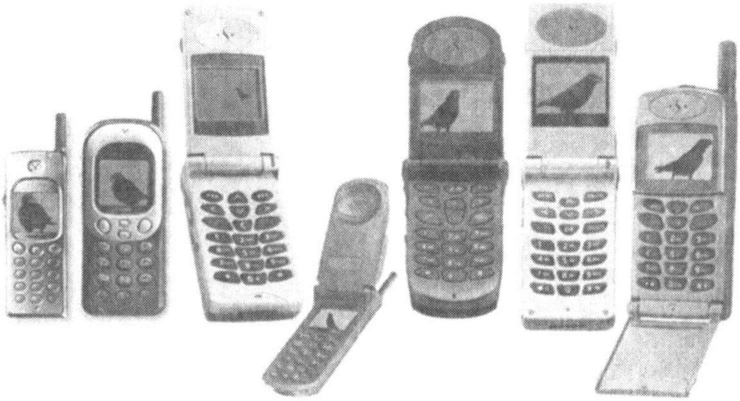

Figure 81: Lazy Crow

MR It is interesting that you chose crows rather than homing pigeons-because that is basically what they are.

FR It was visually nice to see all the crows and to imagine one might be yours. It was the projection in the real world that we thought was the nicest aspect of the project. Those were our favourite ones out of all of them.

MR Now I am glad you explained that one because I didn't understand it at all from the website.

FR Any others you were specifically interested in?

MR Well the satellite one seemed interesting.

FR The one where when you make a journey you received Landsat images of your location?

MR Was that sent to your mobile?

FR Basically you go on a journey and wherever you are, if there is any CCTV camera or satellite image, when you walk into a cell it triggers an image creating an alternative view of the journey. It would be incredibly difficult to do anything of the sort in reality.

MR Haven't you got the CCTV coverage now?

FR You have, but then getting access to material is really problematic. Even under the Freedom of Information Act you have to pay £10 per image (Laughs). You can ask, but it costs-so I think that one was dead in the water.

MR I think these ideas were great and have so many parallels with what is going on now that they deserve a broader exposure. I know people know of your work, but it doesn't seem to turn up in people's thinking on new work, when they are doing very similar projects.

FR Well I guess it was done so long ago and we moved on into other areas.

MR Are you doing anything with mobile technology now?

FR We are, but I am afraid it is all under wraps at present. Its very early days looking at mobile services with 3G.

MR Well we all look forward to seeing what emerges.

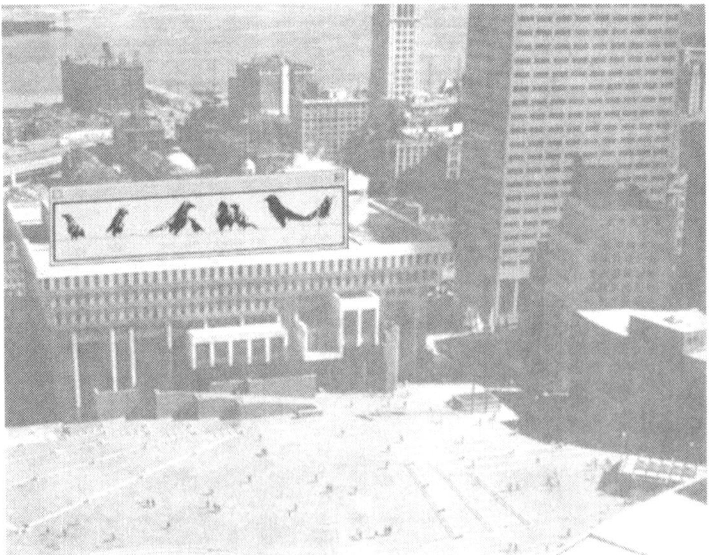

Figure 82: Lazy Crow visualization

Trace, The Choreography of Everyday Movement and Drift

Teri Rueb

(Interview with Martin Rieser Plan ICA London 2nd February 2005)

MR I want to first ask you about the Trace[1] project from 1996

TR 1996-9

MR This was on a trail in the woods near Banff Art Centre. What was significant about your choice of Site? Was there something meditative about the place that inspired you? Was there something specific about the site, which generated this idea?

TR The site was actually very specifically chosen. Firstly, the move away from the screen and keyboard and into real space was actually quite new at that point in digital media. People weren't doing Augmented Reality at the time. That term hadn't emerged in the mainstream. It was kind of a big leap to take people away from the relatively disembodied experience of the screen and keyboard and into an environment where the meaning could be enhanced by the physical effort it required to engage in the work. It really demanded a lot, but we will talk about that more in terms of audience. The space/site was a set of hiking trails in Yoho National Park, which is about an hour away from Banff. It's the location of the Burgess Shale Fossil beds. They are very famous, particularly popularised by Stephen Jay Gould's Book 'Wonderful Life', in which he describes their discovery and interpretation as a pivotal moment in reshaping thought about evolutionary theory.

The Darwinian notion of a linear progression to a cone of increasing diversity was disputed and it was recognised that there are all these contingencies in evolutionary process. In fact in the fossil record there was more diversity in some species at that time than there is now. It was a kind of reversal of assumptions. I wanted to use the site on two levels. It was a site visitors can relate to in a very general and abstract

---

[1] Trace was produced in its first iteration in the summer of 1999. The database of memorials expands over time as interested participants continue to contribute memorial songs, poems and stories. The first launch of the installation featured works by Diana Berry, Susan Davis, Sarah Drury, Stelios Gannoulakis, Joanna Goodman, Henry Israeli, Bruce Ledbetter, Akiko Matsumoto, Lisa Moren, Tim Nohe, Mayumi Reinhard, Thomas Rodebaugh, Vivian Adelberg Rudow, Melora Zaner-Godsey, and Sirpa Vaara. See <http://www.terirueb.net>

way, though I don't like to say universal. But, the idea was was that here is a kind of origin story written in the landscape. The fossils themselves function as a metaphoric record or inscribed gravestone.

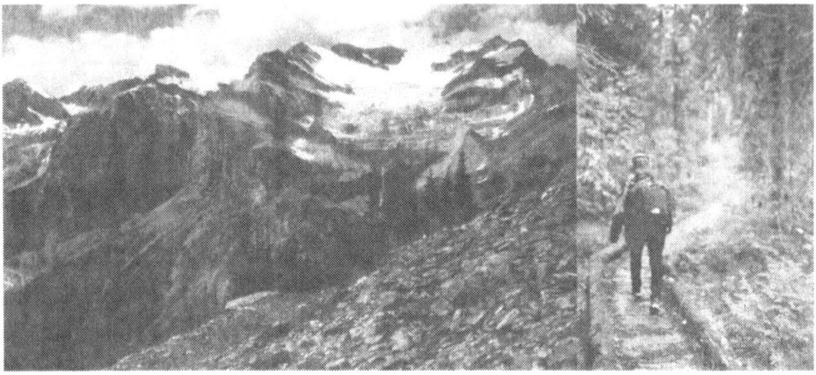

Figure 83: Trace

MR And are they also an ark for all these life forms that have passed away?

TR It's a database of sorts, one that we can relate to on many levels

MR So that was one reason for that location-a very general one, was there another that inspired you to create human memorials, rather than something else?

TR Well, I reallly wanted to point to a larger scale of time. I wanted to take people out into an experience, which would engage them physically and make them feel more alive, even when reflecting on loss. Its a difficult piece, in that respect as it can be emotionally sombre, but at the same time it can be very exhilarating, because you are in this incredibly beautiful landscape which makes you put things in perspective in a different way. It was important to me that people experience the work almost as a kind of healing.

MR In terms of the posting of the different narratives into the space-how was that achieved? Could you give me an example?

TR It was technically very simple, although no one had done this before, (I later learned that Ian Mott had done some similar work), the GPS was just used with a laptop and a software patch in Max / MSP (a sound sampler for musical composition, which was a software used to sample sound among many other things, electronic music composition), but here it was just used as a sound player as you moved along the trail. The poems, songs or stories were spaced very far apart in very large spaces or regions, so it was intended that you hear the entire piece–there was no interruption when you step out of a hotspot. I didn't want people to become interested in that interaction. I didn't want them to be thinking "Oo-what do I do to make it work? What do I do to make it break?" I wanted it to be a much more contemplative experience, I didn't want people to interact with the technology, even to see it. You checked out a knapsack, which was a very natural article to be carrying when you hike and the whole thing happens automatically, though you do wear headphones. I spaced the regions of sound far apart as an aesthetic decision, because I wanted the experience of the landscape to be prior. I wanted people to be able to talk to each other on the hike. Also the sound came through open headphones, so there was a sort of combing for ambient sounds, combined with the sound of the work itself. Although there were some interesting audience responses based on the use of the system.

MR What for example?

TR I imagined they would wear the headphones the entire time because they were comfortable and lightweight. It turned out that they were oftentimes too hot or you wanted the feeling of the wind in your ears, so people would wear them around their neck and as soon as they heard something put them back on. It was an unintentional improvement on the part of the audience.

MR How did you gather the sound pieces?

TR I actually asked people through an international call to submit works based on interest alone. I wasn't concerned to curate or impose a quality control on the work, though the pieces necessarily had to be scaled in first iteration to fifteen works across twenty-two kilometres of trail. The pieces ranged from professional works made by artists and poets and musicians to more amateur or non-specialist recordings that were very, very personal. What was interesting in terms of the audience relationship was that very few people experienced the piece because it was so remote and it required a high-maintenance operation to equip people and get them out on a trail, and it basically took a whole day to experience. Many of the people who had submitted works had never been to that location and never did go. It was intentional in my mind that people would have a relationship to the place which was almost like that of a place of pilgrimage-you may never visit, but its a place you hold in your imagination as an important locale.

MR What kind of feedback did you get from the participants?

TR I did a series of interviews I taped, I had essentially three different responses: for the most part people were drawn into the work and carried along with it without a lot of self-consciousness about how the system itself works-its being new media etc-a lot of the people who experienced it had no relationship to new media. They were either tourists passing through or local people who had seen a flyer in a cafe and were curious to have a different type of hiking experience. So in some sense the piece selected for someone who was already open to it. It wasn't something you just walked by on the street and plugged into, you had to reserve a place. So to say that people had a generally positive response and were open to the work is important to frame, because it was already a very specific audience experiencing it. There were other responses as well, I remember one in particular- a writer, art historian and critic described it as I had hoped; he experienced it as a kind of spatialised narrative. In the sense that there were these very arduous couple of miles that you have to go up: very steep switch-backs at the beginning of the track until you reach altitude and things evened out, and you'd get your wind again. The sounds then kicked in with a little more frequency.

MR So the geography reflected and enhanced the narrative curve.

TR Exactly, also he had a coincidence with the weather and the way he experienced it, by the time he got to altitude the sun had come out, it had been raining prior, so he described this sort of epiphanic moment.

MR Of course you had prearranged the weather.

TR Of course these were all responses, which were culturally specific to the individuals. Another thing he reflected back that I appreciated was that as you come back down you hear the same works, however they are different because it is a second audition and the landscape is different when you are facing the other way and of course you have the prior experience-the leading up, the middle and then the kind of denouement coming down.

MR So on the *Choreography of Everyday Movement* [2] which you did four years ago?

---

[2] This project was originally developed in the context of a group project funded by the Rosenberg Gallery at Goucher College curated by Laura Burns. The project involved an 18-month dialogue between the artist and Denise Tassin (painter), in collaboration with Goucher faculty and students:

TR Yes, 2001 was the first iteration. Framework and process?

MR Yes, how did it develop?

TR This work was originally commissioned as collaboration between three parties-a painter, a choreographer (and her senior capstone studio-so we had dancers) and myself. It unfolded over an eighteen-month period of exchange. We did not know each other prior and we were commissioned by the Goucher College, Rosenberg Gallery by a curator who had a sense of what would happen if you put these three people together. She was also a voice through the process. In that work we each struggled to find common ground. We had very different sensibilities, so in many ways it wasn't collaboration, but rather a dialogue, which went in three different directions in the end.

MR You each ended up doing your own thing?

TR In a sense... yes. Though it was presented altogether as "collaboration". It was very valuable exchange and influence, nonetheless, even without deep collaboration. My awakening through that process was to discover the world of dance and performance, which is distinctly different to the way I conceptualise as a visual artist. It was wonderful to find-an echo, as I was trained as a sculptor-their concern with physicality, with the physical experience. I think some choreographers are much more conceptual, but this choreographer was about the physical nature of the body in space. So what I did was to ask a lot of these young dancers (they were seniors in college) to think about choreography in terms of their everyday experience and that it is in a sense a politicised activity as they move through the space of the city. I asked them to carry a GPS with them on their daily travels and mark their paths, their movements. They were willing to do this; it didn't require a whole lot of technical knowledge. I would download the tracings and transpose them, but they saw a different way of relating to their movement. I will come back to that in a second, I want to outline the larger scope of the project: There were tracings made from their personally recorded journeys, then there were the tracings made live as people would come to a scheduled event when a dancer would go out into the city and we would watch their movements live, transposed as a drawing using a Java

---

Professor Amanda Thom Woodson (choreographer), Andrew Cole (composer) and Kirsten Maurer (dancer). It was presented with the participation of dancers from Amanda Thom Woodson's senior dance studio, Fall 2001. Java Applet programmed by In Choi. See <http://www.terirueb.net>

applet, over the internet. So it was performative and but with layers and layers of insulation or mediation.

MR But all you saw was a line being drawn; there was no figurative content?

TR You understood that that tiny dot represented a person, but it was very unusual for the dancers not to be watched, physically witnessed.

MR So they lacked an audience really?

TR That's right- to respond and feel. The whole aspect of performative work relies so much on that exchange. To be honest I think the dancers were not very interested in what this work was about!

MR Dancers aren't conceptualists...

TR No, and it was a real awakening for me. Anyway those were the two aspects of the work, the drawing archive, which evolved over time, and the real-time performance.

MR A 'white cube' work in fact?

TR Very much so they were layered against each other in a very cool and analytic way, almost as statistical data marking a morphological trace of movement over time.

MR Were you conscious of a minimalist aesthetic process as you put them together?

TR "Minimalist aesthetic?" I was, but I don't have a clearly articulated position in relation to that history. What I found was a progressive move away from the visual as an explicit composition.

MR These were more the trace of process?

TR What I found interesting again was the contrast between that marking and the marking of dancers tracking their own movements. In the real time you are getting a sample every interval, so you get a very smooth contour. In the case of the dancers carrying the GPS they would mark at whatever moment they chose

MR So it was quite jerky?

TR Very angular, very jerky and they also moved in a way that wasn't concerned with making or creating a visual image at all.

MR They were concerned with how they felt in a particular space?

TR That's right and they did things, which seemed very unusual to me, but natural to them, like run in and out of a building many times over, repetitiously.

MR So you got scribbles on the line?

TR That's right, they we're not meant to be graphically expressive. They were indexical. Then in the next iteration (I did a couple) but the one that makes the most interesting contrast was in San Antonio. In this case I had people driving through the city. They were not everyday people; they were familiar with the context of Siggraph and digital artworks as such, but not San Antonio. They would drive, in this case I had multiple movers that were tracked at the time, so it came to be a dialogue between the two lines, an emerging pattern.

MR So they could see the other person's movements on screen as they were driving.

TR Yes, although we discouraged computing when driving, but there were moments when you stopped at a light and so on. Not only did you see a contrast between the way two different people drove through a city, both of whom didn't know that city, so at another level that was not everyday. It was a reflection of how we move in cars through an average mid-sized American city, which tend to have a grid like lay out. But you did see a contrast between the ways in which people explored, some would stay in a relatively small region, explore wander and occasionally try to draw an explicit symbol. That wasn't particularly interesting for me, but it was interesting to see how people are drawn to such activities.
Then there was the question of surveillance in the work. This was something people reflected back sometimes as "This could be for studies" or "This could be used to track how often people go to this Starbucks or that one" That was Siggraph's audience. But not without exception: there were people also who looked and saw distinctions between walking versus driving, and there were people who found a sense of humour in the idea that this little dot you could project into and imagine "Oh they've stopped to buy something" or "They're waiting at a stop light" or "I know that road, I know that bar."

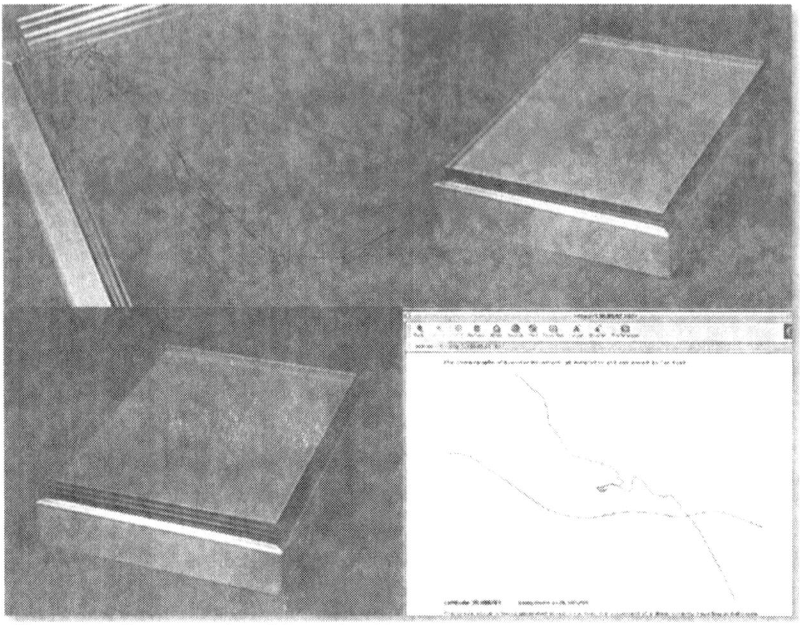

Figure 84: Choreography of Everyday Movement

MR So there was an imaginary that could work at this very minimalist level?

TR Yes, quite a bit. It was important for me to split the line off the map. The map was there on the wall so you could reference it. I wanted people to acknowledge the line in and of itself.

MR If we move on to *Drift* [3], which is only a year or two old...

TR Yes, it was April 2004.

---

[3] Drift, was installed along the Wadden Sea for the exhibition "Ohne Schnur: Art and Wireless Communication." The installation covers a 2 km x 2 km region that is filled with areas of interactive sound. The region moves with the tide such that at low tide all the sounds are out on the Watt, at high tide they flood the town. Sounds play automatically as you wander through these interactive areas with a Pocket PC, GPS and headphones. "Drift" software was created by Computer Science students under the direction of Dr. Zary Segall, Distinguished Professor of Computer Science and Electrical Engineering at the University of Maryland Baltimore County (UMBC) Technical support and implementation in Cuxhaven: Erik Conrad (UC, Irvine) See <http://www.terirueb.net> and <http://www.ohne-schnur.de/>

MR Could you describe how that came to be?

TR It really emerged out of a personal moment. A sense of lifting my head after a certain period in my life and looking around and feeling somewhat adrift, but also having a desire to go off the map. Particularly having worked with GPS for some time and seeing all around me suddenly a proliferation of GPS in artwork beyond GPS drawing, which was the earliest thing we saw. There was a lot of talk at the time about the potential of being able to annotate real space and being able to layer content to specific locations and having different voices framing those environments. Also there was a lot of commercial hype around the potential of a spatialised internet and I think I felt very wary of the idea, the assumption in the dialogue, that we would know where we are or know better where we are through all of the layers of information. GPS of course is supposed to offer that.

So I wanted to invert the tool-and this has been consistent in my work at least since Trace. I wanted people to lose a sense of indexical or concrete relationship to space in these works. I want there to be a suspension of disbelief.

MR So you are creating an imaginary within that environment?

TR That's what I hoped to do

MR Because what struck me about Drift was that you used counterpoint in the sounds to describe, for example the footsteps on different surfaces, rather than on the beach you were actually walking on.

TR Yes

MR I found that a very interesting idea. How did it come to you and what significance did it have for the work?

TR I think I've always been interested to create experiences of space and time that are multiple, overlapping and even sometimes contradictory, so there's a sense of disorientation, but also transport to another place. So I think a novelist does that through narrative, but as a sculptor I come to it through space first. I knew that I would use passages from texts, which addressed themes of wandering, being lost or disoriented. The first inspiration was Dante's opening Canto of the Inferno.

MR Did you use Baudelaire as well or did you keep away from the "Flaneur"

TR No, no Rosseau is very much part of the work's inspiration. We know that there is a whole critique of the Flaneur and it is important in my works, which can at times be understood to be perhaps romanticising. They do take people out into landscapes and they are often solitary works, but I think that for me running up against that tension between the technology, which is in a sense very unexpected in this landscape, is where I get a sense of relief from what might be mistaken as pure nostalgia, or romantic

MR Because you take the romantic notion of landscape and invert it.

TR Yes, which might be interpreted as trying to get people to experience landscape without the constant awareness of the system, which is a big challenge because you have the headphones which are really intrusive. And if other people are around you, you look a little silly or curious-what are you doing, you know? There are these scenes and disruptions, which I still struggle with. In this piece in particular I found I was working explicitly against interaction conventions as they have begun to form now: an expectation of what people demand from interactive works outside of video games, we have at least moved that far. To have people not know where the sounds were; they were always moving. You could not go out and make the sounds work, they had to wash over you.

MR They went in and out with the tide and you had to go with the natural phenomena, which I liked particularly.

TR You had to really experience the landscape first; otherwise you would end up really frustrated. Some people were really frustrated.

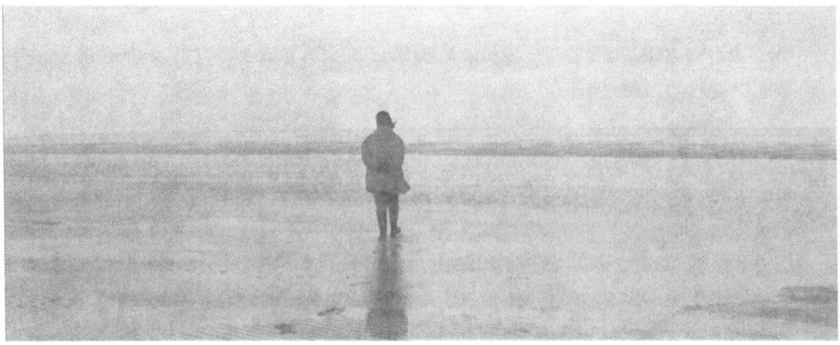

Figure 85: Drift

MR So you got a really diverse experience, some people who could relax into the work and people who fought against the framework?

TR That was very much an integral part of the work-to try to explore how you get people out of a goal-oriented sense of movement and position in space, which the GPS encouraged, and get them to be "inspired."

Blast Theory
Matt Adams
(Interview with Martin Rieser January 15th 2005, Bath)

MR There are a number of issues your work raises, which are to do with the ludic aspects of relationships between real and virtual worlds, I think that might be a summary of what you do. As a group, how did you come to the decision that that was the area of discourse in which you were most interested?

MA We began making work in 1991 and the origin of the group was already a fairly hybrid one. The original members of Blast Theory all worked in the same cinema as ushers and cashiers and bar staff, but we were all either photographers or fashion designers or artists, a hybrid collection, so when we came together to make work we already had a multidisciplinary approach, and in fact the first two or three years we worked together were mainly about establishing the most basic grammar for talking about what we were doing.

MR And the role of technology, did that grow, because you started out more as a performance group, didn't you?

MA Yes , although we had technology in the very earliest work we made, it was more by happenstance than intention. We used projection very much from the beginning, partly because our work was so influenced by club culture and we tried to reinvent that language; but we also used video and computers in the very first piece of work we made-*Government Gunmen Kill Three* to film the audience and make photofits from the moving imagery and so on. In the second one we made-*Chemical Wedding*, we were using large computer graphic projections throughout the entire performance. By 1994, with *Stampede*,[1] we were already using interactive systems with pressure pads on the floor of the performance space. So that even though none of us in the group were technologically savvy, we were always very interested in that approach. The thread that really runs through that early work and where I see continuity with the work we are doing now is an interest in social and political realities, popular culture, and fictional forms. We happened to meet up in 1997 with

---

[1] See *The Independent*: <http://www.independent.co.uk/arts-entertainment/blast-theorys-experiment-in-crowd-control-1614278.html> and <http://www.blasttheory.co.uk/bt/ work_stampede.html>

the Mixed Reality Lab in at Nottingham University and Steve Benford. The work which they were doing there was on the relationship between real and virtual space, so that really gave us a technical sophistication and a platform to do more complex works using technology.

MR There seems to me to be this moment where the two worlds touch that is a poetic that you explore in all these related pieces. What really worked for me in *Desert Rain* [2] was the instant that the real person walks down the beam of the projector and cuts the curtain of water and you are confronted with a human being, rather than a very abstract virtual space. I thought that was a very strong moment, and I thought that playing some of the games-well I've only played *Uncle Roy* [3]— that there was a little bit of a difference between the net-based players and one's experience of actually being in the streets and trying to navigate. When I played it there were messages like "Are you all right?" "Do you need help?" and it wasn't really possible to put into any kind of context how you could get the help, so it didn't quite connect for me in that sense. Obviously I enjoyed playing it and I particularly prized the fact that I found my way to Uncle Roy's place. I know that the next one on-*I like Frank*,[4] was based more on 3G technologies, but did it take that interplay with an observing audience out on the net a bit further?

MA Yes, I think the central challenge of works like *Uncle Roy All Around You* and *I like Frank* is how to combine these two spaces-the online world and the space of the street player out there in the city. About how to deal with the very divergent sets of expectation and experience that those two spaces engender: online Players may only have ten minutes to play, street players may have bought a ticket and have set aside an hour to play. Online players can log off at any time; street players would very rarely do that. Online time is a very peculiar thing, where the time it takes for a link to open can seem like an eternity, when you want to get onto the next page, but you can have played an online game for five or six hours without any sense of the time passing at all. To be grounded in the performative time that the street players experience is a much more familiar and much less malleable form of experience, so when we moved from work on *Uncle Roy All Around You* (with a model which is collaborative, where online players are essentially charged with the task of helping

---

[2] See <http://www.adobe.com/uk/motion/features/blasttheory/> and <http://www.blasttheory.co.uk/bt/work_desertrain.html>

[3] See <http://www.blasttheory.co.uk/bt/work_uncleroy.html>

[4] See <http://www.blasttheory.co.uk/bt/work_ilikefrank.html.

street players to achieve their destination) to *I Like Frank*, we looked at a more competitive… or a structure that's more at cross-purposes.

Essentially the structure of *I like Frank* is that street players are looking for Frank, who is out in the city somewhere. Online players are looking to get a street player to a particular location in the city to retrieve an object for them which they subsequently find is a postcard. There are four places in Adelaide where those postcards are concealed-in the saddlebag of a bike, in a bar and so on. When an online player approaches a street player and they ask "will you go to this bar and retrieve a postcard for me?" they may see that as somewhat at odds with their current role in the game, which is looking for Frank. In fact they are not going to find him until they have helped an online player, so there is a slight slippage there and it combines competitive and collaborative elements. And I do think it was more effective as an experience for online players.
Certainly we are under no illusions as to just how large and complex a task we are trying to achieve in these works. Trying to make something that works as a performance and as an artistic experience; as a game that is a social, but also narratively determined; something that combines two different audiences in two locations using a whole set of interfaces and technologies. We are under no illusions just how far we have to go to deliver on all of the promises that works like that make. But I do think that what we felt about *I like Frank* was that we had a very rich experience for online play. When we came back to do *Uncle Roy All Around You* again, as we had two further performances in the UK in Manchester and West Bromwich, we thought 'great, we just put together the great online experience of I like Frank with the great street player experience of *Uncle Roy All Around You* in London had had'. But the first gig we did was in Manchester and we really found it an extremely difficult experience and it was far from straightforward to do that.

MR Is it hard to get yourself in the mindset of all the players, because when I played it, it was quite frustrating (*Uncle Roy All Around You)* you know you'd be wandering the same few blocks trying to work out where this place was. Some of it was serendipitous; because if you didn't know the location intimately, some of the clues were lost on you-it was quite good that it was embedded in the locale in that way-but I wondered what players who were on the street who never found the installation felt. Did they feel frustrated?

MA It was clearly frustrating. I'm reading a book "The Rules of Play" by Katy Salem and Eric Zimmerman [2003] at the moment, which is a really superb text on the issues of game design, and one of the points they make is that the fundamental aspect of game design is that it has to be an iterative process in which you test the structure of the work as quickly as you possibly can. You play it-the only way to deal with the sinuous complexity of a rule based structure of game play is to immerse yourself within it. A strategy we have long practiced, but with works of the complexity of *Uncle Roy*—effectively the only way to test that work is for it to be up and live. So for the first few days we were there in Manchester we were well aware that it was not gelling in terms of the relationship between all the different components. What we have developed as a group is a process where a team of people working on a piece like *Uncle Roy All Around You* (about 20-24 people), were able to run it for five or six hours a day, then in and around that be developing the work as well, and responding to feedback.

MR I thought that what *I like Frank* seems to suggest is that because you had embedded a lot in the environment—people were going to locations to find actual things—there was a tighter relationship between the physical environment and the virtual, which could mean that you have series of small rewards rather than one big one at the end. Is that a structure you might take further?

MA Yes I think we are at a really interesting moment: *Uncle Roy All Around You* took about eighteen months to two years to develop through to the summer of 2003 and the following twelve months after that to the summer of 2004 we were touring *Uncle Roy All Around You* and developing *I Like Frank* and we really suddenly in the since last autumn we have come into a period of real open time for the first time in three years to really step back and appraise where we are.

MR But you've got a new project in development that follows on?

MA Two in particular we are developing now-one for New York: we are due to spend three months in New York to develop a piece for the area in Lower Manhattan around the World Trade Centre site. We have also begun a three and a half year programme of research that is part of a European project called the Integrated Project for Pervasive Gaming with a number of partners across the EU, predominantly in research labs, but also including commercial partners such as It's Alive, who have done all several the leading-edge mobile phone games, Sony and Nokia.

MR It puts you in a fantastic position

MA Its a great opportunity and fantastic challenge, but I think in terms of our work we felt that *Uncle Roy All Around You* and *I Like Frank* were so complex, as I've intimated already, that essentially every interlocking part relates to every other interlocking part with such enormous reverberation, so that if say we need to change this aspect or structure the clues *Uncle Roy* sends you so that they are more definitive in fact that immediately lowers your dependence on needing help from an online player and the experience becomes more self-generated , so that reduces the amount of traffic and communication between online and street players and then online players' experience is diminishing, because suddenly street players are more self-sufficient. Those kinds of complex inter-relationships are incredibly difficult to deal with, so I think one of the things for us to do is probably to divide the work into two or three different strands and say, "Is it going to be a narrative driven thing? Is it going to be driven by game play or is it going to be a socially structured thing where it is primarily based on social interactions?"

MR And much more communication traffic between players?

MA Yeah, possibly, the other thing we are looking at very carefully is whether you need to have a synchronous relationship between online and the street. Because of our performance background there is this tremendous compulsion to make work that's very much about live presence, it. It was a natural assumption for us that online and street play would be synchronous, but we are now looking to see if there are other possibilities for achieving the pieces via other models. I think the other question is how the complexity and depth of the work is made manifest to others. What we found is that *I Like Frank* and *Uncle Roy All Around You* are very subjective experiences and that within an hour you will have had three of or four people who have had an extremely thrilling experience and have been very moved by what happened and in the same time you can have two or three people who have come back and said "I can't understand it, no one talked to me..." and they leave. In Blast Theory there are three of us and in particular Ju and I have a predominant role merely to observe and try and work out what is going on and what's determining those tremendously varied responses.

MR Just technically-because there is some much technology involved and so much exchange of information (and I know that in cities you didn't want to use GPS because it was too unreliable, so you relied on people polling their positions back) —I mean was it bandwidth? For example if you had mounted a video camera on every player, that could have gone back to the online players and they might have felt much more directly involved with the subjectivity of the experience. Was it purely bandwidth that prevented that, or had you made a decision that that approach wasn't helpful?

MA No, its really that our starting position is that we are artists who make work for today that has to roll out in real time to members of the paying public in a way that would give them a meaningful experience. So although there is a tremendous amount of research attached to our work, and some people see our work in that respect, what we are interested in is in making work that delivers to the public on that day. From that a whole chain of precepts swing into view, which is that we want our work to reach the largest number of people we can at the time and therefore every pound of equipment that has to go on each player limits how many players you can have. In *Uncle Roy All Around You*, you have twelve players and the limitation there is the cost of a PDA and a GPRS card-so you have essentially £500-£600 of equipment for each player multiplied by twelve, and that was the maximum we could possibly afford. It may be very prosaic but essentially those are the kinds of factors that govern whether every one has a video camera or not.

MR But now 3G phones have got video and it's all compressed on the one item, it becomes less difficult presumably?

MA It does, but you still have development issues and our learning experience of working with 3G in Australia is that it is the most unpleasant development environment you could imagine. The devices are not designed to let you do many things; they are specifically fire walled to stop you dealing with certain areas of the functionality of the phone. For example in *Uncle Roy All Around You* people were leaving voice messages on their PDA, which was uploading to online players. It worked very neatly all as a part of the interface of the game, without any problem. When we tried to do that with 3G phones it was a complete nightmare to make voice messages even though the device is a phone. Where the application runs games, it's completely separate from where the phone deals with voice calling. So what you had to do was quit out of the game application to make a voice call and then re-establish back into relaunch the game-and that in itself was several weeks of nightmarish development-and in the end about a week into the I Like Frank experience we cut voice messaging out all together because it was so dysfunctional, it was causing the application to crash and so on.

MR So it suggests that phone developers need to talk much more to the artists at the early stages if they want the content you are developing (and which might be the killer app for this platform) to be actually usable?

MA I think the phone industry is sufficiently large that the people who understand what the possibilities are still marginal to the agendas of the large phone manufacturers and that they who have their eye on a completely different ball and they are dealing with the leading edge of the market, so there is no imperative to deal with some of these things that they would might perceive as very esoteric. Which is an interesting set of confusions, and it's one thing about our work that puts us in a good position because of the game-play element. They do totally understand that games are the critical development for mobile phones, but at the moment there's a real tension between just porting consul games and PC games onto mobile phones and really rethinking where games might be altogether. I think that the fact that the EU have funded the pervasive gaming project, precisely reflects that there is an understanding at a strategic level that here's a real opportunity to think of phones as devices that have a cultural significance, not just as consumer items.

MR Absolutely, you just don't port old media on to them and hope that that will sell them. Can I come back to the art and the concerns of the content, because a lot of your earlier work was involved with things like subversion, surveillance, issues around terrorism I suppose—quite edgy, political stuff. Certainly you looked at the Gulf War and how easily people are coerced into certain behaviours by group interaction. How much of that is carrying over into the new projects? How much is there room for that?

MA Well I think *Uncle Roy All Around You* reflects a different approach to the work that we have made, and it really comes out of our observation of touring *Desert Rain*. We toured *Desert Rain* for about four years and I think there were about two and a half thousand to three thousand performances of *Desert Rain* (not numbers of people) and so we observed this work over four years and it was a tremendously important learning experience for us about the nature of interactive work. Each group of six people went through there and we listened in to what they were doing and observed those dynamics, and what we realised was that interacting with the work itself is such a powerful and important part of the experience of that work, that content is really something you have to think about in a different way.

Tetris, for example, as something that is not played is a very dull thing, if you were engaging with Tetris in anyway apart from playing it, it is a very dull thing. And yet the act of putting those shapes into that order is a surprisingly engaging and complex process and that is both an intellectual process, but it's also a kinesthetic process it has its own linear structure, it has narrative peaks and falls etc. The experience is what's significant. So I think what we took out of *Desert Rain* and into the stages of work that we have made now, is an understanding that the act of the experience is a tremendously powerful part of the work and that therefore there is less need to make content an announced part of the work. We've had some very forceful criticism from people who see the work—*Can You See Me Now* and *Uncle Roy*—as having no content. I think this is a very great misunderstanding of the dynamics of the work.

What *Uncle Roy All Around You* in particular is about is whom you talk to, what you make of the conversations you have with them and what level of trust you ascribe to them. That simple dynamic of a mediated experience that of course culminates in the contract that you are asked to make at the end of *Uncle Roy*, which is you commit to be there for a stranger for twelve months to help them in time of crisis. That segued from a relatively hokey game structure—go and look for *Uncle Roy*—which is not going to challenge an eight year old too much as a simple premise, but then gradually it turns into a social contract. To me, that is an inherently pertinent thread to ask about electronic space, and mobile devices in particular. The kinds of things we are thinking about are: what is happening about the nature of the way we talk to each other; how you build trust; or how you deal with the social nuances of a mobile, online communication. That's partly chat room type spaces, and on one level Uncle Roy provides a chat room situation for online players. You're in there (10-20 of you at a time) and everyone's sharing messages, which are one sphere. In another sphere, those online players are also engaged in one-on-one conversations with street players, so you have these two communication forms happening simultaneously.

It's partly inviting you to choose where you put your value, inviting you to choose what's richest about those different experiences. I was talking earlier on about trying to make some of these things manifest. I think one of the things about *Uncle Roy*, one of its weaknesses as a work, is that some of the very powerful experiences that people have had are not visible in any way to any other players, they are very private. Certainly in *I like Frank* we had people in floods of tears, very overwhelmed by it as an emotional experience, but then we had the next player just coming through going " It was just rubbish and I didn't really like it" and those two groups are really independent of one another. One of the things we are doing now is to look at that that in terms of the work we make next and it also helped to change the structure of *Uncle Roy*, so that the version we did in West Bromwich, the final version ever, the postcards that street players write when they find Uncle Roy's room are now all scanned and online. You can read all of those and they are a very beautiful set of cards. They are filled with tremendously interesting responses to this experience of being alone on the street with nothing in your pocket. This one communications device is the only tool that you have and it's quite a challenging and threatening thing...

The work is also, of course, about the city itself. It is about he city as a site of the banal in which we work and consume versus the city as a vehicle for the fantastical or cinematic. And *Uncle Roy* looks at the line between interior and exterior. There is an intentional blurring between the world of the game and the rest of the city; street players are often confused about who is in the game and who is not. We have many great anecdotes from players who went into office buildings and told the front desk that they have an appointment with *Uncle Roy*. Even when they were met with total bafflement, they persisted. Their attempt to impose their fiction onto the world around them says much about how we cope with the proliferating subjectivities within contemporary cities.

MR Can I come back to this emotion that *I Like Frank* produced again, because I haven't played that one, although I roughly know how it works, but if you could just explain what triggered those emotional outpourings?

MA Well the germ of it really came from a purely practical accident in *Uncle Roy* which was that we wanted every street player to be picked up by a limousine as they came out of Uncle Roy's office and we were running twelve players through an hour, so that meant and we could only afford one limousine, which meant that we had six minutes for each player, assuming that they came at neat staggered intervals, which was of course never going to happen. So what we did then was to build a structure where when you came out of Uncle Roy's office you were told to wait in a phone booth and we had four booths so you could stack four players in a holding pattern. One of my jobs was to be dressed as a businessman hanging out on the street with a mobile phone, observing all four phone booths and logging who went into which one at which time and liaising with the limo driver as to who would be picked up next. Inevitably at very busy times, someone was going to be waiting in a phone box for four or five minutes, sometimes as long as eight minutes, completely isolated, wondering what the hell was going to happen next.

So the strategy we put in place was that I rang those people asked them about their experience, asked them about the postcard that they had just retrieved from Uncle Roy's office, what they had written on the card and invited them to talk about the line of someone changing from being a stranger to being something else. The question on the card was "When can you begin to trust a stranger?" Some of those phone calls were very powerful, I felt. I really enjoyed them and anecdotally a couple of people said "For me that was the best bit" And it was only ever a structural bit just to try and kill time...So when we came to *I like Frank*, we really put that as the apex of the experience and we wanted to use the video call on a 3G phone. So briefly, the culmination for a street player who made it to that stage of I Like Frank was that they received a video call on their phone. When they saw the image come up on the phone they realised it was a shot of them and one of us was across the street filming them on another video phone and they were invited to follow the person who was filming them—when they spotted them. That person then walked into the atrium of a large mirrored office complex, with a sunken tropical garden with palms and a couple of benches for office workers to eat their lunch. In fact the entire complex was almost empty, so this was a deserted area. They would run across the street and follow this person they glimpsed and were invited to take a seat and then a dialogue took place between the performer who was out of sight and this person sat alone on the bench about a whole series of questions. They depended on what was appropriate for the individual player, but they were all dealing with the issue of people who you have left in your past, or lost track of or lost communication with. Some people would find it very overwhelming to talk about someone they think about, but with whom they have lost touch.

So it's really looking at the flip side of the promise of all these communications technologies. We are all in touch everywhere and anyone you want to call you can. What we are interested in is looking at the 'impotence' that arises from some of these new technologies. If you have a technology that allows you to ring anyone at any time but feel unable to for some reason, then your level of impotence is only enhanced. Its a double blow to be sitting looking at a mobile phone with a hundred numbers stored in it and you don't feel you can make the call, so that's really one of the critical underpinnings.

MR That's very interesting, because I suppose there is this urban loneliness that is the concomitant of all these extra technologies and in a sense when you use a mobile, people go out with this bubble around them, they are in a different world. They are cocooned really and in terms of social space it's the very opposite of the agora and direct interaction with people. Do you think that the technologies are going to flip over now that they have this locative element and start bedding people back into real social space, rather than virtual space?

MA Yeah, I think that Sadie Plant makes the point very astutely that mobile phones are about the privatisation of the phone book. There used to be a book where you could find anyone's number. Now your phonebook is entirely private to you and if you lose your phone it is lost and no one else knows your phone book and whom you are talking to at all. So that in itself reflects a very important shift. If you lose a number, you have really lost it. It's a weird token thing to give out your mobile phone number. It has a charge of intimacy and social relationship in its own right.

In terms of how locative media might work, I think it's too soon to say. I think the opportunity to embed things in the local geography is there and I really aspire to that. We've talked a lot about ideas for work that is about your street-even if you just had a virtual model of your street and you could annotate your flat that says 'Martin lives here, I'm a lecturer and I'm interested in mobile technologies'. Someone in your street could go "Oh my God there's a lecturer in mobile technologies lives ten doors down the road." Even that simple form of ambient awareness of who your neighbours are, that kind of thing would be fantastic, so I do think there is a possibility there. The question is 'can money be made from that?' (of which I am not sure) or 'can we make sufficient social capital out of an idea like that that?' means there is an interest in the technology providers delivering it. That is the critical difference between the Internet and Mobile phones—that the level of control and ownership of the mobile world is absolute. We have oligarchs running the mobile phone world and so our interest in where mobile technologies are going is that we are at this point where this language is quite unformed, and there might be an opportunity for a number of Davids to take on Goliath. Let's face it—the Internet is only the way it is because of the Tim Berners-Lees and the various visionaries who were willing to work for free, who were willing to work for the good of the public domain. I wouldn't in any way compare ourselves to that, but we are inspired by the notion that we might possibly be able to contribute to the kinds of languages and kinds of discussions that are evolving in mobile phone companies about what values and ethos they might implement.

MR This whole thing about urban space and the phone companies and "Digital Commons." Are we going to enter an age of total enclosure as happened to the real commons in Britain or is it in the interests of these companies to have open agendas? I expect in the long term it is. It would be a much richer environment, but they may never be persuaded.

MA The irony is that the opportunity now for one of them to break ranks is colossal. If one phone company went "No we're going to go with open standards. We're going to allow anyone to develop for our phones. We're going to throw the doors open." The critical mass that any one of those companies could achieve by doing that I think would be phenomenal. The Internet proves so cogently the value of open systems.

MR Even just allowing sources out. I mean the Windows /Apple thing. At one time in the early Eighties they had equal share of the PC market. But because Apple wouldn't release code and Windows would, they own three percent now of the PC market. I think you could wave that at some of the companies and say this is the sensible way to go.

MA But that is not the same as saying that an organisation of 20,000 people has any way to respond. And if you look at the UK 3G market about £20 Billion for the licenses and about the same again for the implementation, I can understand why they are pretty preoccupied with how they can generate revenue at all costs! I can understand how if you are running a company of that scale in a market of this speed, where the average life of a model of a phone is six months. Even if you get it three months before launch-and you can't get it any earlier-nine months later that model is dead in the water, so you are trying to deal with this incredibly feverish product cycle. Now some manufacturers I would argue are their own worst enemies.

MR After all they developed that strategy

MA That's their model

MR That's not why we buy fixed phones is it-it's about utility?

MA But inevitably that will wind down. With the first few computers I bought, every time I got a new one it would enable me to do something at a quantum leap difference to that I had ever done before. I can remember getting an AV Mac you could put video on-and it was unbelievable. But with phones they are going to run out of innovation-pictures, video, internet connect, that's going to bottom out so I think we are in an interesting moment, where they will have to think more carefully (apart from just getting us to buy more handsets) about the ways to do it. But they have carefully structured the market so that the cost of buying a new handset is mostly hidden. If you are on a decent contract they send you a new handset just for the fun of it. So that is very much in their interest. H, how long they can sustain that, I am not sure...

MR So, coming onto your relationship with the Mixed Reality Lab in Nottingham, you indicated earlier that was a seminal relationship for you in terms of finally getting the technical support you needed. How is that relationship continuing to develop and can you try and quantify the aspects of it that have been the most successful?

MA It has been the single most significant development in our work since we met Steve Benford of the Mixed Reality Lab in 1997. On a whole number of different levels, certainly the first couple of years were we as artists trying to understand the scientific method, the agenda, the kind of rubric under which scientists operated. I think that that was reciprocal. For the Mixed Reality Lab as well there was a huge learning curve. They had worked with other artists before us, and continued to work with other artists subsequently, and I think that question of scientists understanding what artists do and how that might be applicable to research is a steep learning curve. There were certainly periods in that early phase where there was some degree of tension around those differences; around the incredible power relationship that exist between a University department with thirty researchers all on salaries, in a lab that has millions of pounds of equipment and is pulling in six or seven Figure sums every year, and an artists group, where at the time we met them our annual funding was twenty five thousand pounds a year for the four of us to live on. Even getting hold of a PC was still a major challenge for us.

So you had this incredible power imbalance that is really problematic and is surprisingly rarely remarked on. Some people feel it is bad form to mention this, but it is integral to the relationship between scientists and artists. In general artists are approaching scientists asking to get access to some form of knowledge or technique. , I think that is changing and that is a very welcome development. The Arts and Humanities Research Board, for example, is really significant in that respect, and they have funded us several times in our collaboration with the Mixed Reality Lab to sustain and deepen the collaboration.

In terms of what we get from it—we have found that by having a very close relationship with a people with a very profound understanding of programming and design and development issues of complex code, as well as a very broad interest in the social relationships that are created by these new technologies, has been tremendously enriching for us. The real mark of affinity that we have with the MRL is that they believe that if you are engaged in research in the social sciences, it only has meaning in a social context and that therefore to test any form of work in the lab is a relatively unrewarding experience. You must take it out into the public domain. Therefore the issue of the content of the work and what the work says and is about is critically important. That's what I think they feel we are able to bring—to be thinking not just about, say, the act of linking an online world with the streets of the city, and thinking about how to do that—but thinking about what the meanings and resonances are of that, what kinds of questions it raises, and what that might mean to people doing it, rather than just saying I can find my nearest post-box with this device or certain low order things.

To my mind, it is the longest lasting and deepest collaboration between artists and scientists anywhere. This is the eighth year we have worked together pretty much on a weekly if not often daily basis. We have made *Desert Rain, Can You See Me Now, Uncle Roy* and *I Like Frank* together and we are now working on three projects together in the next year. It's been a tremendous privilege to be able to do that. Its because Steve Benford and several of the key staff at the MRL: Ian Taylor originally, Martin Flintham, Dr Adam Drozd more recently, they are all very culturally engaged, fascinated by games, by music, by film, by artistic process. It matters to them very much that people have a good experience, and they really appreciate, and are excited by the act of implementing these works in the real world. That's a tremendously exciting thing for us as live practitioners, that's always been our impetus as well. So that provides a very solid foundation for us to work together.

Bibliography

Salem, Katy, and Eric Zimmerman. *The Rules of Play.* Cambridge: MIT Press, 2003.

Mixed Reality Lab

Steve Benford

(Interview with Martin Rieser Nottingham University 15th March 2005)

MR Steve—the first question is really to go back to the beginnings with Blast Theory, and particularly to ask about how those came about, and then I would like to look at how your concerns with collaborative working and public space merged together and what that meant in terms of evaluating audience behaviours and interactions—so it's a long question, I am afraid!

SB—Ok, well I have been interested in developing new technologies to support collaboration online and new kind of social interaction, I guess really since the late 1980s. Initially that focussed a lot upon technologies at work and co-ordination of office and meetings and all those things, but in the early 1990s I got very interested in collaborative virtual environments, on-line, virtual worlds and new technologies and their possible applications. That led me more and more to thinking about non-work—entertainment, artistic, social 'hanging-out' types of new applications.

In 1996 we sponsored the Now Festival. We staged a poetry slam on-line using VR, through collaboration with rap poets Chocolate Arts, and it was an absolutely eye-opening experience. It was really exciting to take this technology and put it together, and for the first time we created live performance on the web in real time. In many ways it was very exciting and also deeply technically led and the whole experience was phenomenal, chaotic: as could be expected, there was a huge element of chaos. So after that we embarked on a series of projects that were about working with television companies, such as John Wyver's Illuminations TV, to create more coherent forms of multi-user virtual world television titles.

We started to build the technologies to stage-manage live events, we saw that as being very important, and to nudge people into certain positions at certain times and really create proper flow. The whole history of those experiments met, more or less, with success. We published a lot of interesting work, but I don't think that it was both scientifically and artistically really interesting. We never made that connection.
Then in 199, Andrew Chetty introduced us to Blast Theory. At the time I was very interested in the idea of projection screens that you could step thorough as a way of making a transition between being a spectator of something towards being a primary interactor. We'd had this idea for a bit, published papers about it, and come up with a lot of uses, split curtains, sliding partitions, but Andrew mentioned that

they were (as part of a residence at Nottingham Trent University) working with projecting images into water-spray, so I went round to look at that. As soon as I saw it I thought 'yes that's a traversable screen, that's what I'd like to work with', so we formulated this project called *Desert Rain* and it was funded through the European Research Project 'Arena'.

It was a long development process—18 months (completely changing the concept) with long residencies, particularly one at the ZKM in Karlsruhe, but I think that was for 10 days, absolutely essential to get things together, and then *Desert Rain* premiered I guess in 1997 in Nottingham, and then it toured after that!

MR I took part in *Desert Rain* when it came to Bristol, and the thing that really struck me about it was how it really divided into 3 parts, and some were more successful than others. In the 'computer game' aspect of it, I think some people confused by, you seemed to end up entangled in a web. There was a conflict between letting people down and rescuing them, which was okay, but it didn't really move me. What really moved me was the moment when, down the beam of the projector, a real human being appeared through the curtain of rain, and I thought that was one of the most seminal moments I've ever had from a New Media work—it was your initial idea made flesh. How did you feel about the piece? I mean what kind of feedback did you get from the people who took part in it?

SB—Um, so how did I feel about it? I think on the whole it was a quite successful piece; it certainly became refined by that stage. I thought that it was a kind of the key-point, that act of crossing the curtain. I think that the ritualistic briefings you went through and the deliberate confusion that you were often placed in, in the virtual world, were part of preparing you for that moment. They played an important part in psychologically responding to that, in that you are taken and put in a certain state of mind, one that was pretty anxious. I think that that was pretty important. One of the things I learnt as a computer scientist was the reason that that moment works is that the technology is embedded in a whole process, in a whole physical set, and a whole series of rituals that combine to make that moment work.

Now, someone from Performing Arts would say 'yes, yes—really obvious', but to a computer scientist who's used to the 'plug and play' model of technology—sort of 'off you go'—that is quite a revelation. The way we studied that was through a team of ethnographers working with Christian Heath, King's College, London, who observed audiences. They particularly looked at behind-the-scene activities, as much if not more than the actual players' activities. What they wrote about was the way in which Blast Theory would orchestrate those moments. Behind the scenes they would monitor what was going on with a whole bank of computer screens. Also, through using the asymmetric properties of the rain-curtain. One of the most beautiful properties is that when you're standing on the projector side of it, it's pretty much transparent. So you can actually look through it without being seen, and you can observe people's reactions and time your intervention really nicely.

MR—So—a bit like a two-way mirror?

SB—Yeah, has that two-way mirror effect. And that was kind of really important, and basically, the study of *Desert Rain* documented that process of orchestration, and I guess it shapes a lot of our agenda. So, the problematic moments of *Desert Rain*: for me, the biggest problem, which got partially resolved, was always the moment after the rain-curtain. This release of tension as people got together and start talking before they were led on to what was meant to be the final payoff where you swipe your card and it is explained who your targets had been. In the early performances was definitely quite a problem. I think the experience was reengineered, and redesigned to largely deal with that, but I think always that was a difficult moment. It wasn't the moment before that caused the most confusion, it was the moment afterwards where somehow the tension would disappear, and there was this risk that people would start talking about—whatever—the weather or "how did you find that?" It would suddenly revert to this mode of being outside the experience.

MR—You lose the diegesis, and then you're in another place, and I did feel that when we went to the kind of *Desert Rain* ''anteroom'', that it was a bit of an anticlimax by than point.

SB—I think that was a problem, and I think it was certainly problem, it possibly always remained one, but it was certainly a problem in the early Nottingham version, and I think Bristol was the second *Desert Rain*. I think it was still very much a problem, and I think it was kind of recognised as much as possible, worked around and we would separate people much more and make sure that there were moments for the release of tension afterwards.

MR—That's pretty interesting.

SB—But I guess how we leant that (I won't ramble on too much) but it wasn't so much direct feedback from the participants. It wasn't really a process of questionnaires, surveys etc. Obviously we heard what they said afterwards, we did get critical reviews, we'd get e-mails, but it was really a process of observations. A kind of ethnography, and then discussion of that with the Artists about what was going on. What did they feel the issues were? That was where the issues came out.

MR—Right. So, in terms of how that fed into the next project you did with Blast Theory: could you just explain a bit about how it was conceived, and how what you had learnt was reapplied?

SB—So, we really tried to buy *Desert Rain*, we thought first-off, we had created a piece of work that had both genuine artistic payoff, and scientific payoff, so we were very excited, and then the Equator project was funded around that time. That was a 6-year programme, eight universities, a big broad agenda to look at how we could interweave physical and digital worlds for 'everyday life', which involved 'non-work' stuff. You know, not the normal agenda with design—moving away from the workplace towards other things. And within that we knew we wanted to take a road continuing this kind of performance work and breaking into work with Blast Theory as part of that. So *Desert Rain* was really a project where we, I suppose, started things rolling and we wanted to do another one. Broadly speaking to start mixing the physical and the digital.

We were very interested in mobility and location-based stuff, making it into something. And then there were a series of exploratory workshops where we tried out really rough and ready things, with GPS and PDA's, basic experiences in the city centre around Nottingham. We projected stuff onto surfaces and other ideas, and out of that process eventually then came the concept of *Can You See Me Now*, which was Blast Theory's formulation. So yes, the process was a little different there, I would characterise it as having a bit of a 'technology' push. We said 'we'd like to do something in this area, we messed around with technologies in a very practical way for a bit, with no particular agenda, and then there was a moment where it kind of crystallised into Blast Theory and we thought 'we'd like to do this'. It was all done in a great rush, shooting with other artists in Sheffield in December 2001.

When the idea came out eventually the whole thing took less than three months, possibly two months window to make the first version work. It was very tight, and it really was a 'seat of your pants' job, rough and ready, but it had some real energy about it.

MR—Where did it first take place, in London?

SB—In Sheffield, as part of a guest event call 'Shooting Live Artists'. I can't remember if that was the name of the whole festival or not.

MR—And everything worked?

SB—Everything—no, it was a real 'seat of the pants' job, and certainly I don't think the concept worked as seamlessly as you might have hoped. We had this idea that Blast Theory would seamlessly run through the streets and touch people and so on. It turned out in practice, that GPS has all sorts of inherent limitations, so in that sense technically there were lots of problems, but it worked in the sense it had real energy, and when it worked people really responded to it very positively. There were lots of times, which were problematic, but there were moments when it worked and the response from people over e-mail and what we saw was very, very powerful, there was real raw energy there.

MR—So they felt that they had been transplanted from a computer game into the real world, or was it a different kind of feeling they got?

SB—I think—what's interesting about *Can You See Me Now* is a very simple framework that leaves space for quite a range of experiences. So those are some of the ways in which it works—not for everybody—but for some people. Some people seem to tune into the fact that their actions in this almost empty on-line virtual world, actually have a real effect on the streets. If they cross over a line on a map where some poor soul is actually running through the traffic, they may get knocked over. If they run along this virtual line—someone's having to go up a hill, or whatever, then some people understood that and then the game is really close to the edge—you realise you've essentially...

MR—...finally got gravity?

SB—Yeah, you're actually having a real effect and if you could only tune into what the remote person's experience of the city is like, then you could manipulate that in some way. I think for other people, just the being chased is a really powerful feeling.

MR—Yes, very primitive.

SB—Hearing your name said, realising they're onto you gave a real buzz, and I think there were moments also when—I remember playing once with my wife and child on-line: I was Rotterdam and they were in Nottingham and they got dumped at one end of the games place and I at the other, and I was desperately keen to meet up with them, even though only virtually. It proved desperately hard because we got chased and that became a really big challenge for me. I think even simple things like that turn out to be quite powerful.

MR—One of the things that I read about was that people began to use the accidental things like the shadow areas of GPS to hide in. Can you say a bit more about that?

SB—Yeah, we kind of put that into context, again we had a team of ethnographers there capturing video material, some of players where there were public terminals, a fair amount of the runners on public streets and a fair bit of video from the control room and we also logged people's movements in the system and their text messages to try and get a sense of what they were saying to each other, how they were trying to play the game and how they co-ordinated their actions. And from that we observed a number of things. We observed that GPS and WiFi had a profound impact on the game—the inaccuracy of GPS, but it's also the limited coverage of GPS and WiFi. At times the runner is in the city and is just not able to take part in the game, he's just disappeared. not getting a GPS or WiFi fix. But then we observed that people had very different reactions to these things so the online players—very often the structure of the experience hid these things away from them. It's not like a sort of augmented reality experience when you see the physical and the virtual overlaid, and if they're out of line it's obvious.

In *Can you see me Now?* There is more of an adjacent reality structure—your inner virtual world is connected to something physical, but that connection's fuzzy, particularly the audio. So that meant that to some extent they were hidden, but at other times it is clear there is a problem, and as you said one or two players, particularly ones who are quite knowledgeable, would seek to exploit that. You would see in the text messages the odd times people would say if you hide in the shadow of the building there's going to be rubbish GPS they can't catch you, and equally we observed that the runners made use of the GPS—they would lure people into areas where they could make the best us of GPS to get a sudden hit or acceleration to catch them. So people did make use of it tactically, and that was really interesting for us. And the final thing to say about that, I know that's quite a long-winded ramble, the other thing we was that for the runners it was painfully apparent when the two were out of line—you were standing in one part of the city, and your map on your handheld was showing you in the next street, and there was no hiding that, and so for them it became a job of work to learn about how GPS performed, where it would be good, where it would be bad, where they should go to get re-connected, and this became a matter of that they shared their knowledge over the walkie-talkies and with the control-room, about the nature of the particular game-space. They would build this up over a couple of days' play and so by the end, they would have quite a good understanding of GPS and WiFi in relation to one particular place.

MR—So it was a kind of compensatory mental map they built up?

SB—Yeah, in some ways. And actually, after the project, we built a number of visualisations of GPS and WiFi across different game spaces in order to try and help that process in some ways. And there is also a temporal element when one of the interesting features of GPS is its performance in an urban space changes very quickly over time, as satellites move their positions in the skies, so where the shadows are—in five minutes a game could move from being playable to almost unplayable.

MR—Yes, and they had the same problem with *Riot* in Bristol. It worked at different times of the day, very unevenly, for the same reasons.

SB—During the Rotterdam Tour, I remember we had a member of the team who would look up likely satellite positions on the web, and there would be a briefing before each session, that said 'look, we're out there for two hours and at this time we think it's going to be awful', and later on Anthony Steven, UCL. built some visualisation tools that actually tried to predict likely shadows.

MR—Yes, those were shown at PLAN at the ICA.

SB—Yes, they were shown at PLAN.

MR—Excellent. So, moving on, what you took from that I presume went into *Uncle Roy All Round You*, which was the next piece?

SB—Yes, one more thing I've got to add about *Can You See Me Now?* and that is what's interesting about this kind of relationship for us is also there is a trajectory from research through to early performances, where you can really study and a performance get refined, and changed, through to stability through to touring. And touring is quite interesting because as a research lab it's increasingly difficult to justify your involvement in that process. You need to move on to other things and certainly you can't use a research funding for that. And one of the interesting things about *Can You See Me Now* is quite a bit of effort, into finally being able to train up a group of people to really deploy that without us being there.

MR—A travelling team?

SB—Yes, there is now kind of a travelling team of freelancers that Blast Theory can hire, they can run *Can You See Me Now?* in most places and there may occasionally be telephone support in particular fixing this and tackling local issues, but on the whole it's really interesting to see that we learnt a kind of migrating the project into one where we can actually get rid of the technology to some extent and say right its out there.

MR—Just a rider to that, because obviously it tours into different cultures, was there a different behavioural aspect to say in Japan compared to Britain?

SB—It's a bit hard for me to comment, because I didn't go to Tokyo, so I only got that second-hand.

MR—Did you have ethnographers over there, or any one else evaluating?

SB—We didn't, but in feedback, there was some differences, the most notable one seemed to be that the people of Tokyo would play the game on-line and then some of them would come down to the game-space, and run with the runners—I've no idea why, whether it was a particular feature of that game-space that it was easy to do, but I found that to be very interesting. I think we've seen a little bit of that in Barcelona, where we've had a group of kids that ran with runners, but I think it's the first time that people seem to feel quite regularly the urge to then go down and actually run through the streets with them, kind of really nice!

MR—Ok, well I'd like to move on to *Uncle Roy* then-what aspects of the previous project did you take and rework in *Uncle Roy*, what were you research objectives?

SB I think that there were two big things that we wanted to do-one was we wanted to put the public on the streets as well as online. I think *Can You See Me Now* works but there is a sense that the performers get as much, if not more of the fun than the online players. We wanted to get the public on the streets as well and that was a big challenge. Secondly, we did want a deeper experience—and the game we chose is very exciting and it does tap into things that engage—but we did want to embed more in a particular location and to draw on that location, something that was more contemplative, that took on a bigger issue. So *Uncle Roy* was addressed to do that, particularly the notion of trust in' the game, online strangers, bystanders, trust in whatever and it had this structure of following clues and undertaking a somewhat slower-paced journey through the city.

MR How did you evaluate feedback? Did you interview people or observe them?

SB I think the analysis of that was a bit more rounded. There were certainly things like exit questionnaires, from the early ones that got feedback and certainly there was more attempt to get people to e-mail us with their experiences. Some wrote us essays, so I think there was more of that seeking directly the audience's view. But at the same time we also took this observational perspective of seeing how the thing was working. That involved again following street players. We could only follow a few. Part of the problem was that the sheer act of wandering around videoing someone completely violates what *Uncle Roy* was all about, so we had a very limited opportunity to do that. Again there was a lot of videoing in the control room. The control structure for Uncle Roy was even more complex and intense that it had been for *Can You See Me Now*. We also recorded some online players as well and we captured system logs and we made more use of those-text logs and movement. Again there were two issues: the first cut we did was about positioning technologies. In *Uncle Roy* we ditched GPS and used the self-reporting of position and were very interested to see how people used that and how accurate it was.

MR Was it accurate?

SB Well it is interesting, but it depends how you looked at it. If you took a boneheaded view: no. So one of the things we did for the twelve players we followed through the game, one of the things we looked at was we laboriously transcribed where they estimated themselves be according to where they said they were and where they physically were and set those measurements against each other. And said let's call the difference" error" and if you did something crude like that you found the average error is 20 metres which can go up to 50 or 60; quite a bit worse that GPS. But if you then look a peoples reactions to that they really didn't very often notice problems and that's because the positions they give are very plausible- in theory out by a long distance, but semantically absolutely believable they are not in a random location. In fact one of the things they turned out to be was dislocated in time not space. They tended to declare themselves in terms of where they were going-not unsurprisingly they wanted to get information about that, and sometimes where they had been. Sometimes people would pass by very close to where that position was-but 30 seconds later.

MR One of the problems I felt with the piece was what happens if you never find *Uncle Roy*? There is no payoff and you are just wandering the streets, gathering clues, but never orientated. Did you debrief people who had had that experience?

SB I don't think we lavished a lot of attention on debriefing that particular group of people.

MR Do you know what ratio of people succeeded in finding *Uncle Roy*?

SB I could lay my hands on the figures. But I would guess that around 10% didn't make it-the majority were cajoled through to that point. I think its a really interesting issue and one that arises not so much from the structure of *Uncle Roy* but one that arises out of a structure of mixing live performance and games within a formulated set of rules with an outcome-win or lose. And a performance experience is orchestrated with a payoff at the end and those two things sit together in *Uncle Roy* and that can cause a problem if you don't make it to the end you may feel you have not had the full experience.

MR I suppose losing at Poker brings the same disappointments?

SB If you read it as a game or performance you will be reacting differently. The whole issue about *Uncle Roy* is that it is a very subjective experience. I presume considerably more so than going to a conventional piece of theatre. It becomes even more subjective and several people indicted how they had wanted to talk to other people about it. I think one thing all these location-based experiences have in common is a point of reflection, a need to relive and discuss it with other people.

MR If we move on. This having one aim and the need to share with other people-did this feed into *I like Frank* where you had multiple destinations?

SB I think *I Like Frank* historically sat in the middle of *Uncle Roy*-it followed early *Uncle Roy* at ICA and preceded the two later versions. So it wasn't really an opportunity to revisit all the issues. I don't think we really got our hands on all the feedback issues, which came at the end. What we were more concerned with was this whole relationship between street and online which is difficult fragile territory. Obviously *Can You See Me Now* and *Uncle Roy* were concerned with what relationship you could establish between someone at street level and someone online. The answer is when that works it can be very powerful, but it is very hard to set up and very fragile. The technology is very fragile so even if you establish that relation it can be destroyed very quickly. It was rare in early *Uncle Roy* to get that kind of payoff where someone really felt they had managed to guide someone. I did it myself and it was very powerful when it worked, but it was extremely hard work. In *I Like Frank* we tried to make a tighter connection between the street and the online player. In particular we experimented with getting the online player to do more with the street player in the city. Retrieve a postcard from a location, whatever, for which there would be a further payoff at the end. So I think one of the main things in *I Like Frank* was trying to make those connects. It was one of the things that proved really difficult and I don't think we've yet quite got to grips with how that online/street relationship can be made to work.

MR When I spoke to Matt Adams about *I Like Frank* he said there was a cathartic moment when a player would sit on a bench and talk with a performer and there were some deep confessional moments, which were very different from *Uncle Roy*. Did you observe those?

SB I don't think the performer sat on the bench with them. The way it worked-there wasn't the big office and limo as in Uncle Roy, the increasing crossing of boundaries: going to an unknown office (you shouldn't do that), getting into a strangers car (you certainly shouldn't do that)-*I Like Frank* took a different trajectory. You went to a fairly quiet place in the city to sit where you got a video call on a mobile phone from a performer who would ask a series of questions. I think those moments were extremely powerful and I think one of the things has come out of that and informed on going projects now is that one of the most useful things about a telephone is that you can make phone calls with them, it might surprise you to know! One shouldn't forget that in the hurry to put games and interfaces and maps on them. Ultimately the ability to speak to someone, for a performer to speak to someone is incredibly powerful.

MR And the nuance of the human voice...

SB And the ability to call live to someone. I think that was one of the things, which were very successful in *I Like Frank*, and in some ways was something we had to adapt to because of the lack of resource to do anything else and that has lived on in the current project. It has very much got the notion of the telephone as a voice device at its heart.

MR Let's move on to the new project about which you talked obliquely, which I assume uses 3G properties in more integrated ways, or is that a misreading?

SB It's not clear that it has to be 3G and the stage that the new project is at, very much like the early stages of *Can You See Me Now* is 'messing around' with technologies and it is very likely that what eventually emerges will ditch all of the ideas which we have tried so far. It is very hard to talk about it in a concrete way because pretty much anything goes and anything can be thrown out. We are very concerned with the issue of scale in terms of the number of participants. One of the big bottlenecks with the pieces we have done so far has been is it takes an awful lot of performer and orchestration effort to deliver an experience to a relatively few people, and to some extent geographical and temporal scale are also an issue. We have been restricted to a half-mile square of the city or an hour's experience all feel like boundaries to be pushed. So we are looking to build an experience with more people and longer time scale, possibly a larger area and by necessity will need to involve your own phone. That will be a challenge in lots of ways, but I won't guarantee we will do that, but to make the scale change you couldn't give devices out to people they would have to use their own stuff.

MR When I talked to Matt Adams he mentioned how difficult it was when you did *I Like Frank* how hard it was to get into the mobile software, which was all proprietary and hidden away and not amenable to some of the things you wanted to implement. Have you found a way to get round that yet?

SB It was a real problem with *I Like Frank* and I think part of it was the particular combination of phone and network. A completely new 3G rollout in Adelaide, there were only a few of the handsets with 3G and we didn't get one until late in the process and so on... blah blah blah. That made it more difficult than normal, but there still is a bit of an issue about phones really being able to access their functions in an integrated way. So it is still very difficult, as far as I know, to send data while making a voice call. To have these things seamlessly managed. But I think the frightening issue is actually much more about how non-standard things appear to be.

MR Yes I think you have put your finger on it... all these companies in competition, not prepared to share.

SB I think that is the problem if you want people to use their own phones-which phones and exactly how much effort are you prepared to put into porting things and possible is that when you develop projects up to the 24th hour.

MR Do you think that is going to shake out, that the companies will define standards?

SB I don't know. I don't want to predict what they will do.

MR Because it is like the early days of personal computing where they all did their own thing and were in cutthroat competition. In the long term it's not the way to do it.

SB It may shake out if there is enough pressure, but it isn't something we could rely on in the next year.

MR And is the informing idea for the new project quite different to the old ones? Or does it remain with the game format.

SB Not clear yet-sorry! I think it will have some game relation, but it is not clear what that might be at the moment.

MR I think that covers the trajectory pretty clearly from my point of view. Excellent, well—thanks a lot Steve.

The Politics of Mobility
Drew Hemment

MR In your writing and the *Loca* Project, you have focused on the politics of the new mobility. Why was this?

DH In 2001 I curated an exhibition on surveillance called *Broken Channel*, and a lot of the interests, curiosities, concerns and contacts have remained with me since then. In 2003 I had the opportunity to undertake some research and map out a project in this area, and because mobile media threw up so many new questions about issues of surveillance and control I decided to make the focus mobile media and surveillance. At this time I was looking at the Celldar system, which is similar to radar, but which uses mobile phone emissions and infrastructure, as well as more every day phenomena such as GSM tracking.

MR What was your entry point into "Locative Media"?

DH I first met the locative crew—Ben Russell, Marc Tuters et al—in the flesh at *next5minutes* in Amsterdam during 2003. In many ways for me at this time, locative media was defined as much by the characters that populated it as by anything else. I was doing this research on mobile media and surveillance, and encountered a set of people who were looking at the same set of technologies, and largely the same set of applications of those technologies, but accentuating the opposite: the empowering or creative possibilities. On the occasion I first met them, I was given the last slot in the collaborative cartography session, and after what I had heard decided to give my presentation the title 'Anti-mapping' (I am not so sure anyone else got the joke).

What inspired me was that here was a group of people who were engaging in the social possibilities of the technology in a far more imaginative way, without at that time considering the more troubling, less salubrious side. To this day it is the ambiguity between all the exciting creative and social potential and the sinister possibilities that are present <u>at the same time</u> that has kept me interested in this field. The term locative media is a new term, and at times can be quite hotly contested, in ways that are not always helpful. I try to interpret it loosely, and to distinguish locative media as a set of technological devices or artifacts from the technological, artistic and social field or movement. Understanding the term in a way that is inclusive rather than exclusive does at times risk a lack of distinction from other mediated engagements in spatiality, but better that than pinning the field down too soon.

MR What has informed your curatorial approach to Locative Media?

DH For many years I have been interested in spatiality, perception and the body. This comes from my past experience in sound and music. I first became involved in music as a DJ and promoter of acid house events in the late 80s, and the participatory, sensory spaces of dance culture have stayed with me as a metaphor when I think about art, installation, interaction etc. In 1999/2000 I did a project called SenseSonic looking at sound and space, which had an online component featuring the likes of David Toop, Maryanne Amacher, Rolf Gehlhaar, Kaffe Matthews and Ansuman Biswas, plus an 'ambisonic club tour' featuring Cristian Vogel.

So with this background my first thought of locative media was what does it feel like, and of course the answer is, pretty cold and distant. But at the same time it offered a space where these old interests of mine could come alive again. Much of the excitement surrounding locative media has come from thinking through the possibilities of connecting locative, networked devices to social software environments, or of the social impact of user-led, collaborative mapping.

Not so much work has been done on tactility, the haptic, or emotive dimensions—in the way that *Crank the Web* (2001) by Jonah Brucker-Cohen or *Mobile Feelings* (2002-03/2004) by Christa Sommerer & Laurent Mignonneau have done for the Web. This is why projects such as *Sound Mapping* (1998) by Iain Mott, or *Biomapping* (2004) by Christian Nold (which emphasise the body within or over-and-above visual representation) interest me (notwithstanding the fact they both have 'mapping' in the title!). Likewise, it was for this reason that in the Mobile Connections exhibition at Futuresonic in 2004, projects that are not thought of as locative media, or at least not in a narrow definition of the term, such as *Oscillating Windows* (2003) by Katherine Moriwaki and *Sonic Interface* (1999) by Akitsugu Maebayashi were placed alongside projects that are.

MR How do the Artistic and the Political connect here?

DH There is a very immediate answer to this question, then, of course, there are bigger issues about art, politics and activism that could fill many books. The direct answer is that because this field parallels and intersects the world of surveillance and control in so many ways, it becomes very hard to claim a neutral or disengaged position. And, conversely, for this reason there is an ambiguity here, which art is in some ways best equipped to encounter.

Quite often there is an assumption, not just in locative media, that the act of appropriation is enough—in other words, that taking technology developed by the military and doing something different or creative with it is inherently subversive. However, to my mind it is not clear that this can be taken for granted. Firstly, cultural practitioners can often prepare the ground for the more widespread introduction of something that would previously have been considered objectionable. And, secondly, locative media projects themselves—in their (obsessive or inadvertent) documentation of the world—contribute to a process whereby the world is progressively analysed and mapped. A fundamental distinct lies in the extent which knowledge is open, transparent and shared. And yet nevertheless there are no clean lines between grass-roots projects, and Total Information Awareness.

Clearly whenever you have ambiguity like this, simple messages cannot to justice to the complexity involved. If you are too forthright with your criticisms then people will ignore you, because they can see that there is a positive side. And if you only focus on the positive side (when developing technology, making art, or creating social projects) then you are exposing yourself to the risk that your project will have unintended consequences that you would detest.

Art here can play a role in exposing this ambiguity without attempting to close it down with simplistic statements. In this sense the art is playing a utilitarian role, in the service of a social issue. Conversely, the political ambiguity offers an opportunity to make some very interesting art.

For many people this would raise danger signals, both because art "in the service of" anything is problematic for many people, not least because this often leads to bad art, and because political action often requires simple statements that people can mobilise around, not cosy meditations on ambiguity.

For me these twin dangers, which are very real, are the motivation that keeps me on my toes.

MR The role of the artist seems to be on the one hand pointing up the dangers and contradictions of the medium in terms of civil freedoms and on the other acting as unwitting ambassador for the uncritical penetration of the media as a "cool" artefact into consumer consciousness (something levelled at Blast Theory in the past)

DH Yes, and of course this can be said of projects that highlight issues of control and surveillance just as much as it can of projects, which ignore these, issues altogether. It is so easy for the culture industry to eat up any critical distance and turn it into edgy 'cool', which can ultimately neutralise the very point the artist is trying to make.

This is actually an issue that is very live in the *Loca* project. We are attempting to flag up the way that new forms of pervasive surveillance operate, and ways in which they can be said to be peer-to-peer or synaptic. In a way we are only showing what is going on already, rather than inventing new surveillance techniques or environments. But the danger is that by causing a small irritation or thrill, the project introduces a form of control to people, and helps them accept it, before they see the more troubling cases that could really touch their lives (e.g. a friend or relative arrested at a protest and convicted under anti terrorism legislation through the use of location data, migrants blah, intrusive marketing or gonzo marketing. The other danger is that it will just seem like gonzo marketing.

Ultimately I do not think that there are any easy answers for how to avoid this. There has been a long tradition of attempting to produce art that cannot be incorporated, but the ability of contemporary culture to eat everything in its path has always won out of the artists' attempts to evade it. Only on a few occasions… Moira Hindley, can the shock be said to outweigh the ... value... and pushed to that extreme the danger is of course that nothing other than the shock itself remains.

The approach with *Loca* is very different. We do want to give people pause for thought, but we are not attempting to shock, and are careful not to suggest that the phenomena we expose are all negative. There is an element which is simply about raising awareness and debate, but that is probably not sufficient on its own. Ultimately the aspiration is that the project might have an impact in another way, either by contributing to discussions of policy and legal safeguards, or, perhaps more importantly, but also more ambitiously, by influencing the emergence of the protocols of the future. With Bluetooth, for example, it is bad for privacy management because it supports anonymous scanning. This is not the result of a conspiracy, but because its current use was not foreseen when it was developed, and so safeguards were not built in. Once a platform is commercially deployed it is not economically feasible to withdraw it, and it is futile to attempt to argue this should be done. What can be done though, is to attempt to make an economic argument that companies should invest in privacy preserving measures. This economic argument can only be made if there is a demand, and there is more likely to be a demand if it is possible to show people the double edged sword of today's network technologies—and this has to be an argument that takes place outside the comfort zone of art galleries.

MR I am interested in pushing the relation between mapping and power a little further, after all maps were the primary tool of both the military (ordnance survey) and the colonial. Possession of knowledge for exploitation of resources is part of the process of ownership and dispossession of the indigenous. The history of the Americas and Africa has much to do with this process. When it also extends into the personal realm, as with locative media, are we about to see a different kind of dispossession, which goes beyond state knowledge and control?

DH There is a broader relationship between knowledge and power, and when this knowledge becomes spatial and involves the ability to locate individuals (or the devices they carry) in real time this does raise many questions. The classic model of control in modern, industrialised societies has been the Panopticon, which Foucault wrote about extensively. Already here there was an internalisation of control, where we start to police ourselves. More recently Deleuze has argued that control has shifted further from traditional instruments of State power, and is now an internal mechanism of consumer society. The choice of the word dispossession is an interesting one, as the argument would run that, on one level, the control increases as we access more services: dispossessed in the midst of all our possessions.

Quite often there is an assumption, not just in locative media, that the act of appropriation is enough—in other words, that taking technology developed by the military and doing something different or creative with it is inherently subversive. However, to my mind it is not clear that this can be taken for granted. Firstly, cultural practitioners can often prepare the ground for the more widespread introduction of something that would previously have been considered objectionable. And, secondly, locative media projects themselves—in their (obsessive or inadvertent) documentation of the world—contribute to a process whereby the world is progressively analysed and mapped. One distinction lies in the extent to which knowledge is open, transparent and shared. A key question therefore becomes whether this is enough to put air between the grass-roots projects and Total Information Awareness, or whether these exist on a continuum where many of the aspects that can be found troubling remain as a constant. If not then can we say

these grass roots projects are increasing the efficiency and resolution of the system rather than disturbing its operation?
In many ways we are seeing a fundamental shift, and a whole new environment we need to orient ourselves to. Central to this is the fact that we ourselves become the drivers of the surveillance society because it is a secondary effect of so many things we otherwise value. Equally we are marching us towards a situation in which more and more facets of our lives become in principle knowable, and yet practically unknown simply because there is too much information for it ever to be processed. The basis of the Panopticon was that we would not know when we were being observed or not, and so would come to act as if we were being observed at all times. With network technologies there are a whole new set of variables that govern this same scenario, plus some wholly new issues such as the fact that we now leave data trails behind us that will last potentially forever, meaning that we need not only to think of who or what is observing us in real time but also for at any point during the rest of our lives or even after our death.

MR How do you understand the politics of the new mobility?

DH Part of the problem today is that the only option for people concerned about some of the implications of new network technologies is to switch them off or never to start using them in the first place. The politics of the new mobility will be found somewhere in the space between switching on and switching off.

4.2 Wearables

Memory-Rich Garments and Social Interaction

Joey Berzowska

(Interview with Martin Rieser, May 2005, London)

MR Right Joey, we are going to try and explore this concept of the "Memory-rich garment" which is I think your current obsession and one of the subjects of your coming research. It's one thing to have memory and to use memory, but whose memories are being used? Are we really just talking about signalling between people in the way that cuttlefish signal their emotions through their surfaces? If that is the case, how do you see it being used effectively?

JB The starting point for this current fascination is the word "memory." It has always been used in many different ways, but is currently being used extensively in the media to refer to the many portable devices that are supposed to help us remember. With the proliferation of all these artefacts associated with the "memory industry," the meaning of the word "memory" has changed. Human memory is a very personal and interpretative thing. When computers were being developed and scientists were first working on them, they needed to find a name to refer to the storage component. They thought it was analogous to human memory, so they named it the same, but it's actually a very different thing, because computer memory is really only a repository-a box into which you throw bits and the bits stay there. Perhaps if you have a virus those bits will change over time or if your hard drive fails something may happen to those bits, but basically the bits just sit there. Whereas in human memory we accumulate experiences and we forget them very quickly or reinterpret them. We recreate what we think we remember based on our current experiences. Its a much more fluid, more personal and as I said, more interpretative process. So the starting point for this project was really frustration with how the computer-based meaning of the word "memory" is now being used to describe human memory. For example PDAs (Personal Digital Assistants) are referred to as "memory aids," or picture phones take photos that can be instantly uploaded into your blog or that can record everything that happens to you during the day as a means of storing memories. Those are very different processes. So OK whose memory?

MR The reason I am putting the question is because with fashion “wearables” they are signalling to other people. You wear them as an expression, but you are actually communicating to other people. So, if they are repositories of your memories, how does that square with the communication function?

JB Whose memory? I have always considered that it is the memory of the garments. Because **you** are wearing it, it has a very close relationship to **your** memory. But it is really the memory of the garment. For example if you take the shirt I am wearing right now, I spilt some coffee on it this morning. Sometimes I sweat or cry or wipe something on my sleeve, so my garment is building up its own memory of things that happened to it during the day. So I felt that the garment as this interface between the body and the outside world in many ways has a lot of potential to both record these things (that may seem too intimate to record) and then to maybe modulate them in some way. To display them very much as a statement about what we really want to remember. To record what we feel we want to show. What is public and what is private. A lot of these portable technologies are very invasive, especially the ones that record things, like cell phones with GPS. They always record where you are when you make a call. So a major aspect of this project is really art, trying to lift some of these issues to the surface.

MR So it’s the traces of life that these garments exhibit. In the past you have done projects where you used a sense of humour. They would show for example where you had been groped as a visual joke. Are you going to take that much further into something that’s essentially serious or is the fun aspect of what you are doing, a primary motivation?

JB Well on the one hand looking at fashion—a lot of it is not serious. It’s about play, showing things and hiding others. What I do is not exactly fashion, but being playful doesn’t necessarily mean it’s not also serious. Because fashion is a very serious thing-it makes a lot of money, but remains playful. Part of the reason I enjoy these more humorous applications is because i think that more serious ideas are not necessarily more meaningful.

MR So that’s a methodology for exploration with which you feel comfortable?

JB Because I am based in a Fine Arts Faculty I feel I have the freedom to explore all these ideas in more artistic ways.

MR To go back to what motivates you. A lot of your projects are very physical and tactile. Do you think that imperative is what underlies the way you interpret wearability? When Katherine Moriwaki interviewed you for Rhizome she quoted how you were motivated "by the poetic and the personal," but I also get the impression that what you are interested in is the tactile and whole physical body?

JB Yes that's very true and comes from my previous experience of working in this field, particularly working at MIT Media Lab. A lot of the scientists who work with the traditional definitions of wearable computing (as in wearing a computer on your body) are actually very afraid of the body. They don't like to admit that things, which are on our bodies, have a very intimate relationship to the body. A lot of wearable computing and a lot of the wireless aspects of wearability brings us closer in many ways. We can send e-mails, send images to each other and have conversations, but at the same time it allows us to be further apart physically. I am not making a value judgement, but it is one aspect of these technologies. So again in a playful fashion I want to bring back the idea of physical interaction. To create projects where people are compelled to have physical contact, whether it's sexual (obviously all these ideas flirt with sexuality) or otherwise; but to really bring the idea of physical intimacy to the surface. There are also a lot of references to ideas of embodied or tacit knowledge and to disembodied knowledge and to gender theory, which come into this.

MR Could you give a couple of concrete examples of where that has been a central concern, where you have looked at those issues in depth?

JB I can think of two projects we have been doing in the last year with our research group. One of the projects we have been working with was called "Intimacy maps." It was just a graphical representation of different zones of the body that might be considered more or less intimate, and we have been using this fantastic textile printer that prints on a 3-4 foot wide roll of fabric. So we have just been printing dozens of dresses. We print the dress pattern with the design right on it, cut it out and stitch it together. We have been doing this very rapid prototyping of different interpretations of "intimacy zones." For example one of the series are called finger dresses because they have hand prints all over them, but they are on different parts of the body. So it's not only the fact that some are on the shoulders or the hips others on the belly or more on the chest area. It's not just the areas. Some are positioned in such a way that it looks like I was just putting my own hands on the dress. Others are positioned in such a way that it would have been impossible for me to place them in that direction. They obviously hint at having been touched by somebody else. So I am trying to map out this whole territory of what it means to be touched and then to wear that data. Then we have also been looking at the different ways of representing that data. With the handprints it's obviously a very direct representation, there is no metaphor there at all. We have also used scientific visualisations and heat maps and the idea of a road map where some areas are more densely populated and some less. We have been printing dozens of these dresses and having my research assistants wear them and I photograph them and we think about it (Laughs)

MR What technologies would you apply to these after the initial research? Thermographic or pressure sensitive fabrics? Where are you going with this?

JB Almost any technology can be used. Because this part of my research is about developing textile-based display technologies, I also want to spend a lot of time thinking about "once we have the technology, what are we going to do with it." So these sketches, these intimacy maps don't refer to a particular technology, they are more to see where the boundaries of representation might lie, in particular, with physical data.

MR That's very interesting because very few people in the industry are doing this primary research. One of the questions I wanted to ask is: since Fashion is about human behaviour (and a lot of fashion is about very simplistic behaviours), I think it would be very interesting to see if the new technologies alter behaviour patterns and maybe reinforce more subtle kinds of interaction between people and whether that subtlety is on of your goals?

JB On the one hand portable technologies are changing behaviour. The way we arranged to meet for example. People don't do that anymore, naming a place and time and both making sure to be there at the right time. People don't do that at all since their ideas of time have become much more fluid. We both have mobiles and say see you later and then call each other. Our behaviours are definitely changing. I also feel that a lot of these technologies are developed in an environment that is far away from fashion. In fashion these questions of what this clothing does-to hide and reveal—these words come up over and over again and a lot of it symbolises power, status and sexuality. But these portable technologies are developed in a vacuum more in the consumer electronics industry. There is almost a bubble where those things don't matter (although having the biggest i-pod confers status) but I am really serious in emphasising that the body is a space for these intimate conversations and communications. If we put a message on the shoulder and then on the elbow then put the same message on the breast that is going to be a completely different message! So that is where the "intimacy maps" project came from. I also want to force people to consider how we are redefining intimacy and to bring back this idea of the body and locations on the body.

MR Moving on to the other of the two projects which you mentioned-that of shape memory, part of the research money allocation to XL Labs. What exactly are you doing in that complicated area of emergent technology?

JB The way that I think about "shape memory" is that it is just another kind of display technology like the other stuff we work on. Like colour change or integrated light-emitting components for textiles, treated in an efficient and "soft" way (where components illuminate or change colour). But with shape memory we use different kinds of shape memory materials, in particular Nitinol, which is a shape memory alloy. They eventually make something change shape which is another way of displaying information, whether its flat or bumpy (Laughs). So it's part of our display research, except it obviously has this whole other dimension.

MR My understanding of shape memory is that it is about materials that want to return to their original shape. How does it hold memory, isn't that more like a kind of erasure of deformation?

JB That's one application of Nitinol-to weave it into fabrics to make them wrinkle-free. The way we have been using it has been knitting and weaving and stitching Nitinol threads, and shape setting the wires to give them a particular shape. The way you create the shape is to bake it at a precise temperature for a specified time. It becomes a specific shape and then we can control the shape change using currents of very low voltage. We are doing a very different project because the technology has been a complete nightmare-we've had so many fires and short-circuits and things melting or malfunctioning. It is especially problematic if you are wearing it on the body.

MR Do you give it to a model or do you act as guinea pig?

JB We employ all our scientific ingenuity to make it as safe as possible. One of the things we are creating is a dress with a neckline, which has flowers on it. The flowers can open and close. When they close the petals move down so the neckline plunges. Hanna Soder is working on it and it's really so gorgeous. At the same time as developing these flowers we discovered this flower in Africa called "Caprice des dames" which translated means 'woman's whim' because this flower opens in the morning and closes at night-it only lives for one day and is very beautiful. So on the one hand there is this decorative element and on the other the closing flowers extended your neckline to reveal a little bit more of the body.

MR What control does the wearer have over that change?

JB If you look at our experiments previously such as in the "Pure play" garment, there is a very rich space for discourse there. Is it electronically controlled? Is it controlled through body data? The "Pure play" garment had these alternating elements of colour: one half controlled anatomically and the other half controlled by a switch. It raises the question of what is voluntary or involuntary control. So with the shape-change flower dress (it doesn't have a formal name yet) we are going to be looking at two systems: one will be having something you can control voluntarily with a sewn on switch and the other option will be to just have a programme on a chip that will almost be controlling it in a random pattern. So we are going to implement and evaluate both.

MR You said that fashion was as much about concealing as revealing. If it transparently reveals your true desires and emotions through say galvanic skin response, you might have real problems?

JB There is still something ambiguous there, because the conductivity of your skin can change in various ways if you are excited or just sweaty from running up the stairs.

MR The other question I wanted to address was that of multidisciplinary. How you make projects succeed where you are working in complex collaborations with designers, artists and technologists. What happens at XS Labs. Do you recruit for particular skill sets or do you have a network of people you use for particular bits?

JB We try to have a mix of both specialists and generalists. All the grants that I have received are government-based research grants that carry a very strong component of training new researchers. So I work with students, and most of them are undergraduate rather than graduate students. I have a couple of grad students. I definitely hire them from a wide variety of backgrounds so I have an electronic engineer, designers, programmers, textiles, students—from our programme in computational arts—and throw them all together.

I am very much guided by the personality of the student, so when I interview them and when I look at their work I am seeking a sense of humour and curiosity. Everyone who works at XS labs has to know a little bit about everybody else's field, so everybody takes a sewing workshop. I give an electronics workshop to everybody, because a lot of that knowledge can't be communicated from a book. It's what is called tacit knowledge: what happens if you stitch a conductive thread? If you a re working with wire you can imagine what will happen, but I you are working with a thread what happens when you stretch it?

Maybe the resistance will change. They learn the basics of everything. Then I think of myself as the person who makes the-I've described it a certain way in the past—where everyone is a specialist of a certain thing but for real innovation to happen those people who are specialists need to look at the question from a different perspective. So I think of myself as the person who kind of digs the trenches between the perspectives, like makes the water flow from one perspective to another.

MR An irrigator?

JB Yes and then of course I always have my master plan (laughs), which is whatever bizarre idea I have dreamt up between waking up in the morning and getting in the shower. Intimacy maps or memory rich garments.

MR These come at five in the morning?

JB So I throw them out to the students. It is very important to me that everyone has fun and is learning and I learn a lot from them and they learn a lot from me. But it is definitely the very fun atmosphere, which opens up a lot of avenues for discovery, as well as making mistakes.

MR That's very very different from the normal research lab atmosphere.

JB Yes

MR I tend to agree that play is an absolute motivator especially given the subject.

JB Exactly

MR The only other question I want to get to grips with is the way we are changing as a species through the use of technology. I mean if you look at say the Middle Ages, they had the art of rhetoric. People would be capable of learning a whole book off by heart using memory mapping or theatre of memory techniques. Are we moving into an era where people no longer remember, where they are so dependent on technology that memory is something out there rather than internal. If that is the case does that diminish us rather than increase our humanity?

JB It's an excellent question. Some people argue that other abilities might flourish while others might asphyxiate slowly. The other aspect of that question though is that it is not only the devices that we have that help us record and remember things that we no longer have to try to remember, but its also all this research in ubiquitous or pervasive computing where every object will remember or sense and be written to a hard drive somewhere, so its a very different way of thinking about the world, where we almost give up some of our agency to the inanimate objects around us.

MR Well that's the question-behind every inanimate object recording there's a human programmer who is deciding on priorities and I think we touched earlier on the double-edged nature of the technology. It faces both ways: it invades our privacy and improves our communication. But there is human agency behind the decision-making of these devices and that really is the question., because it is a political question. You are inventing these things. How much control will you have when these children of your mind go out into the world?

JB And who's going to control the children? Now that's very scary and very complex. There are two things: one is access and the other is control. Those are the two things that I find most worrisome about all of these technologies. Access because who will have access to it is a very political question. We will, because we are both academics working in this field. But will people with less money in our countries have access? What about those from other countries who might be perceived a some kind of political threat? Who will control these technologies, monitor them-and that will be different for different people?

MR It seems to me that all the technologies we have been talking about-GPS and wearables all derive originally from military research budgets. You are obviously riding a tiger because they come from that provenance. But on the other hand the internet developed from military research and is now more on the side of the angels as an open source area. Do you think we could move into that situation with these newer technologies?

JB I think it is very different because the Internet is more of an infrastructure whereas more of these technologies which monitor, sense, record and save are designed with a very particular purpose in mind. Sure there are people doing amazing artworks using these technologies. Social artwork using social commentary, but I believe that those technologies are more-you know people say technology is neutral, but then some technologies are less neutral than others. Like these ones are some of the least neutral technologies I can think of, because, though there might be great medical or safety applications, they are really predicated on monitoring and recording data about things that happen. The student I was talking to at Goldsmiths this morning wanted to have this camera in your eye. We were talking about this fictional corporation that would sell these. I asked her "How do you know they are going to sell it? Maybe their business model will be to rent these things so that they always retain ownership of these fictional products." We have no idea of what the business models will be, and if these things are only rented to us, the same as cable boxes in North America, anything that happens to be recorded by that device is actually property of the corporation or whoever rented it to you. So there are lots of problematic issues around any device with the ability to sense and record.

MR I think the optimistic point is that many of these debates are very current and we keep growing more aware of these issues. I think the essence of a democratic society is that you don't baulk at discussing these issues and actually build agendas around them. Thank you Joey.

Heart on Your Sleeve

Annie Lovejoy

(Martin Rieser Interview, Bristol, February 2005)

MR So are wearables visible aesthetics, visible data?

AL I think aesthetic visualisation of data has potential for wider shared experience. Locative media can move beyond cartographic documentary style narrative. My work is generated through interactions with sites or situations, technological tools are party to this, but not central. I'm currently attempting to articulate a non-media specific overview of this approach for a web space. *"Interfaces of location and memory"* is a helpful phrase: *interface*—as a point of connection between things, *location*—the site, situation, context or position of something and *memory*—as experiential, collective, associative, knowledge base, data and archiving.

MR Where shall we start—do you want to talk through chronologically or shall we talk about Millennium Square first?

AL OK, the Millennium Square project presented a chance to explore 'context-aware' technologies within a research environment at Bristol University's Mobile and Wearable computing group. The resulting 'prototype' offered an immersive experience: an audio led compositional journey dependent upon moving around the square. The project hurled me into a world of wearable technology issues, to do with encumberment ('bearables'), elitism and invasiveness.

MR Invasive in terms of the way it feels when you wear it or the way people react to someone wearing it?

AL More like the nature of GPS, the business of being locatable and gridlocked, although it's not always totally accurate. Mobile phone networks demonstrate this pervasiveness.

MR But are you talking about the political implications of tracking people? That's invasive.

AL Yes, and discovering the types of applications that the Wearables Group at Bristol was producing, such as "where is the next pub?" Or ideas like, if your friend you haven't seen for 20 years is around the corner you will know—and you can decide if you want to meet them. So my big question is: where does spontaneity go, where does surprise go, when everything is data driven?

MR It derives out of 'happenstance'?

AL We would miss so much in our lives if we had no serendipity.

MR Are you saying we all become Stepford Wives in our predictable suburbs?

AL Too much information and too cerebral, and the fact that it's embodied—you and the data. In Millennium Square the number of wearables available limited the experience of the audio. Issues of access and security were highlighted when suggestions of a deposit for use of the jackets arose, such as a credit card or passport.

MR So—it was very limiting in terms of public participation

AL Yes, an experience restricted to evidence of identity! For me, the project was quite a learning curve, I felt uncomfortable with these limitations. The kind of thing I was suggesting at the time, was to bring in a visual component to the jackets. The audioscape contained maverick moments such as a skateboarder crossing the square. I was interested in introducing a performative element, a skateboarder wearing a jacket programmed to display graphic LED representations, which changed (dependant on the location); to extend the experience beyond the plugged-in participant to other passers by.

MR That's interesting because in *Riot!* in Queen's Square and some of the other (Mobile Bristol) projects I felt that if there had been some kind of indicator on the ground, then spectators of those walking around with headphones would have had better engagement with the project

AL That's the gap, which led me to consider that the environment might be a better situation for responsive technologies than the encumbered body, freedom from clutter, not having to carry anything.

MR It's a richer experience?

AL The bus stop might give you poetry

MR Before we move away from Millennium Square, just give me some idea of how the sound bites were intended to work.

AL I can show you some documentation, this is the indoor ultrasound rig at HP Research Labs (also used by Liz Milner for *A Walk in the Woods* [1])—it's totally out of context, a diagrammatic exercise, and a technology demo.

MR One of the things we have discovered is that when you get an ambient sound which coincides with a sound within the piece you get a very strong moment of reinforcement or counterpoint and all the sonic accidents become quite profound when read from within the piece. An indoor installation leaves that out.

AL For *Millennium Square* [2] composer Roger Mills and I collaborated on a conceptual mapping of audio content based on the physical attributes of the site. For instance, in proximity to the large mirror ball structure of the Imaginarium the audio-led experience reflected multi faceted and multi ethnic aspects of the city, sourced from adhoc interviews conducted in the streets, cabs, shops etc.

MR And the transitions of each sound overlap with the next on their borders.

AL The sound was spatialised, so in the central area you would be led by audio streams enabling you to compose vocal and instrumental elements as you walked. This area of the square was based on a musical collaboration with singer Keti Nickoladze from Bristol's twin city of Tbilisi (Republic of Georgia).

MR Were all the cities represented in the work?

AL No, just Tbilisi, as a result of the *Here nor There* [3] project, an international collective of musicians and artist working across geographical and cultural boundaries.

MR So there is quite a wide fall off-it's on a large scale across, say fifty feet. Because in *Riot* if you moved out of one sound you couldn't go back, they didn't stay there.

---

[1] See <http://www.dshed.net/digest/04/content/week4/wired_woods.html> and <http://wearables.cs.bris.ac.uk/public/wawiwo.htm>
[2] See <http://www.herenorthere.org/msquare/index.htm>
[3] See <http://www.herenorthere.org/>

AL You could return to the sound, the spatialisation was great but there were loads of other glitches and cut offs. Apparently the compasses didn't work well in the square, due to the presence of the underground car park and magnetic field. But isn't it interesting that even when it's full of glitches people have this excitement, this urge for new and imagined applications. I found this very evident in the Mobile Bristol events which were incredibly useful for HP Invent 'end user' research but for the artist quite frustrating! I guess that's the same story for any industry led 'emergent' technology.

MR You struggle with the limitations. Let's now deal with wearables proper. I know that after *Pillow* , you worked with Cliff Randall and made a number of designs for garments to be used in different contexts.

AL The customized jackets I produced have been used by the Wearables Group for various applications.

MR They had no aesthetic?

AL Aesthetics hadn't come into the agenda, and wasn't an aspect of application interest or development. Their previous jacket applications were consumer-led: the tourist jacket, shopping jacket etc. Let me ask you something, when you refer to a "Mobile Audience," what do you mean?—it implies that technology is essential to experiential being; surely mobility is inherent to life, the audience of the everyday. Technology is just part of that.

MR Not necessarily in an artwork-I know we have had public art and people have been walking around sculptures for thousands of years, but I think when you add interaction of the kind we are familiar with from computer-based artwork, then you do have something new.

AL that depends on what you mean by "interaction"—referencing computers originally meant the ability to open another programme from within the one you were running. Now that makes sense, because the problem with binary architectures of information is that they invite designed interaction, which is very different from human understandings of the term "interaction."

MR But artists have to bridge between the two-that's the whole problematic.

AL We are interacting now in a far more sophisticated way, gesturally and face to face, than will be gleaned from the recording or transcript you end up with—that's interaction.

MR To get back to wearable's—and your position in the group?

AL Having produced *Pillow* (presented by HP Invent at the Tomorrow's World 2000), which used intricate embroidery to combine working and redundant technology, I was invited to work with the Wearables Group. I was keen to explore the merging of circuitry and fabric. The Wearables Group needed an artist (or maybe a seamstress!) to sort the ergonomics and invisibility of the hardware. Their previous jackets housed the tech in ripped linings stuck together with gaffer tape. The jackets held various components; CPU, GPS compass, aerial, accelerometers and heavy camcorder batteries. The lack of opportunity for aesthetic visualisation of data led me to deconstruct the hardware as far as I could without destroying the connections.

MR So you started exposing all the wiring connections?

AL I started to, but then held back as the project was slowing down due to the magnetic field problems. The ideas channeled into a more recent collaboration with Wearable Group technologist Cliff Randell.

MR A sort of Richard Rogers's approach where what's inside is made visible and external?

AL A making transparent of the technology.

MR So how did all this lead to "Your Heart on my Sleeve"?[4]

AL I had always been fascinated by various e-broidery projects, particularly the Sensatex vest that merged sensors and fabric—you know, the one, which tells you when you've been shot!! Cliff knew I was keen to experiment with electroluminescence and when the opportunity to show work at Siggraph arose, he invited my input.

---

[4] See <http://wearables.cs.bris.ac.uk/cyberfash.htm>

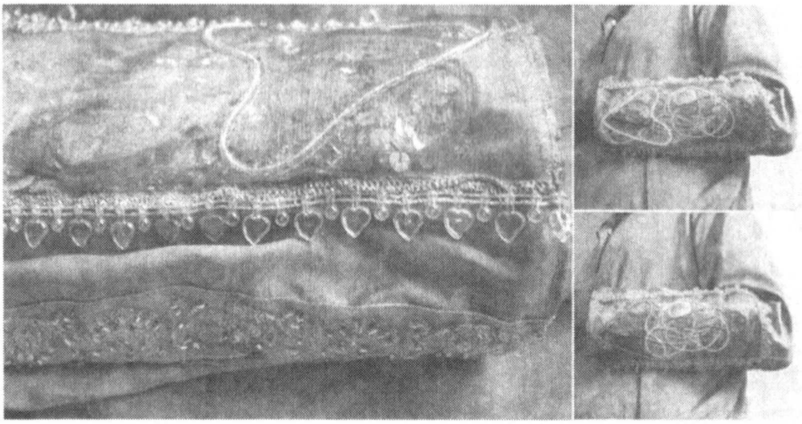

Figure 86: 'Your Heart on My Sleeve' by Annie Lovejoy and Cliff Randell ; presented at Siggraph 2004 Fashion Show Los Angeles, USA (Image: © Annie Lovejoy 2004)

MR Unfortunately so much of the technology is driven by military imperatives

AL But that's nothing new. Here's another jacket I made for a performative demo with Martin Dupras (University of the West of England). The sensors allow gestural input to generative sounds. Here, light is achieved with reflective bands.

MR So this brings us on to "Your Heart on my Sleeve" which appears to be really elaborate embroidery. The threads are illuminated-are they optic fibre?

AL It's electroluminescent wire, which looks like neon, but is flexible. It's a bit gimmicky so I've worked it down, made it subtler by combining it with various transparent and reflective materials.

MR So there is a pressure monitor on the chest that transfers the data through a computer to the lights—is that how it operates?

AL A Polar heart rate monitor wirelessly transmits data—the heartbeat, to another person wearing the sleeve, hence: *Your heart—on my sleeve.*

MR Ah, I had read it as the other way around

AL Yes, you may be confusing it with Cliff's "Heart on my Sleeve," which is a small screen displaying a pulsating heart animation. This one goes a little further… you are separated from your lover and but their presence is reflected on your clothing.
MR Laurie Anderson talks about the danger in her performances of the technology becoming a "Cool Tool" and that overriding the intentionality of the artwork. Isn't that the major problem-How do you stop the technology dominating the meaning of the artwork and what you want it to say to people.

AL Absolutely. You see this often with technology driven artwork, the seduction of the medium. I'm terrible when I go to an exhibition, I find myself figuring out the technology before considering the art. So there has to be a big leap for the work to become meaningful. We need to find ways to work with technology that are emotive and experiential. It's often too defined by the software or hardware, too cerebral, too gimmicky or too info based.

MR That brings us on to a whole other argument that are we now part of the mainstream and is technological art something that happens in the galleries anyway, so why are we bothering with these metaphors? We know the debates around these technologies are very important but are they exhausted? So we appear to be in a transitional period.

AL I think it's a tricky time for institutions, particularly galleries. What were once defined disciplines have become hybrid, trans-disciplinary, experimental and collaborative. There's a proliferation of networks that exist without needing to be building based, and which are not driven by a curatorial overview. 'Emergence' is a continuum, yet there's still this obsession with the 'new'—innovation, new and surprising ways, new contributions to knowledge, new audiences—buzzwords for funding criteria, in this sense the debates don't really change.

MR Well not since the 60s and EAT-we seem to be continually revisiting those debates.

AL Artists making experiential work in response to context, situation and social relations have always made use of existing and emerging systems. The pervasiveness of technologies within society is reflected in cultural practice—networked projects, time-based interventions, social interactions, performic events and architectural integration. Ubiquity is here to stay; these technologies exist, be they intimate, useful, dangerous, invasive or prone to glitches. They permeate the everyday and artists will, as they always have, find potentials for emotive communication.

Contributor Biographies

Adams, Matt

Co-founded Blast Theory in 1991. With a background in professional theatre since the early 1980s, Matt Adams has created dozens of live events for nightclubs, theatres and the street. He has taught and presented papers from Cairo to Riga, from Sydney to Toronto. He specialises in scriptwriting, graphic design and video editing.

Beloff, Laura

With an acclaimed international reputation as an artist, Laura Beloff's works can be described as peculiar wearable objects, programmed structures and participatory, networked installations. In her pieces she combines technology fluently with various media ranging from video to textiles, from sound to sculpture and organic materials. Many of her works deal with individuals in the global society, trying to adapt to highly complex technologically enhanced world, which is becoming increasingly mobile. Collaboration with other artists, musicians and computer scientists has been one of the features typical for her working methods. Beloff has exhibited widely in various museums, galleries and major media-art events in Europe and worldwide, recently in Vienna (2011), in Russia and Brazil (2008) and in the Venice Biennale (2007). She has received various grants, residencies and awards: 2009 Prix Ars Electronica Honorary mention in Next Idea-category, 2002 Vida 5.0 honorary mention/Spain, 2000 File-festival the 3rd prize/Brazil, etc. 2002-06 she was Professor for media arts at the Art Academy in Oslo, Norway. 2007-11 she was awarded a five-year grant by the Finnish state. 2009-2011 she has been a visiting Professor at The University of Applied Arts in Vienna (AT). Currently (2011) she is on the verge of finishing a PhD within Planetary Collegium, University of Plymouth.

More information on her works: http://www.realitydisfunction.org

Benford, Steve

Co-founded the Mixed Reality Lab in 1997. As a computer scientist, Professor Benford has received international acclaim for his pioneering work at conferences such as SIGGRAPH and CHI. He has made work for TV, for schools and for the heritage sector with partners such as British Telecom, Channel 4, Illuminations Television and the Royal College of Art.

Berzowska, Joey

Joanna Berzowska is an Assistant Professor of Design Art and Digital Image/Sound at Concordia University in Montreal. Her work and research deal primarily with "soft computation": electronic textiles, responsive clothing as wearable technology, reactive materials and squishy interfaces. She is the founder of XS design studio in Montreal. She is the founder and senior design advisor of International Fashion Machines in Boston, where she developed the first electronic ink wearable animated display and Electric Plaid, an addressable color-change textile. She received her Masters of Science from MIT for her work titled Computational Expressionism. She worked with the Tangible Media Group of the MIT Media Lab on research projects such as the musicBottles. She directed Interface Design at the Institute for Interactive Media at the University of Technology in Sydney. She holds a BA in Pure Mathematics and a BFA in Design Arts. Her art and design work has been shown in the Cooper-Hewitt Design Museum in NYC, SIGGRAPH, Art Directors Club in NYC, Australian Museum in Sydney, NTT ICC in Tokyo and Ars Electronica Center in Linz among others. She has lectured about the intersections of art, design, technology and computation at SIGGRAPH, Banff New Media Institute in Canada and Interaction Design Institute Ivrea in Italy among others. Website: http://www.berzowska.com

Bossoli, Arianna

Arianna Bossoli recently worked as a Research Fellow in the Human Connectedness group at Media Lab Europe, the European research partner of the MIT Media lab. Her main interest lies in researching the possibility to use wireless technologies and mobile devices to better connect people on a local scale. Coming from a mass media background, she is now very fascinated by the potential that peer-to-peer applications have to decentralise the production and distribution of multimedia content, and to support grassroots phenomena.

Before starting the tunA project, she worked on a project called WAND (Wireless Ad Hoc Network for Dublin), a collaboration between Media Lab Europe and Trinity College Dublin, whose aim was to create a testbed network for new ad-hoc applications and services that could integrate with the interests and needs of the local communities. For this project she also conducted a local study to inform the application development, of which tunA can be considered as an output. about tunA: tunA is a mobile wireless application that allows users to share their music locally through handheld devices. She is currently a PhD candidate at the London School of Economics and Political Science, in the Department of Media and Communications. Her main research interest is the design of mobile proximity-based applications, technologies that support communication and data sharing among co-located people. Her work has been sponsored by the EU funded project Bionets.

Brucker-Cohen, Jonah

Johah Brucker-Cohen pioneered the mix of art, science, new technology and digital music as director of festivals such as Now in Nottinghamand FutureSonic in Manchester. As Head of New Media at the ICA in London he worked with partners such as Sony and Motorola. Currently he is artistic director of The Public in Birmingham.

Website: http://infamia1.infamia.com/coin-operated.com/

Cubitt, Sean

Sean Cubitt was born in Lincolnshire of Irish parents. He studied at Queens' College Cambridge and McGill University, Montreal. In the 1980s he worked freelance in art schools, video workshops, journalism and at the Society for Education in Film and television. He spent the 1990s in Liverpool, where he was involved in developing the Foundation for Art and Creative technology (FACT). In 2000, he moved to New Zealand with wife Alison and dog Zebedee, and now holds dual nationality with New Zealand and the UK. Sean Cubitt is Professor of Screen and Media Studies at the University of Waikato, New Zealand. Previously Professor of Media Arts at Liverpool John Moores University, he is the author of Timeshift: On Video Culture (Comedia/Routledge, 1991), Videography: Video Media as Art and Culture (Macmillans/St Martins Press, 1993), Digital Aesthetics (Theory, Culture and Society/Sage, 1998), Simulation and Social Theory (Theory, Culture and Society/Sage, 2001), The Cinema Effect (MIT Press, 2004) and EcoMedia (Rodopi, 2005); and coeditor of Aliens R Us: Postcolonial Science Fiction with Ziauddin Sardar (Pluto Press 2002) and The Third Text Reader with Rasheed Araeen and Ziauddin Sardar (Athlone/Continuum, 2002). He is the author of over 300 articles, chapters, papers and catalogue essays on contemporary arts, culture and media. A member of the editorial boards of Screen, Third Text, The International Journal of Cultural Studies, Futures, Time and Society, Journal of Visual Communication, Leonardo Digital Reviews, Iowa Web Review, Cultural Politics, fibreculture journal, International Journal of Cultural Politics, Public and Vectors, he has lectured and taught on four continents, and his work has been published in Hebrew, Arabic, Korean and Japanese as well as several European languages in publications from Latin and North America, Europe, Asia and Oceania. He has also curated video and new media exhibitions and authored videos, courseware and web poetry.

de Souza e Silva, Adriana

Adriana de Souza e Silva is an Assistant Professor at the Department of Communication at North Carolina State University (NCSU), and the director of the Mobile Gaming Research Lab. She is also a faculty member of the Science, Technology and Society Program at NCSU. In 2004/2005, she was a Senior Researcher at the UCLA Graduate School of Education and Information Studies (GSE&IS) at CRESST (Center for the Study of Evaluation). Dr. de Souza e Silva holds a Ph.D. on Communication and Culture at the Federal University of Rio de Janeiro, Brazil. From 2001 to 2004 she was a visiting scholar at the UCLA Department of Design / Media Arts. Dr. de Souza e Silva's research focuses on how new media (mobile) interfaces change our relationship to space and create new social environments via media art and hybrid reality games. She holds a Masters degree in Communication and Image Technology at the Federal University of Rio de Janeiro, Brazil.

Dovey, Jonathan

Jon Dovey is Professor of Screen Media and Director of the Digital Cultures Research Centre at the University of the West of England, UK. Before becoming an academic he worked in media practice producing documentary and experimental works located within the independent film movements of the 1980s, most notably founding the original mash up scratch video team Gorilla Tapes in 1984. This work has recently been revived in shows at FACT Liverpool (2003), the Tate Modern (2004) and the ICA (2007).

Since returning to UWE in 2008 he has been working with the Pervasive Media Studio in Bristol and working on setting up the Research Centre with colleagues in the Faculty of Creative Arts at UWE. He previously taught at the University of Plymouth and at the University of Bristol where he was Head of Subject in the Department of Drama: Film, Theatre and Television.

He was a Knowledge Transfer Fellow at the The Pervasive Media Studio, Bristol 2010-2011. This project was a research programme into the Aesthetics and Value of Pervasive Media innovations. His practice-based work is with Rik Lander and Greg Browning in the media innovation company 'You Are Here'. His research focus is on technology and cultural form and he has worked collaboratively with members of the UWE School of Culture, Media and Drama on a number of projects. He is a founding member of the Play Research Group.

His books include:

Game Cultures Dovey J & Kennedy H W, Open University Press 2006
New Media - A Critical Introduction Lister M, Dovey J, Giddings S, Grant I and Kelly K, Routledge 2003
Freakshows - First Person Media And Factual TV Pluto Press 2000
Fractal Dreams: New Media In Social Context, ed. Dovey J, Lawrence and Wishart 1996

Fang, Victoria

An Interaction designer and new media artist with over 10 years of experience in interactive media, exploring the ways in which technology can enhance play, evoke wonder, and innovate upon traditional experiences. Senior Interaction Designer at *frog design* from January 2007 to the present. Engaged in user research, competitive analysis, concept development, information architecture, experience design, documentation, and user testing for online, mobile, and real world interactive experiences. As a New Media Artist since 2004, she has engaged in the exploration, development, and creation of interactive spaces and objects that are designed to challenge traditional forms of play and storytelling. She exhibited at SIGGRAPH, Transmediale, Kapelica Gallery, BOFFO, Leipzig Games Convention. Nominated for the Transmediale.05 award. She is recipient of a 2006 Rhizome commission. She is a graduate of Parsons School of Design 2002-2004

Fleuriot, Constance

Constance Fleuriot is a Research Associate at the Digital Cultures Research Centre, part of the University of the West of England that is based in the Pervasive Media Studio. She is working with Jon Dovey on an AHRC funded Knowledge Transfer Fellowship investigating the language, aesthetics and value of pervasive media. She founded *featherhouse*, an independent content agency specialising in digital, social and pervasive media, which in 2010 launched *fashion popcorn*, a new network for creative people working in film, fashion and digital media. She was also a founding member of staff at the Pervasive Media Studio, Bristol, working on public engagement projects for the studio, attempting to draw in people of all ages and backgrounds.

As a Research Associate at Bristol University, Constance was one of the principal investigators on the Mobile Bristol Project investigating the social impact of emerging pervasive and mobile technologies. Since then she has been working as a freelance consultant in user research and locative media design, and facilitating location-based mixed-media projects with a variety of user communities.

She worked on the development of the Futurelab *createascape* website, a resource for pupils and teachers on how to use HPlabs *mscape* software in an educational setting. She still runs *mscape* workshops, mainly with schools, but has experience in teaching *mscape* to people of all ages.

Her interests encompass showing people how to be creative producers with pervasive technology and not just consumers. She is also interested in exploring what might be embedded or hidden in pervasive technology applications, whether deliberate or due to the assumptions of the designer. She is a Fellow of the Royal Society of Arts.

Graham, Beryl

Is an arts worker and educator with a special interest in new media art. Her professional work over 15 years has included curating the *Serious Games* exhibition of international interactive artwork, and co-editing the CRUMB website - a resource for curators of new media art. She has a PhD concerning interactive media artwork and is currently a Research Fellow at the University of Sunderland.

Having lived and worked worked in the UK and North America, she has lectured internationally, including Banff Centre for the Arts, and the Slade School of Art, London. Her writing includes a book for Heinemann, chapters for Routledge, and articles in *Art Monthly*. Her artwork comes from a background in photography and video, and ranges from interactive media installations to low-tech games.

Website: http://seacoast.sunderland.ac.uk/~as0bgr/cv/index.htm

Haque, Usman

Usman Haque designs interactive architecture systems and researches how people relate to each other and their spaces. He has created responsive projection environments, interactive installations, digital interface devices and choreographed performances. He is a former partner in architecture practice Pletts Haque and teaches at the Bartlett School of Architecture. He has been artist-in-residence at the International Academy of Media Arts and Sciences, Japan; a researcher at the Interaction Design Institute Ivrea, Italy, and has also worked in USA, Malaysia and UK. His current research interests include: The application of 'open source' strategies to architectural systems;Softspace technologies (e.g. smell outputs, temperature zones, wireless networks);The non-tangible, non-physical aspect of spatial interactions.

His work has been exhibited at the Institute of Contemporary Arts (London), Ars Electronica (Austria), the Hillside Gallery (Tokyo), the Tokyo Metropolitan Museum of Photography, Belluard Bollwerk International (Switzerland) and Fabrica gallery, (Brighton). He has presented at Doors of Perception, The Design Council London, OneDotZero, The Tate Modern and SIGGRAPH 2004, among others. His interactive and telecommunications projects have appeared in several magazines and journals including Archis, Art & Architecture, Science, Wired (online), Architects' Journal and RIBA Journal.

Website: http://www.haque.co.uk

Hemment, Drew

Director of Future Everything, a non-profit creative community interest company. In 1995 established Futuresonic International Festival, currently in its 10th year. As AHRC Research Fellow at University of Salford developed Loca, a collaborative arts-based project on mobile media and surveillance, accepted for 'Interactive Cities' at ISEA2006. Founder member of PLAN - The Pervasive and Locative Arts Network. Involved in early UK electronic dance culture as DJ and event organiser. Completed an MA (Distinction) at the University of Warwick, and a PhD at University of Lancaster.

Website: http://www.drewhemment.com/

Huhtamo, Erkki

Erkki Huhtamo is a Finnish media researcher, writer and curator. He is Professor of Media History and Theory at the University of California Los Angeles (UCLA) , Dept. of Design | Media Arts. He has published extensively on media archaeology and media art, curated exhibitions and directed television programs. He was born in Helsinki, Finland and received his Masters of Arts with honors from the University of Turku in 1987. He has worked for more than ten years as a writer, researcher, lecturer and curator in the field of audiovisual media culture. Huhtamo has lectured at the University of Art and Design in Helsinki, The Academy of Fine Arts in Helsinki, The Universities of Turku, Tampere, Oulu, Lapland, and was formerly professor of media studies at the University of Lapland in Rovaniemi, Finland.

Hull, Richard

Since joining HP Labs in 1985, Richard Hull has worked, managed and published in a broad range of research areas including knowledge-based management systems, manufacturing planning, handwriting interfaces, situated computing, and appliance communications. His current research interests are in the use of pervasive mobile computing in experiential applications. Within Mobile Bristol, he has responsibility for the Application Framework and is particularly active in developing the framework architecture and core software components. He also leads the Mobile Bristol contribution to the Savannah project and is active in many of the other application projects. RIchard gained a BSc in Electronic and Electrical Engineering from the University of Surrey in 1979 and a DPhil in Computer Science from the Universtiy of Sussex in 1984. He has also held research and lecturing posts at Philips Reseach Labs, Standard Telecommunications Laboratories and the University of Sussex.

Jacobs, Margot

Margot is an American interaction design researcher focusing on playful, emotional incorporation of technology in everyday life. She has worked at Interactive Institute DESIGN Göteborg (previously RE:FORM and PLAY) studio for the last three years, and she holds a deep interest in developing innovative design methods and experimental prototypes for social interventions in public space. Her previous experience includes a year as research fellow at the interaction Design Institute Ivrea, Italy. Margot has a bachelor's degree in Industrial Design from the Georgia Institute of Technology in the USA and a master's degree from the Interactive Telecommunications Program in New York. Margot takes a great interest in reaching the public and other audiences. Outreach activities include organizing local creative new media collective [fringe], lecturing, tutoring postgraduate students, and creating a gallery space for community expression. Margot is currently living in America.

Jakob, Anke

A Researcher at Kingston University London, her research interests are the Synthesis/fusion of digital image and physical surface/materiality; digital display technologies and theirapplication for surface design; vision and visual perception of space; application of light and projection within architecture and spatial design

Based on her background as trained textile designer and the particular interest in digital media technology, herresearch explores conditions, which lead to the fusion of physical surface and digital display - evoking the confusion and interference of the real and the virtual. In this process, her investigation focuses on visual ambiguities and illusions resulting from the interaction between digital image, material and light influencing visual perception and sense of space. In 2007, Anke was awarded her PhD for her practise-based, interdisciplinary research project 'Blurring the boundaries between the real and the virtual' at Bath School of Art & Design, Bath Spa University, where she experimented with printed textiles/surfaces, animated patterns and projection technology.

She holds an MA in textile design from Central St Martin's College of Art & Design (1996) studied digital graphics and animation at the Academy for Media & Arts, Cologne, and textile and fashion design at University of Art & Design in Halle, Germany.

Besides her research activities, she is partner at Architecture and design practice leit-werk Ltd since 2001, engaging in projects with rather interdisciplinary character within the realm of architecture and urbanism and focusing on the application of digital media and ornament. Further, her professional background includes working in the textile industry as well as in commercial textile design practise for many years.

Jaschko, Susanne

Susanne Jaschko is a Berlin based independent curator and researcher in contemporary visual and electronic art. From 2000-2004 she was deputy director and curator of "transmediale – international media art festival berlin". Within the structure of the festival she was mainly engaged in curating exhibitions of electronic and digital art and building up international co-operations. Recent and upcoming curatorial projects include a traveling show entitled "Open House. Intelligent Living by design" that critically reflects future home designs in an era of extensive application of technology, urbanism and an overaged society. This show is produced by the Vitra Design Museum, Germany, and Art Center, College of Design, Pasadena, USA.

In addition, she currently curates SCAPE Biennial 2006 dedicated to art in public space, in Christchurch, New Zealand and an international series of exhibitions entitled "urban interface" which opened in Berlin in spring 2007. In 2005 and 2006 she has taught at the Academy of Applied arts in Potsdam in the department of European Media Science. In 2005 a grant for researcher in residence at the Daniel Langlois Foundation in Montréal enabled a research about "The temporal and spatial design of early film and video-based installation art, its perception and effect on the visitors' action".

In 2004, she twice visited the Statens Kunstakademie in Oslo, Norway, as a guest teacher. In 2003 she was a guest teacher at the Leipzig University, Germany, where she taught in the department of media science and communication. From 1997 – 1999 she worked as a programme assistant, later as programme curator at the transmediale. Beside this freelance work for the festival she co-ordinated the conference "monomedia: value" on new values in media design at the University of the Arts, Berlin in 1999/2000. She wrote her doctorate thesis about "Self-portraiture and self-understanding in the painting of the GDR from 1945 to the 80ies" at the Rheinisch-Westfälische Hochschule Aachen, Germany.

During her studies she casually wrote for newspapers and magazines about art and culture. Moreover she assisted in two Berlin galleries, the "Projektraum Berlin" and "Dogenhaus Galerie Berlin", where she initiated and organised contemporary art shows. In 1995 she curated and organised the show "Entgrenzungen" presenting young artists from East Germany, for the Ludwig Forum, Aachen. As a jury member she participated in a number of national and international competitions in media art visual arts and digital culture. Since 2003 she is on the advisory board for the official web portal of Germany, <www.deutschland.de.> Moreover she lectures internationally and gives presentations on media art and digital culture.

Lander, Debbi

Following a degree in Humanities (Anthropology and Media Communications) and two years professional work in Advertising (Bartle Bogle Hegarty of Levi 501 fame), Debbi joined shinkansen as co-partner in 1991 (research & production unit working with dance, performing arts and media technologies). She is also a director of Future Physical Ltd (a non profit cultural organisation working with responsive environments and technologies).

Website: http://futurephysical.org

Lovejoy, Annie

Is a media artist based in Bristol, whose work is often site-specificand uses a variety of technologies including location aware software and sensing, her works include:2004 'your heart on my sleeve' sensor, an led lightwork using wearable technology shown at Siggraph Los Angeles, ISWC Washington, USA; and 2002 millenium square a GPS / wearables urban situated technologies project with Mobile Bristol (Bristol University / Hewlett Packard Research Labs); 2002 - reverberations: responses to 2 locations in Swansea, Wales, the Helwick Lightship & the River Tawe; Locws International 2002 - pillow: (HNT) Kibla Multimedia Center, Maribor, Slovenia

Website: http://www.annielovejoy.net/

Moriwaki, Katherine

Katherine Moriwaki is an artist and researcher investigating clothing and accessories as the active conduit through which people create network relationships in public space. Currently a Ph.D. Candidate in the Disruptive Design Team of the Networks and Telecommunications Research Group at Trinity College Dublin, her work has appeared in IEEE Spectrum Magazine, and numerous festivals and conferences including numer.02 at Centre Georges Pompidou (02), Break 2.2 (03), Ubicomp (03,04), eculture fair (03), Transmediale (04), CHI (04), ISEA (04), and DEAF (04). She is a 2004 recipient of the Araneum prize from the Spanish Ministry for Science and Technology and Fundación ARCO.

Website: http://www.kakirine.com

Nisi, Valentina

Valentina is a researcher and media artist working on the overlapping possibilities of cinematic narratives and place, enabled by mobile wireless technologies. Her fascination lies in the poetic potential of bringing stories out into real space and merge architecture, environment and landscape with the narrative experience. Her research is centred on designing and producing such experiences for public spaces.

Valentina is currently writing her Doctoral thesis on 'Place based narratives' describing her interactive location based installations (Hopstoryand the Media Portrait of the Liberties) and their relations to the place and the audience. Valentina's thesis was supervised by Dr. Mads Haahr.

Since September 2005 She has been a Researcher in the Distributed Systems Group, at the Computer Science Department of Trinity College Dublin, on location aware and mobile content project such as Context Aware Multimedia Environment (CARMEN). From January 2001 till August 2004 she has been a researcher at MediaILabEurope with G.Davenport and the Storynetworks group investigating the potential of wireless mobile technologies in cinematic non-linear narratives, designing and producing distributed location based stories for public spaces.

Her work has been presented and published in different festivals and conferences both in Ireland and internationally with projects such as "Weirdview", "Hopstory" and the most recent "Media Portrait of the Liberties" Valentina has an MSc in Multimedia Systems from TCD. She holds a First class Honour Fine Art Degree from the Albertine Academy of Fine Arts in Turin.

Raby, Fiona and Dunne, Anthony

Dunne and Raby design products and services as a medium to stimulate discussion and debate amongst designers, industry and the public about the social, cultural and ethical implications of emerging technologies. Many of their projects are collaborative, working with industrial research labs, academia and cultural institutions to design both speculative products and services. Projects include: Hertzian Tales, a combination of essays and design proposals exploring the aesthetic meanings of electronic objects (published 1999); Weeds, Aliens and Other Stories (with Micahel Anastassiades), a collection of psychological furniture for the home and garden (published 2000); FLIRT, a European Union funded research project investigating location-based services for mobile phones (published 2000); and Placebo, a collection of electronic objects which explore mental well-being in relation to domestic electromagnetic fields (published 2001).

BioLand, their current research project investigates how a critical design approach can be applied to the field of biotechnology. Anthony and Fiona were founding members of the CRD Research Studio at the Royal College of Art (1994-2002), they currently lead studios in the Design Product and Architecture & Interiors departments. Design Noir: The Secret Life of Electronic Objects was published by August/Birkhauser.

Website: http://www.dunneandraby.co.uk/content/projects

Reid, Jo

She joined HP Labs in 1994 having worked in Research and Development at Texas Instruments on expert systems and prior to that for Rank Xerox on interface design prototyping and testing. Jo contributes to the Mobile Bristol program as a project manager and as a researcher. Her research interest is in the area of Experience Design and as a firm believer in the value of user participation she has been involved in the design, construction and project management of a number of field trials in the area of appliance communications and Experience Design. Jo recently received an MBA with the Open University.

Projects that she has been involved in are Schminky, Queen Square, Interactive Building project, Zap Scan, Walk in the Woods, John Cabot Technology College Workshop, St Gregory School Workshop, Reflective Practitioner Study and the creation of the Mobile Bristol web site.

Rieser, Martin

Publications include: Spatial Narratives for the 21st Century Refresh New Media History conference, Banff New Media Institute 2005 and Network Performance Siggraph Panel 2005; Beyond Mapping Altered States, University of Plymouth 2005; Interactive Poetics Interactivity of Digital Texts University of Munster, Germany 2005; Mobile Audience PLAN ICA London; Space Narratives, UCE / Vivid Birmingham 2004; Spatialised Narratives at Intermedia University of Oslo 2004; The Spatialisation of Narrative, Multimedia Histories Conference, Exeter University, 2003; New Screen Media Cinema / Art /Narrative Rieser, M, Zapp, A (Eds.), BFI / ZKM book and DVD, 2002; Interactive Installation: From Gallery to Street in Computers and Art, Mealing, Stewart (Ed.), Intellect Books, UK 2002; Interactive narratives; Convergence Magazine v 3.1 Spring Issue 1997; New structures for Interactivity,

Oberhausen Film Festival Proceedings, Germany 1997; Public Art, Architecture and Interactivity ISEA97 Proceedings Chicago 1997.

He has curated a number of International New Media exhibitions (The Electronic Print-Computer Images and the Fine Artist, exhibition and catalogue, Arnolfini, Bristol, 1989, and The Electronic Eye: European Computer Art exhibition and catalogue, Watershed, Bristol/University of Exeter 1986), and involvement in the curation of a variety of international shows (CADE 2001 / Arcade 3, Glasgow School of Art, Grays School of Art, New Greenham Arts Centre, London Guildhall University, 2001; CADE 99, Middlesborough,1999; CADE 97, Derby, 1997; 2000 Losing the Plot-Ship of Fools, Interactive installation, Pittville Pump Rooms, Cheltenham Literature Festival). His Interactive installations include: The Street (Heat: The art of Climate Change, RMIT Gallery, Melbourne 2008), Secret Door Invideo Milan 2006; Hosts Bath Abbey 2005, Triple Echo (AHRB funded three- screen pilot) 2003; Understanding Echo, ISEA2002, Nagoya, Japan and Watershed Media Centre, Bristol, 2002; Labyrinth, F-Stop Media Station, Bath and Bridport Arts Centre 1998; Screening the Virus, multimedia screening and website, The Watershed, Bristol 1997, From Silver to Silicon, Media Myth and Mania, CD Rom and Exhibition, Watershed, ICA, Focal Point, Milia (1994).

Website: www.martinrieser.com

Rowe, Anthony

Anthony Rowe is a founder member of interactive design agency squid s o u p. Squid s o u p 's work focuses on 3D interaction and combining sound and virtual space in intuitive ways. They have won numerous awards, including an International EMMA [Online art 2000], a BAFTA nomination [Interactive Art, 2002] and their work has been shown at dozens of events worldwide, ranging from ISEA [Japan 2002] and SIGGRAPH [LA 2001] to SONAR [Barcelona 2001]. *Presentation at VSMM (Xi'an, China) 18-20 October 2006* Research paper based on the findings of Come Closer – "Come Closer: Encouraging Collaborative Behaviour in a Multimedia Environment". Awarded BEST PAPER, VSMM 2006 *Artful Gaming (Dana Centre, London) 2-6 October 2006* squidsoup show Ghosts and take part in discussions on the potential creative futures of gaming. *Bloom Festival (Dorset UK) 11-13 August 2006* Ghosts projected onto the roof of a tent at the 24 hour cafe at Bloom, again courtesy of Cybersonica... *Cybersonica (UK) May 2006* Freq2 launched at Tate Britain, Dana Centre, Phonica Records and b.tween; and we do a talk with Soundtoys at the Cybersonica Conference

Website: http://www.squidsoup.org/

Rueb, Teri

Rueb's large-scale responsive spaces and location-aware installations explore issues of architecture and urbanism, landscape and the body, and sonic and acoustic space. In 1999 she launched "Trace", an interactive GPS-based sound installation set along a network of hiking trails in the Canadian Rockies (funded by the Banff Centre for the Arts). She recently launched "Drift" an interactive sound installation set along the tidal flats of the Wadden Sea in Cuxhaven, Germany <www.ohne-schnur.de>. She is currently working on an installation to be installed throughout Boston Common (commissioned by turbulence.org).

She lectures and exhibits world wide at venues including Transmediale (Berlin, 2004), SIGGRAPH (San Antonio, 2002), The International Symposium on Electronic Arts (Nagoya, 2002; Paris, 2000; Helsinki, 2004), Consciousness Reframed (Perth, 2002), The New Museum of Contemporary Art (New York), the Corcoran Gallery of Art (Washington D.C.), The Banff Centre for the Arts (Banff), Bell Laboratories (Holmdel), Interval Research Corporation (Palo Alto), and The Fraunhoefer Institute/GMD (IRCAM, Paris, 2002; Glasgow, 2001).

Rueb's work has been featured and reviewed in diverse publications including "Information Arts: Intersections of Art, Science and Technology", edited by Stephen Wilson, MIT Press, 2001. She holds a

B.F.A. in Art and Literary and Cultural Studies from Carnegie Mellon University and a master's degree in Interactive Telecommunications from the Tisch School of the Arts, New York University. She is currently professor in the Graduate Department of Digital Media at the Rhode Island School of Design.

Sermon, Paul

Professor of Creative Technology, Adelphi Research Institute for Creative Arts and Science, The University of Salford. Born 1966. BA Hons Fine Art, Gwent College of Higher Education, 1988. MFA, The University of Reading, 1991. Awarded Prix Ars Electronica "Golden Nica" Linz, Austria, 1991. Artist-in-Residence, ZKM Centre for Art and Media Karlsruhe,Germany,1993. Awarded Interactive Media Festival "Sparkey Award" Los Angeles, 1994. Dozent for Telematic Arts at the HGB Academy of Visual Arts in Leipzig, Germany(1993 to 1999). Guest Professor for Performance and Environment, University of Art and Industrial Design Linz, Austria, 1998 to 2000. Since June 2000, based at The University of Salford in the research field of immersive and expanded telematic environments.

Sommerer, Christa /Mignonneau, Laurent

Internationally renowned media artists, working in the field of interactive computer installation. They are Professors at the University of Art and Design in Linz Austria where they head the Department for Interface Culture at the Institute for Media. Sommerer and Mignonneau previously held positions as Professors at the IAMAS International Academy of Media Arts and Sciences in Gifu, Japan and as Researchers and Artistic Directors at the ATR Media Integration and Communications Research Lab in Kyoto Japan. They also were Visiting Researchers at the MIT CAVS in Cambridge US, the Beckmann Institute in Champaign Urbana, IL, USA and the NTT-InterCommunication Center in Tokyo. Sommerer originally studied biology (botany) at the University of Vienna and modern sculpture and art education at the Academy of Fine Arts in Vienna (masters degree) Mignonneau studied modern Art and Video Art at the "Ecole des Beaux Arts" in Angouleme, France where he received his masters degreeSommerer and Mignonneau completed their PhD degrees from CAiiA-STAR, University of Wales College of Art, Newport, UK and the University of Kobe Japan, respectively.

In 1992 Sommerer and Mignonneau met at the Institute for New Media at the Staedelschule in Frankfurt where they teamed up and started their collaboration in the area of interactive computer installations. Mignonneau and Sommerer's artworks have been called "epoch making" (Toshiharu Itoh, NTT-ICC museum) for developing natural and intuitive interfaces and for often applying scientific principles such as artificial life, complexity and generative systems to their innovative interface designs. Mignonneau and Sommerer created pioneering interactive computer installations such as "Interactive Plant Growing" (1992), "Anthroposcope" (1993), "A-Volve" (1994), "Trans Plant" (1995), "Intro Act" (1995), "MIC Exploration Space" (1995), "GENMA" (1996), "Life Spacies" (1997), "Life Spacies II" (1999), "HAZE Express" (1999), "VERBARIUM" (1999), "Industrial Evolution" (2000) and "PICO_SCAN" (1999/2000) and "Riding the Net" (2000), "The Living Room" (2001), "The Living Web" (2002), "Nano-Scape" (2002), "Mobile Feelings" (2003) and "Eau de Jardin" (2004).

These works have been shown in around 150 exhibitions world-wide and are permanently installed in media museums and media collections around the world, including the Media Museum of the ZKM in Karlsruhe, Germany, the NTT-ICC InterCommunication Center in Tokyo, the Cartier Foundation in Paris, the Millennium Dome in London, the Tokyo Metropolitan Museum of Photography in Japan, the AEC Ars Electronica Center in Linz, Austria, the NTT Plan-Net in Nagoya, Japan, Shiroishi Multimedia Art Center in Shiroishi, Japan and the HOUSE-OF-SHISEIDO in Tokyo. Sommerer and Mignonneau have won mayor international media awards, for example the "Golden Nica" Ars Electronica Award for Interactive Art 1994 (Linz, Austria), the "Ovation Award" of the Interactive Media Festival 1995 (Los Angeles, USA), the "Multi Media Award '95" of the Multimedia Association Japan and the" World Technology Award" in London (2001).

Sommerer and Mignonneau have published numerous research papers on Artificial Life, interactivity and interface design and they lectured extensively at universities, international conferences, and symposia.Sommerer has organized workshops and invited sessions at various international conferences, such as Dagstuhl Seminar (Germany, 2003), SCI2001 (Orlando, 2001), KES2001(Osaka, 2001), AlifeVII (Portland, 2000), KES2000 (Brighton, 2000) and ART-Science-ATR (Kyoto, 1997). Sommerer has also organized several student exhibitions (Ogaki Biennale, 2003 and the IAMAS Campus Exhibition at Ars Electronica 2004 in Linz) and she serves on many conference committees (Siggraph 2005, ACM 2004, NIME 2004, ICAT 2004 etc.) and is also an active jury member at various international competitions (IP Impulse Program 2005 of the Austrian Business Association, Ars Electronica Jury 2005, UNESCO Digtital Arts Program, VIPER 2002 etc.

Sommerer is an International Co-editor for the LEONARDO Journal, MIT Press and in 1998, together with Laurent Mignonneau, she edited a book on the collaboration of art and science called "Art@Science," published by Springer Verlag Vienna/New York.

Stead, Lisa

Lisa Stead completed her PhD studentship, "Fashion and Technology: Future Functions for 'The Emotional Wardrobe'," funded by the AHRB as part of a collaborative project entitled 'Fashion and Modernity'. The aim of the PhD study was to marry fashion aesthetics with material science and electronic technology to provide interactive personalisation for the wearer. Research into The Emotional Wardrobe has culminated in collaborations with the Electrical Engineering Department at Imperial College, London and Elumin8, an electroluminescent panel company in Dorset. The garment developed with Imperial College is based on the principles of affective computing and changes its aesthetics in reaction to a change in inferred emotional state. It has shown great potential for further development, which could also be used for communicative and therapeutic benefit to autistic, and deaf, dumb and blind teenagers to create inclusive emotional aesthetics. Lisa has presented this work at 2AD conference in Bristol and has co-authored a paper, which will be published in Computers and Graphics later this year. She has also appeared on 'Go Digital', a technology programme on the BBC World Service. For the last 8 years she has lectured on various fashion courses at Foundation Art & Design, ND, HND and MA level. She has also held a course leader position for 'Preparation for Design', during which time she successfully revalidated the course.

Struppek, Mirjam

Mirjam Struppek works and lives as urbanist in Berlin. Since three years she is developing the information-platform www.interactionfield.de about new interactionelements for the urban public space, which she has presented at various events and conferences. After working at PLAY_gallery for still and motion pictures, Berlin, she founded in 2004 Urban| |Resesartch, with a special focus on the livability of urban space, public sphere and its transformation and acquisition through new media. The Institute of Network Cultures and Lectorate of Art in Public Space, Amsterdam, currently commission her to prepare the conference "Urban Screens 05 - discovering the potential of outdoor screens for urban society". She is as well in preparation of starting her PHD at Bauhaus University Weimar. The aim is to connect her background experiences of her academic training and the focus on environmental issues, with her new interest into new media studies and urban art. She is holding a degree in Urban- and Environmental Planning from University of Kaiserslautern, where she also was assistant teacher in Geometry for Architects and Planners and worked as research assistant in the Institute for Mobility and Transport during her studies. In 1999 she spent two research semesters at the Graduate School of Environmental Studies at Nagoya University, Japan.

Vaucelle, Cati

Her challenge as researcher, visual media artist and interaction designer is to explore, design and implement transformable interfaces; personal objects, computational products, audio and visual media tools, collaborative installations that allow creators, by publishing a piece, to engage others in discussing the technological space they live in. The transition from individual ideas to exposure of those in a final expressive piece is the space between improvisation and publication for which she creates technology to support the creative metamorphosis that happens.

Her exploration into transformable interfaces began in 1996 in the Art, Philosophy, Aesthetic department at Paris VIII University, specializing first in Photography and Fine Arts, then in Graphic Design, Video, Computer Sciences, MST'99, to finalize a post graduate degree in Arts and Technology, DEA'00. In parallel I studied Computational Linguistics, Psychology, Mathematics and Economics, BS'98. I have continued this theme at the Massachusetts Institute of Technology as a research assistant at the MIT Media Laboratory from 2000-2002, graduating from the Tangible Media group lead by Dr Hiroshi Ishii, studying Research Methodology, Electrical Engineering, Computer Sciences, and Industrial Design, and as a research associate at the Media Lab Europe conducting independent research in the Story Networks Group from 2002-2004 led by Glorianna Davenport in parallel of researching at the Interactive Cinema Group/Media Fabrics Group at the MIT Media Laboratory. Since February 2005, a Post Doctoral Researcher at the University of Trinity College Dublin working with Brendan Tangney on video/music multimodal interfaces.

Zapp, Andrea

Andrea Zapp is a German media artist currently based in Manchester, working as a postgraduate Research Fellow at the Institute for Research & Innovation in Art & Design MIRIAD, Manchester Metropolitan University. Her projects and installations focus on networked models of narration. They have been shown widely in international exhibitions and on the net. She has been lecturing extensively in international art institutions and Universities and on media conferences. She edited the book and DVD "New Screen Media. Cinema/Art/Narrative", published in 2002 by the BFI London and ZKM Karlsruhe.

Website: http://www.azapp.de

Glossary

Avatar

An avatar, named after the Hindu Sanskrit word for a deity in a visible form, is a graphical representation of a human being on the Internet or elsewise in VR. Avatars may be either two or three-dimensional. They may also be created either entirely as computer graphics, or else by mapping digital photographs of a real human being onto a virtual 3D model.

Bandwidth

A measure of the intensity of the data travelling (or that can travel) across a network connection, as usually measured in Kilobytes per second. High bandwidth data connections (such as those available over ADSL) permit Internet users to download files far more rapidly and to view high-quality, real-time video. For users, the basic message is that the more bandwidth they have available to them, the better.

Blog

Blog: an online diary or journal, typically documenting the day-to-day life of an individual. Derived from "weblog."

Bluetooth

A standard for short-range wireless communication between computing devices and associated peripherals, including laptop and mobile computers, PDAs, and mobile phones.

Bot

Also known as software agents, bots are "software robots" that work for human being across computer networks, undertaking tasks such as search operations.

GPRS

General Packet Radio Service: the technology standard soon to be used by 2.5G mobile phones to provide up to a 56Kbs always-on data connection.

Broadband

Referring to a high bandwidth communications link, most usually over the Internet. At the present time ADSL technologies provide the highest bandwidth broadband connections for most Internet users over landlines, whilst 3G UMTS technologies provide the highest bandwidth wireless links..

Browser
A program or "viewer" that allows Internet users to access pages available across on world-wide web.

Cache
A "cache" is an area of computer memory (either held on a chip or else on a hard disk) that is used to temporarily store data files in order to increase the speed of access to them.

Chat Room
A chat room is a virtual "place" on the Internet in which members can exchange text messages in real-time.

Compact Flash
A standard for small-sized, non-volatile memory cards and other peripherals used in mobile computers, PDAs, digital cameras, and some MP3 players.

Firewall
A device, mechanism, bridge, Router, or gateway which prevents unauthorized access by hackers, crackers, vandals, and employees from private network services and data.

Geosynchronous Orbit
Orbit taken by satellites where the satellite's orbit velocity matches the rotation of the earth, causing the satellite to remain stationary relative to a position on the earth's surface. Geosynchronous orbit demands a position about 23,000 miles above the earth's surface over the equator.

GIS (Geographic Information System)
Generally refers to a database of geographical data. In some circles, it refers to Graphics database .

Encryption
The processing of data under a secret key in such a way that the original data can only be determined by a recipient in possession of a secret key

FTP
File Transfer Protocol: the communications standard used for transferring data files across the Internet. Whilst dedicated FTP programs are used by web designers to "upload" content to a web site, most Internet users access FTP facilities transparently when they download files, as FTP capabilities are now integrated into all web browsers.

Host
Any computer, although typically a mainframe, mid-sized computer, mini-computer, or LAN server, servicing users and their processing at the centrally-based processor but distributing the results to terminal based or client connections.

HTML
Hypertext Mark-up Language: The open coding standard (middleware) in which pages are written for the world-wide web.

HTTP
Hypertext Transfer Protocol: the standard used by browser programs for network communications across the world-wide web.

IrDA (Infrared Data Association)
A group of wireless infrared product vendors that promotes serial infrared linkages and interoperability between vendor products.

GLS (Global Locationing System)
A triangulation system used to locate a vehicle and convey that information to a central management facility.

GPS (Global Positioning System)
A satellite-based triangulation system used to ascertain current location.

GSM
The Global System for Mobile communications: the technology standard currently used by second generation (2G) mobile phones including WAP phones

3G
Third Generation (3G mobile phones offer high-speed Internet connectivity

MAN

(Metropolitan Area Network) A network that spans buildings, or city blocks, or a college, or corporate campus.

MP3

Motion Picture (experts group) 3: MP3 (or MPEG audio layer 3) is the now widespread compressed file format in which digital music and other sound files are commonly transmitted over the Internet for playback on personal computers, PocketPCs, and dedicated MP3 players

Mbps

Megabytes per second-speed of data transmission

MMS—Multimedia Message Service

It is an evolutionary upgrade EMS which is an upgrade from SMS. Short Message Service is a two-way text-based message service on primarily GSM network in Europe.

MPEG

A compressed digital video standard developed by the Motion Pictures Experts Group. Whilst MPEG is a contender to QuickTime as one of the two most popular video file formats used upon the Internet, high quality MPEG files consume a great deal of file space and often require dedicated decoding hardware to replay them in real-time at full-screen video resolutions.

PDA

Personal Digital Assistant: A pocket-sized computer, typically operated via a touchscreen stylus rather than a keyboard.

RFID Radio Frequency Identification

A method of identifying unique items using radio waves. Typically, a reader communicates with a tag, which holds digital information in a microchip. But there are chipless forms of RFID tags that use material to reflect back a portion of the radio waves beamed at them.

Roaming

The ability to access a network anywhere and move freely while maintaining an active link through a wireless connection to a network. Roaming usually requires a handoff when a node (user) moves from one cell to another. The ability to access a network anywhere and move freely while maintaining an active link through a wireless connection to a network. Roaming usually requires a handoff when a node (user) moves from one cell to another.

RFID tag
A microchip attached to an antenna that picks up signals from and sends signals to a reader. The tag contains a unique serial number, but may have other information, such as a customers' account number. Tags come in many forms, such smart labels that are stuck on boxes; smart cards and key-chain wands for paying for things; and a box that you stick on your windshield to enable you to pay tolls without stopping. RFID tags can be active tags, passive tags and semi-passive tags.

Router
A device that interconnects networks that are either local area or wide area. A device that interconnects networks that are either local area or wide area.

Spam
Spam refers to junk e-mail, most usually received in the form of unsolicited advertisements.

Symbian
A consortium, founded and led by Psion, developing software tools to run on PDAs, handheld and mobile computers, and mobile phones.

SMS
Short Messaging Service: the technology standard used for text-messaging on current second generation (2G) mobile phones.

SMTP
Simple Mail Transmission Protocol: one of the two communications standards used across the Internet for the transmission of e-mail messages

Sniffer
A computer program used by the criminally minded to attempt to discover the passwords of legitimate users in order to forge access to a computer system.

TCP/IP
Transaction Control Protocol/Internet Protocol.

TTS
Text to Speech: referring to the use of software to read electronic content (most notably web pages) back to a user as audio.

UMTS
Universal Mobile Telephone System: the technology standard to be used in third generation (3G) mobile.

Ubiquitous Computing ("UbiComp")
The concept of building computers into our everyday working and living environments to such an extent that data, rich media and network access become constantly, frictionlessly and transparently available.

URL
Uniform Resource Locator: the name given to the unique combination of a network communications protocol (for example HTTP or WAP) and Internet address (domain name) that a browser can use to access a particular web site.

VRML
Virtual Reality Mark-up Language: A coding standard used for writing virtual reality (3D) web sites.

WAP
Wireless Application Protocol: A standard for data communications that provides a relatively low speed (typically 9.6Kbs) Internet connection to a second generation 2G mobile phone via GSM.

Wi-Fi
Wi-Fi is short for "wireless fidelity." The Wi-Fi Alliance is an industry alliance to promote wireless networking arrangements based on the IEEE 802.11 specification.

Windows CE
A diskless operating system (with bundled applications packages) invented by Microsoft and used on handheld computers and mobile computers.

WML
Wireless Mark-up Language: The coding standard used for writing web pages viewable on wireless application protocol (WAP) mobile phones and compatible devices.

XML
eXtensible Mark-up Language: allowing for the automated exchange of data between integrated applications across the Internet.

Selected Bibliography
Books and Articles

Aakhus, M. (ed.) (2002) *Perpetual Contact: Mobile Communication, Private Talk, Public Performance.* Cambridge: Cambridge University Press.

Alam, Shahidul (2004) *Migrantsoul: an exploration of transitions across borders.* <http://www.migrantsoul.net/>.

Albert, Saul (2004) "Locative Literacy", in: *Metamute,* M28: Summer/Autumn 2004 (6.07.04). <http://www.metamute.com/look/article.tpl?IdLanguage=1&IdPublication=1&NrIssue=28&NrSection=10&NrArticle=1390&search=search&SearchKeywords=locative&SearchLevel=0>.

Albert, Saul (2004) *Re: [Locative] Locative Gaming: Dawn of the Zombie Cyborgs*. New-Media-Curating Discussion List. <http://www.jiscmail.ac.uk/lists/new-media-curating.html>.

Arns, Inke (2005) *Dispersed Moments of Concentration.* <http://www.hmkv.de/dyn/d/>.

Ascott, Roy (1998) "Gesamtdatenwerk. Connectivity, Transformation and Transcendence", in: *Ars Electronica: Facing the Future.* Linz: MIT Press.

Balnaves, M., O'Regan,T. and Sternberg, J. (eds) (2002) *Mobilising the audience.* St. Lucia: University of Queensland Press.

Bambozzi, Lucas (2004) *Spio.* <http://www.http.uk.net/docs/exhib4/press_release_spio.htm>.

Banovich, Tamas (2001) "Tamas Banovich ", in: Presentation at Curating New Media Seminar, May 10-12, 2001. Gateshead. <http://www.newmedia.sunderland.ac.uk/balticseminar/banovic.htm>.

Barry, A. and Kimbell, L. (2005) "Pindices: Personal Political Indices ", in: *Atmospheres of Democracy. March 20-August 7. ZKM.*

Bassoli, A., Cullinan, C., Moore, J. and Agamanolis, S. (2003) "TunA: a Mobile Music Experience to Foster Local Interactions", in: *Adjunct proceedings of UbiComp'03.*

Beaton, J. and Wadjman, J. (2004) *The impact of the mobile phone in Australia : social science research opportunities.* Report. The Australian Mobile Telecommunications Association Conference.

Benjamin, W. (1970) *The Work of Art in the Age of Mechanical Reproduction.* London: Illumination.

Best, Anna (2002) *Ice cream van convention.* <http://www.annabest.info/cv_ice-cream-van-convention/>.

Black, Ilze (13 April 2004) *Re: April Thought For The Day: Exhibiting Locative Media.* New-Media-Curating Discussion List. <http://www.jiscmail.ac.uk/lists/new-media-curating.html>.

Bloch, E. (1986) *The Principle of Hope.* Cambridge, MA: MIT Press.

Bourdieu, Pierre (1986) *Distinction: A Social Critique of the Judgement of Taste*, transl. R Nice. Cambridge, MA: Harvard University Press.

Bourriaud, Nicolas (2002) *Relational Aesthetics.* Paris: Les Presses du Réel.

Brown, Green, B. N. and Harper, R. (eds) (2002) *Wireless World : Social and Interactional Aspects of the Mobile Age.* London: Springer-Verlag .

Bull, M. (2000) *Sounding out the city – personal stereos and the management of everyday life.* Oxford, UK: Berg.

Bunting, Heath (2003) *BorderXing Guide.* <http://www.tate.org.uk/netart/borderxings/>.

Calvino, I. (1997) *Invisible Cities.* New York: Vintage Classics.

Canetti, E. (1984) *Crowds and Power.* New York: Farrar, Straus and Giroux.

Canonge, Hector (2004) *Ciudad Transmobil: Creating Participatory Spheres through Communication Technologies.* <http://www.ciudadtransmobil.net/>

Cardiff, Janet (2001) *The Telephone Call.* <http://010101.sfmoma.org/>.

Castells, M. (1996) *The rise of network society.* Blackwell, Oxford.
CitizenShift. http://citizen.nfb.ca/onf/info.

Castells, M., Fernandez-Ardèvol, M., Qiu, J. L., & Sey, A.(2007) *Mobile communication and society: a global perspective.* Cambridge, MA: MIT Press.

Clarke-Dickson, P. (2005) "UK cellcos, broadcasters step up initiatives in mobile and iTV", in: *Mobile Media Analyst.* Vol. 6, no.10.

Cook, Sarah (2001) *Multi-Multi-Media: An interview with Barbara London.* <http://www.newmedia.sunderland.ac.uk/crumb/phase3/ilondon.html>.

Cook, Sarah, and Graham, Beryl (2002) "Curating new media: Net and not net", in: *Art Monthly* (Nov), pp. 44-45.

Cook, Sarah (2005) "Immateriality and its discontents", in: Christiane Paul (ed.) *Curating New Media.* Berkeley: University of California Press.

Cook, Sarah and Beryl Graham (2004) "Curating new media art: Models and challenges", in: Lucy Kimbell (ed.) *New media art: Practice and context in the UK 1994-2004.* London: Arts Council of England.

CRUMB, (2001) *New-Media-Curating Discussion List.* <http://www.jiscmail.ac.uk/lists/new-media-curating.html>.

Deleuze, G. (1992) "Postscript on Societies of Control", *OCTOBER* 59, pp. 3-7.

Deleuze, G. and F. Guattari (1987) *A Thousand Plateaus: Capitalism and Schizophrenia II.* Minneapolis: University of Minnesota Press.

Depocas, Alain, Ippolito, Jon and Jones, Caitlin (eds) (2003) *Permanence Through Change: The Variable Media Approach.* New York: Guggenheim Museum. <http://www.variablemedia.net/e/preserving/html/var_pub_index.html>.

de Souza e Silva, Adriana (2007) "Cell phones and places: The use of mobile technologies in Brazil", in: Harvey Miller (ed.), *Societies and cities in the age of instant access* (pp. 295-310), Dordrecht: Springer.

Dietz, Steve (1999) "Why have there been no great net artists?", in: *Through the Looking Glass: Critical texts.* <http://www.voyd.com/ttlg/textual/dietz.htm>.

Eagle, N. and Pentland, A. (2005), *Reality Mining: Sensing Complex Social Systems*, Personal and Ubiquitous Computing.

Ebner, Jorn (2004) "Locative ". New-Media-Curating Discussion List. <http://www.jiscmail.ac.uk/lists/new-media-curating.html>.

Eduardo Kac, (1992) "Aspects of the Aesthetics of Telecommunications, in: John Grimes &, Gray Lorig (eds), *Siggraph Visual Proceedings,* New York: ACM, pp. 47–5.7 http://www.ekac.org/Telecom.Paper.Siggrap.html.

Fletcher, Harrell and Miranda (2004) *Learning to Love You More.* http://www.learningtoloveyoumore.com/.

Foreign Office Architects (2003) *Phylogenesis: foa's ark.* Barcelona: Actar, and London: ICA.

Freeman, Peter <http://www.peterfreeman.co.uk/home1.htm >.

Fritz, Darko (2002) *Migrant Navigator.* <http://darkofritz.net/>.

Gees, J. (2005) *Hello World Project.* <http://www.helloworldproject.com/>.

Gergen, K. (2002) "The challenge of absent presence", in: J.E. Katz, *Perpetual contact.* New York, NY: University Press.

Gillman, Clive (2001) *Re: Too Interactive: Nov theme of the month.* New-Media-Curating Discussion List. <http://www.jiscmail.ac.uk/lists/new-media-curating.html>

Graham, Beryl (2002) *An interview with Benjamin Weil. CRUMB.* <http://www.newmedia.sunderland.ac.uk/crumb/phase3/iweil.htm>.

Graham, Beryl (2002) *Curating new media art: SFMOMA and 010101.* <http://www.newmedia.sunderland.ac.uk/crumb/phase3/append/sfmoma.htm>.

Graham, Beryl (2004) "Webexclusive: Exhibiting Locative Media", in: *CRUMB discussion postings.* Metamute, 20.05.04 . <http://www.metamute.com/look/article.tpl?IdLanguage=1&IdPublication=1&NrIssue=24&NrSection=5&NrArticle=1373&ST_max=0>.

Graham, Beryl (2005) "Taxonomies of new media art: Real world namings", in: J. Trant and D. Bearman (eds.) *Museums and the Web 2005: Proceedings.* Toronto: Archives & Museum Informatics. <http://www.archimuse.com/mw2005/papers/graham/graham.html>.

Graham, Beryl (2005) "Redefining digital art: Disrupting borders", in: Sarah Kenderdine (ed.) *Theorizing Digital Cultural Heritage. Cambridge*: MIT.

Graham, Beryl (2009) "The ironic heirs of serendipity: British new media art 1980s to now", in: Paul Brown, Charlie Gere, Nick Lambert, Catherine Mason (eds) *White Heat and Cold Logic: British computer arts 1960-1980. An historical and critical analysis.* Cambridge, Mass.: MIT.

Graham, Beryl and Sarah Cook (2001) "A curatorial resource for upstart media bliss", in: David Bearman and Jennifer Trant (eds) *Museums and the Web 2001: Selected papers from an international conference.* Pittsburgh: Archives & Museum Informatics. Also available from URL: <http://www.archimuse.com/mw2001/papers/graham/graham.html>. 197-208.

Grether, R. (2005) *Directory of Mobile Art and Locative Media.* <http://www.netzwissenschft.de/mob.htm >.

Grundman, Heidi (ed.) (1984) *Art Telecommunication.* Vancouver: A Western Front Publication, p. 140.

Haraway, D. (1992) "The Promises of Monsters: A Regenerative Politics for Inappropriate/d Others", in: *Cultural Studies*, L. Grossberg, C. Nelson, P. Treichler (eds), pp. 295-337.

Harding, Anna (2003) *Republicart-interview on Interference, with Raimund Minichbauer*. Republicart, July. <http://republicart.net/art/concept/interview-harding_en.htm>

Harwood, Graham and Matt Fuller (2001) *TextFM.* <http://scotoma.org/TextFm/>.

Hemment, Drew (forwarded by Sarah Cook) (2004) *Fwd: April Thought For The Day: Exhibiting Locative Media.* New-Media-Curating Discussion List. http://www.jiscmail.ac.uk/lists/new-media-curating.html.

Hemment, Drew (forwarded by Sarah Cook) (2004) *Re: locative,* in: New-Media-Curating Discussion List. <http://www.jiscmail.ac.uk/lists/new-media-curating.html>.

Husain, Yasir (2004) *Re: mobile media, education*. New-Media-Curating Discussion List. [Online]. <http://www.jiscmail.ac.uk/lists/new-media-curating.html>.

Isis Arts (2000) *The Big M.* <http://www.isisarts.org.uk/>.

Katz, J.E., Aakhus, M. (eds.) (2002), *Perpetual Contact: Mobile Communication, Private Talk, Public Performance.* Cambridge: Cambridge University Press.

Keating, Barbara (2004) *Fountain.* <http://www.commissionsnorth.org/showcase/portfolio/223>.

Kester, Grant (2004) *Conversation Pieces.* Berkeley: University of California Press.

Koh, Germaine (2005) *Relay.* <http://www.balticmill.com/html/viekoh.html>.

Koolhaas, Rem (ed.) (2004) *Conten.* Köln:Taschen.

Krysa, Joasia, in partnership with Tate Modern (2005) *Curating, immateriality, systems*: A conference on curating digital media, 4th June 2005. [Conference]. London: Tate Modern.

Lacan, Jacques (1989) *The Mirror Stage, Écrits: a selection.* London: Routledge (First published: *Écrits.* Paris: Éditions du Seuil, 1966).

Lacy, Suzanne (ed.) (1995) *Mapping the Terrain: New Genre Public Art.* Seattle: WA Bay Press.

Landow, George P (ed.)(1994) *Hyper/Text/Theory.* Baltimore: Johns Hopkins University Press.

Lanham, Richard A. (1993) *The Electric Word: Democracy, Technology and the Arts.* Chicago: University of Chicago Press.

Latour, B. (1999) *Pandora's Hope: Essays on the Reality of Science Studies.* Cambridge, MA: Harvard University Press.

Latour, B. and Weibel, P. (curated) (2005) *Atmospheres of democracy.* Karlsruhe: ZKM. <http://on1.zkm.de/zkm/e/ >

Levine, J., Vinson, A. and Wood, D. (1973) "Subway behavior", in: A. Birenbaum and E. Sagarin (Eds) *People in places: the sociology of the familiar.* London: Nelson.

Lichty, Patrick (2004) *Curating Locative Media.* New-Media-Curating Discussion List. <http://www.jiscmail.ac.uk/lists/new-media-curating.html>.

Lichty, Patrick (2001) *[Re] distributions.* <http://www.voyd.com/ia/>.

Lichty, Patrick (2005) "Re: Gere : Linked References and Open Systems". Tate Online Events*: COdE Of practice: Online Panel Discussion,* June 13 - July 18, 2005. <http://www.tate.org.uk/contact/forums/onlineevents/>.

Lichty, Patrick (2005) "Participation/Curating/knowledge production ". Tate Online Events*: COdE Of practice: Online Panel Discussion* June 13 - July 18, 2005. <http://www.tate.org.uk/contact/forums/onlineevents/>.

Ling, R. (2004) *The mobile connection: the cell phone's impact on society.* Amsterdam: Morgan Kaufmann .

Lozano-Hemmer, Rafael (2001) *Too interactive (belated).* New-Media-Curating Discussion List. <http://www.jiscmail.ac.uk/lists/new-media-curating.html>.

Lunenfeld, P. (ed) (1999) *Snap to Grid: a user's guide to digital arts.* Cambridge, MA: MIT Press.

Lunenfeld, P. (2004) "Media design: new and improved without the new", in: *New Media and Society*, Vol. 16, No. 1, pp. 65-70.

Lyon, D. (2004) "Surveillance in the City", in: *The Cybercities Reader*, S. Graham (ed.), pp. 299-305.

Mack, Gehard and Liebermann, Valerie (2000) *Architecture Landscape Urbanism 3: Eberswalde Library Herzog & de Meuron*, London: Architectural Association.

Mann, S. (1997) "Wearable Computing: A First Step Toward Personal Imaging", in: *Computer*, Vol. 30, No. 2, February 1997.

Manovich, Lev (2003) "The poetics of augmented space", in: John Caldwell and Anna Everett (eds) *New media: Practices of digitextuality.* New York: Routledge, pp. 75-92.

Manovich, Lev (1996) *The aesthetics of virtual worlds: Report from Los Angeles.* Ctheory, Global Algorithm 1.3. <http://www.ctheory.com/e42.html>

Manovich, Lev (1996) *The Death of Computer Art.* <http://www.thenetnet.com/schmeb/schmeb12.html>.

Matustik, M. B. (2004) "Between hope and terror: Habermas and Derrida plead for the im/possible", in: *Epoché*, Vol. 9, Issue 1, pp. 1-18.

Mcdonald, Jillian [curator] (2002) *Location is Everything.* [online exhibition]. <http://www.rhizome.org/>.

McGuigan, J. (2004) *Towards a sociology of the mobile phone.* Agora Center, University of Jyvaskyla.

McKee, A. (2005) *The public sphere: an introduction.* Cambridge: Cambridge University Press.

Micallef, S., Roussel, J. and Sawhney, G.(2003) *[murmur].* <http://www.murmurtoronto.ca>

Mirzoeff, Nicholas (ed.) (1998) *Visual Culture Reader.* London: Routledge.

Mol, A. and Law, J. (1994) "Regions, Networks and Fluids: Anaemia and Social Topology", in: *Social Studies of Science*, Vol. 24, No. 4, pp. 641-72.

Moores, S. (1995) *Interpreting Audiences: The Ethnography of Media Consumption.* London and Thousand Oaks, CA: Sage.

Nardi, B. (2004) [no title] <http://www.darrouzet-nardi.net/bonnie/ >

Nedkova, Iliyana (2001) Presentation at *Curating New Media Seminar,* May 10-12, 2001. Gateshead: <http://www.newmedia.sunderland.ac.uk/balticseminar/nedkova.htm>.

Nedkova, Iliyana and Troy, Maria (eds) (2000) *Crossing Over.* Columbus, Ohio: Wexner Centre for the Arts.

Nokia. (2005) *Culture of Mobility.* <http://www.culture.nokia.com/hasflash.jsp?section=0&division=0>

Oskman, V. and Turtianen, J. (2004) "Mobile communication as a social stage", in: *New Media and Society*, Vol. 6., No. 3, pp. 319-339.

Papadakis, Alexandra & Andreas (eds) (2003) *Innovation – from Experiment to Realisation.* NewArch 07, London: Papadakis Publisher,

Papadakis, Alexandra and Andreas (eds) (1995) *Innovation – from Experiment to Realisation.* Vienna: Passagen Verlag.

Patterson, Zurkow, S. M. and Bleecker, J. (2003) *PDPal.* <http://www.pdpal.com/>.

Paul, Christiane, "Public Forum and Openness". Tate Online Events: *COdE 0f practice: Online Panel Discussion* June 13 - July 18, 2005. <http://www.tate.org.uk/contact/forums/onlineevents/>.

Paulos, E. and Goodman, E. (2004) "The familiar stranger: anxiety, comfort, and play in public places", in: *Proceedings of ACM CHI 2004* , pp. 223-230

Plant, S. (2002) *On the mobile: the effects of the mobile telephone on social and individual life.* http://www.motorola.com/mot/doc/0/234_MotDoc.pdf .

Pope, Ivan (2004) *Re: April Thought For The Day: Exhibiting Locative Media*. New-Media-Curating Discussion List. <http://www.jiscmail.ac.uk/lists/new-media-curating.html>.

Pope, Nina and Guthrie, Karen (1996) *A Hypertext Journal.* <http://www.somewhere.org.uk/hypertext/journal/home.html>.

Pope, Nina and Karen Guthrie (1997) *Homespun.* http://www.somewhere.org.uk/homespun/index.htm.

Pope, Nina and Guthrie, Karen (2001) "On being curated - a critical response", in: Presentation at *Curating New Media Seminar,* Gatershead, May 10-12, 2001. <http://www.newmedia.sunderland.ac.uk/balticseminar/popgut.htm>.

Pope, Simon (2005) "The Shape of Locative Media." *Metamute,* M29, Feb (9.02.05). <http://www.metamute.com/look/article.tpl?IdLanguage=1&IdPublication=1&NrIssue=29&NrSection=10&NrArticle=1477>.

Pope, Simon [curator] (2002) *Art for Networks.* [exhibition]. <http://irational.org/artfornetworks/>.

Popper, Frank (1993), *Art of the Electronic Age*, New York: Harry N. Abrams, Inc., Publishers, pp.122-139.

Raven, Arlene (1989) *Art in the Public Interest.* Ann Arbor: UMI Research Press.

Reid, D. (2004) *Conceit.* <http://www.litencyc.com/php/stopics.php?rec=true&UID=213.>

Richair, (2005) *Let's Do Lunch.* <http://richair.waag.org/>.

Roberts, Spencer and Anneke Pettican (2004-) *[Inch].* <http://www.anthropo.org.uk/gpsweb/>.

Schleiner, Anne-Marie (2003) *"Curation fluidities and oppositions among curators, filter feeders, and future artists",* in: *Intelligent Agent,* 3 (1). <http://www.intelligentagent.com/archive/Vol3_No1_curation_schleiner.html>.

Schleiner, Anne-Marie (2004) *Operation Urban Terrain (OUT).* <http://www.opensorcery.net/OUT/>

Schwarz, Ullrich (ed.) (1995) *Peter Eisenman, Aura und Exzeß – Zur Überwindung der Metaphysik der Architektur.* Vienna: Passagen Verlag.

Spellman, Naomi (26 Apr 2004) *Re: locative.* NEW-MEDIA-CURATING Discussion List. <http://www.jiscmail.ac.uk/lists/new-media-curating.html>

Stenslie, S. (1996) "Wiring the Flesh: Towards the Limits and Possibilities of the Virtual Body", in: *Ars Electronica' 96. Memesis. The Future of Evolution.* Vienna/New York: Springer Verlag.

Svalsen, S. (2005) Interview. mo:life 5 September. <molife@lists.s7digital.com>

Tosa, N., Hashimoto, H., Sezaki, K., Kunii, Y., Yamaguch, T., Sabe, K., Nishino, R., Harashima, H., Harashima, F. (1995) *Networked Neuro-Baby with robotics hand* (An automatic facial expression synthesizer that responds to expressions of feeling in the human voice and handshake), demonstration presented at Siggraph'95, Los Angeles.

Treadwell, S. (2004) "Earthquake Weather", in: S. Cairns (ed.), *Drifting: Architecture and Migrancy*, London: Routledge, pp. 203-233.

Tschumi, Bernard (1994) *Event Cities - Praxis*, Cambridge, MA: MIT Press.

Tuters, Marc (2004) "locative forbearers." *New-Media-Curating Discussion List.* <http://www.jiscmail.ac.uk/lists/new-media-curating.html>

Tuters, Marc (forwarded by Sarah Cook) (14 Apr 2004) "Re: locative? tracked!" in: *New-Media-Curating Discussion List.* <http://www.jiscmail.ac.uk/lists/new-media-curating.html>.

Walker, John A. and Chaplin, Sarah (1997) *Visual Culture – an Introduction*, Manchester and New York: Manchester University Press.

Urry, J. (2000) *Sociology Beyond Societies: Mobilities for the Twenty-First Century.* New York: Routledge.

Umble, Diane Zimmerman (2003) "Sinful Network or Divine Service", in: Lisa Gitelman and Geoffrey B. Pingree (eds.), *New Media – 1740-1915*, Cambridge, MA: MIT Press.

Urry, J. (2003) *Mobile Cultures.* <http://comp.lancs.ac.uk/sociology/030ju.html>

Virilio, P. (2003) "The Overexposed City", in: A. Hoete (ed.), *ROAM: Reader on the Aesthetics of Mobility.* London: Black Dog Publishing, pp. 264-297.

Venturi, Robert, (1966) *Complexity and Contradiction in Architecture* (Museum of Modern Art papers on architecture, 1).

Williams, R. (1974) *Television: technology and cultural form.* London, Fontana.

Williams, R. (1968) *Communications.* Harmondsworth, Penguin.

Wilson, Stephen (2002) "Telecommunications", in: *Information Arts: Intersections of Art, Science, and Technology*, Cambridge: MIT Press, pp. 457-601

Yudicé, George (2003) *The Expediency of Culture: Uses of Culture in the Global Era.* Durgam, NC. Duke University Press.

CPSIA information can be obtained at www.ICGtesting.com
Printed in the USA
LVOW06*1957281013

358948LV00007B/473/P

9 789042 031272